Papers on Fuchsian Functions

Henri Poincaré

Papers on Fuchsian Functions

Translated by J. Stillwell

With 39 Illustrations

Springer-Verlag
New York Berlin Heidelberg Tokyo

Translator
J. Stillwell
Department of Mathematics
Monash University
Clayton, Victoria 3168
Australia

AMS Subject Classification: 30-03, 30F35

Library of Congress Cataloging-in-Publication Data
Poincaré, Henri, 1854–1912
Papers on Fuchsian functions
Bibliography: p.
1. Functions, Automorphic. I. Stillwell, John.
II. Title.
QA353.A9P652513 1985 515.9 85-20800

Printed and bound by R. R. Donnelley & Sons, Harrisonburg, Virginia.
Printed in the United States of America.

9 8 7 6 5 4 3 2 1

ISBN 0-387-96215-8 Springer-Verlag New York Berlin Heidelberg Tokyo
ISBN 3-540-96215-8 Springer-Verlag Berlin Heidelberg New York Tokyo

TRANSLATOR'S INTRODUCTION

by John Stillwell

I. General Remarks

Poincaré's papers on Fuchsian and Kleinian groups are of great interest from at least two points of view: history, of course, but also as an inspiration for further mathematical progress. The papers are historic as the climax of the geometric theory of functions initiated by Riemann, and ideal representatives of the unity between analysis, geometry, topology and algebra which prevailed during the 1880's. The rapid mathematical progress of the 20th century has been made at the expense of unity and historical perspective, and if mathematics is not to disintegrate altogether, an effort must sometime be made to find its main threads and weave them together again. Poincaré's work is an excellent example of this process, and may yet prove to be at the core of a new synthesis. Certainly, we are now able to gather up some of the loose ends in Poincaré, and a broader synthesis seems to be actually taking place in the work of Thurston.

The papers I have selected include the three great memoirs in the first volumes of **Acta Mathematica**, on Fuchsian groups, Fuchsian functions, and Kleinian groups (Poincaré [1882 a,b,1883]). These are the papers which made his reputation and they include many results and proofs which are now standard. They are preceded by an unedited memoir written by Poincaré in May 1880 at the height of his creative ferment. (I shall refer to this as Poincaré [1880], though it was only published posthumously in 1923.) This records the initial development of his ideas and is quite disorderly, but the passage where the geometry emerges is worth waiting for. Also included are two **Comptes rendus** notes of February 1881 in which Poincaré summarizes his results. Finally, I have included the fourth Acta paper (Poincaré [1884]) because it is his most ambitious, and still relevant to present day research. Inserted in the text are the footnotes of N. E.

Nörlund in his edition of Poincaré's works, together with a few notes of my own. These are distinguished from Poincaré's own footnotes by "N.E.N." and "Translator's Note" respectively.

In the comments which follow I have concentrated mainly on Poincaré's geometric ideas. This was unavoidable because of my lack of expertise in analysis, but in any case I believe that the geometric side of Poincaré's work is the most original, and probably most in need of explanation. I have therefore attempted to trace the earlier geometric ideas which influenced Poincaré — he is unfortunately very lax in citing sources — to pick out his key insights, and to indicate how loose ends in his work were tied up by later authors. At the end, I have provided annotated tables of contents of the four Acta papers, so as to provide technical summaries of them in terms familiar to the modern reader. Some books that will be valuable as background reading are Ford [1929] and Lehner [1964] for automorphic functions, and Magnus [1974] and Beardon [1983] for Fuchsian and Kleinian groups. The general history of automorphic functions is nicely sketched in the first chapter of Lehner, and is to appear at greater length in a book by Jeremy Gray.

II. Automorphic functions before Poincaré

Poincaré's theory of automorphic functions is a spectacular example of the sudden crystallization of a mathematical theory from ideas that were "in the air". Poincaré expressed it in his own words as follows:

> "For fifteen days I strove to prove that there could not be any functions like those I have since called Fuchsian functions. I was then very ignorant; every day I seated myself at my work table, stayed an hour or two, tried a great number of combinations and reached no results. One evening, contrary to my custom, I drank black coffee and could not sleep. Ideas rose in crowds;

I felt them collide until pairs interlocked, so to speak, making a stable combination. By the next morning I had established the existence of a class of Fuchsian functions, those which come from the hypergeometric series; I had only to write out the results, which took but a few hours.

Then I wanted to represent these functions by the quotient of two series; this idea was perfectly conscious and deliberate, the analogy with elliptic functions guided me. I asked myself what properties these series must have if they existed, and I succeeded without difficulty in forming the series I have called theta—Fuchsian.

Just at this time I left Caen, where I was then living, to go on a geological excursion under the auspices of the school of mines. The changes of travel made me forget my mathematical work. Having reached Coutances, we entered an omnibus to go some place or other. At the moment when I put my foot on the step the idea came to me, without anything in my former thoughts seeming to have paved the way for it, that the transformations I had used to define the Fuchsian functions were identical with those of non—Euclidean geometry. I did not verify the idea; I should not have had time, as, upon taking my seat in the omnibus, I went on with my conversation already commenced, but I felt a perfect certainty. On my return to Caen, for conscience' sake I verified the result at my leisure.

Then I turned my attention to the study of some arithmetic questions apparently without much success and without a suspicion of any connection with my preceding researches. Disgusted with my failure, I went to spend a few days at the seaside, and thought of something else. One morning, walking on the bluff, the idea came to me, with just the same characteristics of brevity, suddenness and immediate certainty, that the arithmetic transformations of indeterminate ternary quadratic forms were identical with those of non—Euclidean geometry."

This passage is from Poincaré's essay "Mathematical Creation" in Poincaré [1918], translated by Halsted.

Of course, we do not know what ideas were actually in Poincaré's head, but we do know the ideas which existed in the literature of the time, and specifically in the literature which Poincaré acknowledges to have influenced him. This literature concerns elliptic functions, modular functions and triangle functions.

The theory of elliptic functions began with integrals such as

$$\int \frac{dx}{\sqrt{(1-x^2)(1-k^2x^2)}}$$

Hence the nomer

which arise, for example, in finding the arc length of an ellipse. These integrals were a stumbling block in the early history of calculus because they cannot be expressed in terms of elementary functions. They were subject to certain identities, however, such as Euler's addition formula (Euler [1768])

$$\int_0^a \frac{dx}{\sqrt{g(x)}} + \int_0^b \frac{dx}{\sqrt{g(x)}} = \int_0^c \frac{dx}{\sqrt{g(x)}}$$

where $g(x) = (1 - x^2)(1 - k^2x^2)$ and

$$c = \frac{b\sqrt{g(a)} + a\sqrt{g(b)}}{1 - k^2a^2b^2}.$$

Eventually it was realized, by Gauss around 1800, and for the first time in print by Abel and Jacobi in the 1820's, that these identities became much more intelligible in terms of the **inverse** functions of the integrals. It is completely analogous to the situation with integrals such as

$$\int_0^a \frac{dx}{\sqrt{1 - x^2}} = \sin^{-1} a$$

where the inverse function, sine, has much simpler properties. In particular, it is single—valued and **periodic**: $\sin(\theta + 2n\pi) = \sin\theta$ for $n \in \mathbf{Z}$. Indeed, as Gauss, Abel and Jacobi discovered, the elliptic functions have **double periodicity** as functions of a complex variable z. Associated with each elliptic function F are complex numbers ω_1, ω_2, with $\omega_1/\omega_2 \notin \mathbf{R}$, such that

Periods ω_1, ω_2 not linear over $\mathbb{R}$

(1) $$F(z + m\omega_1 + n\omega_2) = F(z) \quad \text{for } m,n \in \mathbf{Z}$$

Geometrically, this says that F is periodic with respect to the following tessellation of the complex plane (Figure 1)

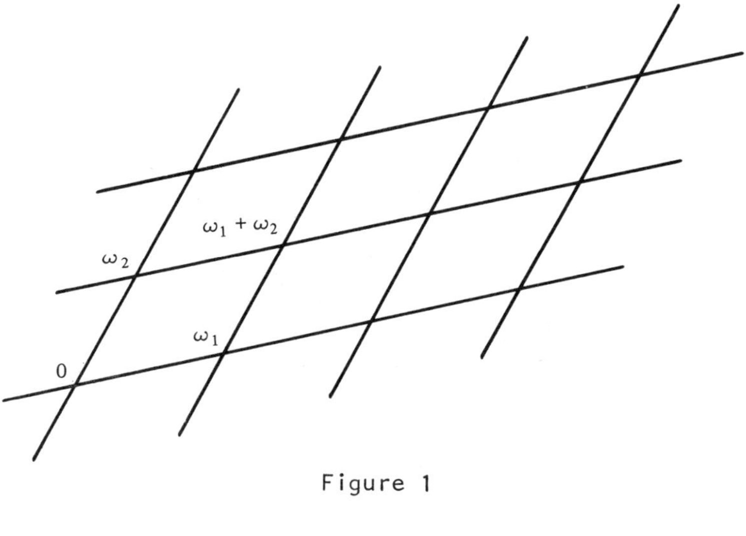

Figure 1

The condition $\omega_1/\omega_2 \notin \mathbf{R}$ says that the parallelograms of this tessellation are non-degenerate.

The reason for the double periodicity became completely clear with the introduction of Riemann surfaces by Riemann [1851]. The integrand f(z) of an elliptic integral has a two-sheeted Riemann surface with four branch points, which makes it topologically a torus (Figure 2).

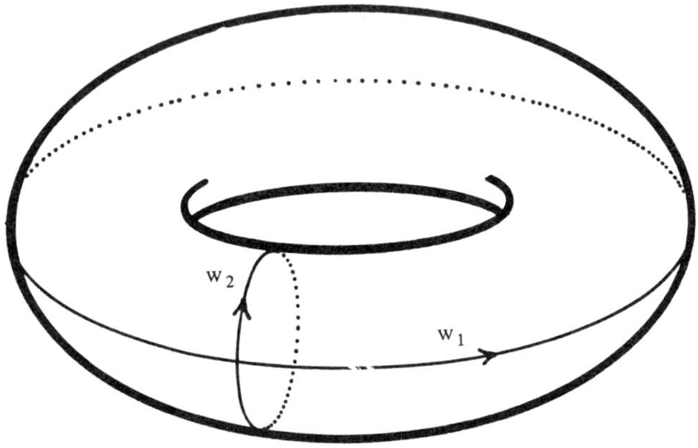

Figure 2

Integrals around closed paths on the torus are linear combinations of the integrals along the paths w_1, w_2, hence

$$\int_0^a f(z)dz$$

will be a "many-valued function" whose values are ambiguous up to a term $m\omega_1 + n\omega_2$, where $m,n \in \mathbf{Z}$ and ω_1, ω_2 are the integrals along w_1, w_2 respectively. The inverse function, F, therefore has the periodicity property (1).

One could not ask for a more satisfying explanation of the mysteries of elliptic integrals. It is as if an analyst innocent of geometry had begun with

$$\int \frac{dx}{\sqrt{1 - x^2}},$$

discovered sin and cos, and finally discovered the circle.

In fact, the intuitive topological torus we have used can be replaced by a precise complex manifold defined by the tessellation in Figure 1. It is just the quotient of the complex plane **C** by the group

$$\Omega = \{m\omega_1 + n\omega_2 \mid m,n \in \mathbf{Z}\}$$

or, less formally, the result of identifying opposite sides of any parallelogram in the tessellation.

Each elliptic function F is said to be **automorphic** with respect to its group Ω. The circular functions sin and cos are automorphic with respect to the even simpler group $\{2n\pi \mid n \in \mathbf{Z}\}$. However, one does not invent the automorphic function concept, or even notice that groups are involved, when faced with such simple examples. Before 1880, a handful of other functions with some kind of periodicity were known, and with hindsight one can see that they are automorphic with respect to certain groups. These are the elliptic modular functions and triangle functions. A modular function F, defined on the upper half plane, is automorphic with respect to the **modular group** of transformations

$$z \mapsto \frac{az + b}{cz + d}$$

where $a,b,c,d \in \mathbf{Z}$ and $ad - bc = 1$. If a,b,c,d satisfy these conditions then

$$F\left(\frac{az+b}{cz+d}\right) = F(z).$$

The modular group arises when considering equivalence of binary quadratic forms, and was known to Gauss in this context. It also arises when investigating the dependence of elliptic functions on the **moduli** ω_1, ω_2 (hence the name), because moduli ω'_1, ω'_2 generate the same group Ω as ω_1, ω_2 if and only if

$$\begin{bmatrix} \omega'_1 \\ \omega'_2 \end{bmatrix} = \begin{bmatrix} a & b \\ c & d \end{bmatrix} \begin{bmatrix} \omega_1 \\ \omega_2 \end{bmatrix}$$

where $a,b,c,d \in \mathbf{Z}$ and $ad - bc = \pm 1$. We shall explore this situation further in Section 8 in order to illustrate Poincaré's approach to spaces of groups.

Modular functions became fairly familiar after their use by Hermite [1858] to solve the fifth degree equation, and were intensively investigated by Dedekind and Klein in the 1870's. Klein [1878], in particular, emphasizes the role of groups and tessellations, though still with only a small range of examples at his disposal.

Triangle functions arise in connection with the hypergeometric differential equation of Gauss, as Gauss himself seems to have discovered. This can be guessed from Figure 3, which is an undated fragment found in his Nachlass (see Gauss [**Werke, Vol. 8,** pp 102–105]). The figure is a tessellation of the unit disc by curvilinear triangles with angles $\pi/4$, and it reflects the periodicity of the inverse to a certain quotient of solutions to the hypergeometric equation. The group in this case consists of certain linear fractional transformations which leave the unit disc invariant. Invariance under linear fractional transformations arises quite generally from second order linear differential equations as follows. If y_1, y_2 are two linearly independent solutions of

$$\frac{d^2y}{dx^2} + p(x)\frac{dy}{dx} + q(x) = 0 \tag{2}$$

then any other solutions y'_1, y'_2 are linear combinations of y_1, y_2:

$$y'_1 = ay_1 + by_2$$

$$y'_2 = cy_1 + dy_2$$

hence the quotient of the new solutions is related to the quotient of the old by a linear fractional transformation:

$$\frac{y'_1}{y'_2} = \frac{a\eta+b}{c\eta+a}$$

where $\eta = \frac{y_1}{y_2}$.

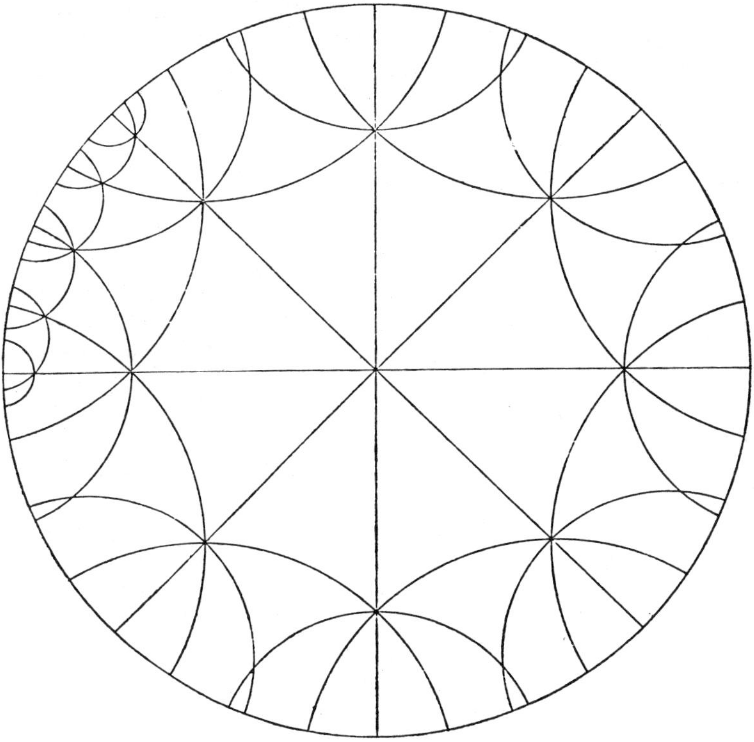

(Gauss: **Werke, Vol. 8**, p. 104)

Figure 3

By inverting the quotient of solutions one obtains a function automorphic with respect to a certain group of linear fractional transformations. In the case of the hypergeometric equations, one is led to groups of automorphisms of certain triangle tessellations of the unit disc. The Gauss figure is one such tessellation. Others were found by Riemann in lectures of 1858–59 (discovered only in 1897 and published in Riemann [1902]), and the general theory was worked out independently by Schwarz [1872]. Figure 4 is from Schwarz' paper,

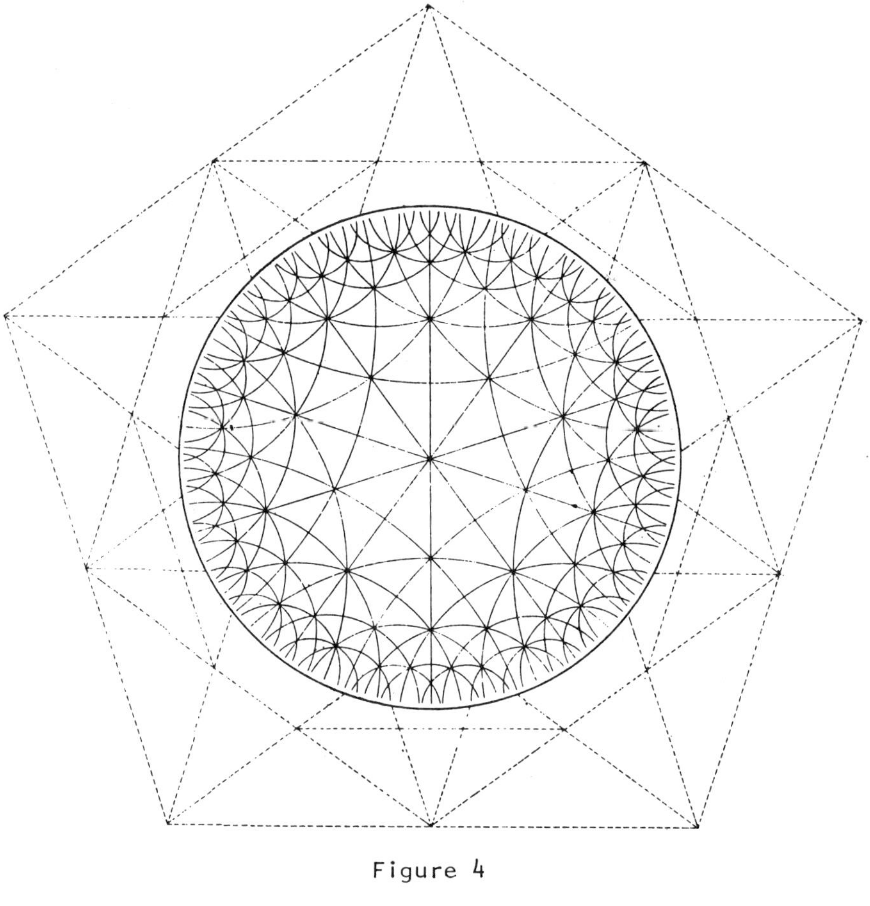

Figure 4

and is the first of this beautiful family of tessellations to appear in print. It is also almost unique in showing the construction lines outside the unit circle, which relate to the inversive properties of the tessellation. Each circular arc α in the figure divides the tessellation into two halves which are inverse with respect to the circle α.

Schwarz was familiar with the inversion symmetry of these tessellations from a previous encounter with them in the theory of conformal mapping. According to the famous mapping theorem of Riemann [1851], any plane region bounded by a simple closed curve can be mapped conformally onto the unit disc. Finding explicit mapping functions, however, is another matter, and the first non–trivial mapping, from the square onto the disc, was constructed by Schwarz [1869]. It uses an elliptic function, which is not surprising for, as Schwarz points out, one can equally well map the square onto the half plane, then use the Riemann–Schwarz reflection process to extend the mapping to successive reflections of the square in its sides. The mapping function must therefore be doubly periodic. In the same paper, Schwarz constructs conformal mappings of triangles bounded by circular arcs. The same reflection argument then shows that the mapping function is automorphic with respect to the group generated by inversions in the sides of the triangle.

Since inversion in a circle can be regarded as a generalization of reflection in a straight line, Schwarz came within a hair's breath of attributing geometric symmetry to the triangle tessellations. What he lacked was a notion of length which is invariant under the inversions. Such a notion had already beeen invented by Beltrami in his work on non–Euclidean geometry, but it was left to Poincaré to observe its invariance under suitable inversions and linear fractional transformations.

III. Non–Euclidean Geometry before Poincaré

The story of non–Euclidean geometry has often been told in books on geometry and the history of mathematics. I shall therefore assume that the reader is familiar with the story up to the time when the logical consistency of Bolyai–Lobachevsky (i.e., hyperbolic) geometry became generally accepted, say with the Beltrami [1868a] pseudosphere model. Most writers in fact stop there, having only a logical axe to grind. However, the pseudosphere model was not Beltrami's deepest contribution to hyperbolic geometry. Beltrami [1868b] gives the first constructions of the projective and conformal models of hyperbolic spaces, of all dimensions, relating different models and dimensions in a masterly way which supplies almost all the geometric results needed for the theory of automorphic functions. These models are usually attributed to Klein and Poincaré, and the only account of hyperbolic geometry giving adequate credit to Beltrami seems to be Milnor [1982]. Since Milnor's discussion is in n–dimensional generality, I shall briefly sketch Beltrami's results in 2 and 3 dimensions, the range which is relevant to Poincaré.

The starting point is a **hemisphere model** consisting of the open hemisphere

$$x_1^2 + x_2^2 + x^2 = a^2,\ x > 0 \tag{3}$$

in 3–dimensional (x_1,x_1,x)–space, with the line element

$$ds = R\,\frac{\sqrt{dx_1^2+dx_2^2+dx^2}}{x} \tag{4}$$

One could check that the geodesics for this metric are vertical sections of the hemisphere, however, this also follows from the second model, the "Klein disc", obtained by orthogonal projection onto the plane $x = 0$ (Figure 5). The model is now the open disc $x_1^2 + x_2^2 < a^2$, and Beltrami shows directly that its geodesics are straight line segments.

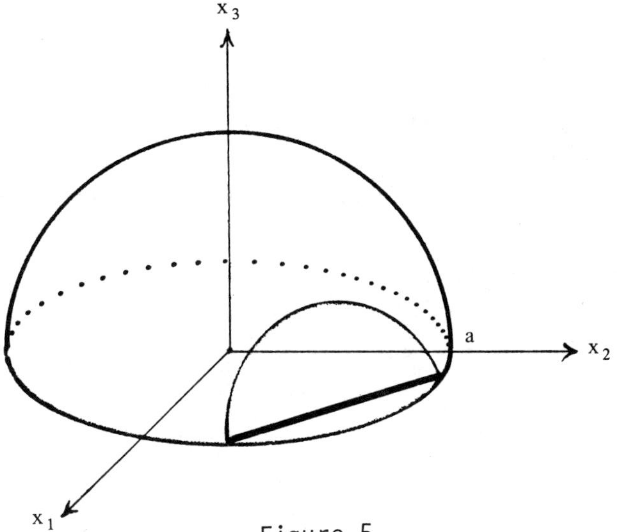

Figure 5

The third model is obtained by stereographic projection of the hemisphere onto its horizontal tangent plane (Figure 6)

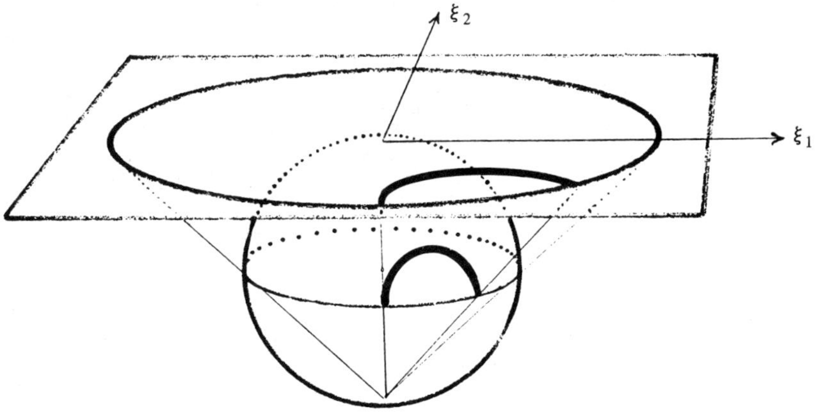

Figure 6

and the line element is computed to be

$$ds = \frac{\sqrt{d\xi_1^2 + d\xi_2^2}}{1 - \frac{1}{4R^2}\left[\xi_1^2 + \xi_2^2\right]} \tag{5}$$

a metric which was stated by Riemann [1854] to be of constant curvature. We observe that (5) is a multiple of the Euclidean line element $\sqrt{d\xi_1^2 + d\xi_2^2}$ in the (ξ_1, ξ_2)–plane; the multiple of course varies with position but not with direction, hence angles are preserved. Thus we have the conformal, or "Poincaré disc" model, 14 years before its appearance in Poincaré [1882a]. Presumably Riemann would have noticed the conformal property of his metric, but he missed the hyperbolic geometry, and hence was unable to exploit this geometry in complex analysis as Poincaré did. It is tempting to suppose that Beltrami is the missing link between Riemann and Poincaré.

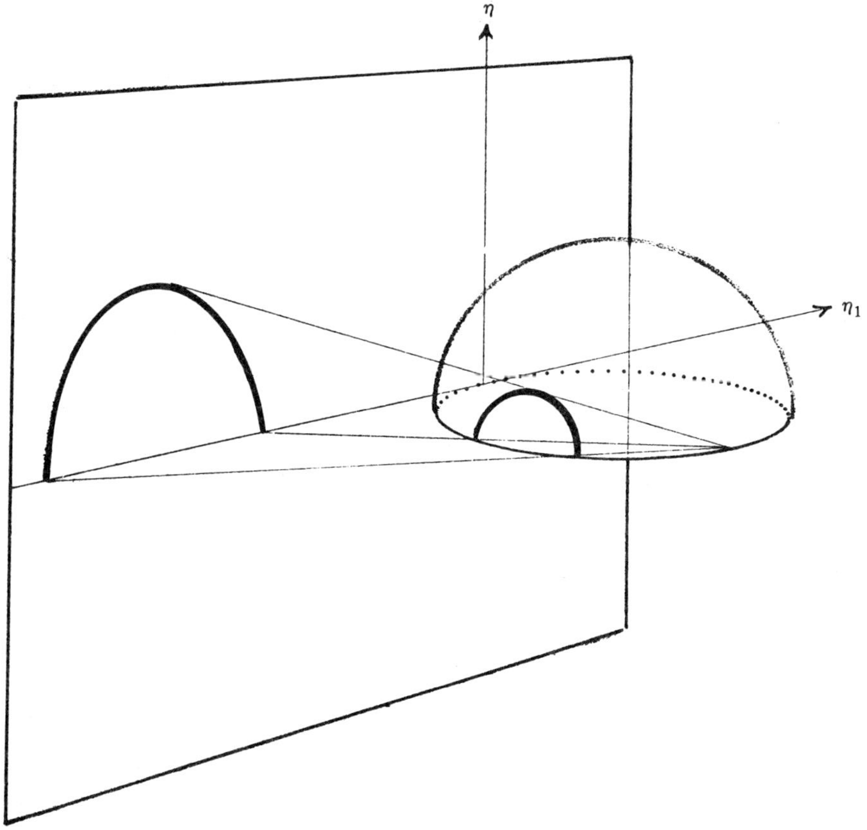

Figure 7

The fourth model is obtained by stereographic projection of the hemisphere onto (the upper half of) a vertical plane (Figure 7). This gives the line element

$$ds = R\,\frac{\sqrt{d\eta_1{}^2 + d\eta^2}}{\eta}$$

for the "Poincaré half plane", which Liouville [1850] had previously obtained by an ingenious transformation of the line element for the pseudosphere. Like Riemann, Liouville missed the hyperbolic geometry. The n–dimensional generalization of this line element, for the **half–space model** of n–dimensional hyperbolic geometry

(6) $$ds = R\,\frac{\sqrt{d\eta_1{}^2 + \ldots + d\eta_{n-1}{}^2 + d\eta^2}}{\eta}$$

is particularly interesting when n=3, because if we write (6) with $x_1 = \eta_1$, $x_2 = \eta_2$, $x = \eta$ we get back the line element (4) we started with. As Beltrami points out, this means that (3) and (4), his hemisphere model, can be viewed as the restriction of the half–space model of hyperbolic 3–space to a hyperbolic plane.

This then was the state of non–Euclidean geometry before Poincaré. As mentioned in section 3, his contribution was to note the invariance of the conformal metrics (5) and (6) under inversion, and to deduce interpretations of the rigid motions of 2– and 3–dimensional hyperbolic geometry as linear fractional transformations. This brought non–Euclidean geometry into the mainstream of mathematics for the first time, and also put complex analysis on a new geometric basis.

IV. The 1880 prize essay (unedited memoir)

As N.E. Nörlund writes in his note at the beginning, this essay helps us to see Poincaré's ideas as they first developed, and also the speed with which he worked. Poincaré wrote it as part of his entry in the 1880 Paris Academy competition, the object of which was "to perfect, in some important point, the theory of linear differential equations of a single variable." As he explains, it arose from his reading of Fuchs [1880] paper on the functions which result from inversion of integrals of linear differential equations. The essay was completed less than a month after he read Fuchs and shows some signs of haste, for instance the incorrect Figure 1. Instead of correcting it, Poincaré simply asks the reader to ignore the incorrect part! Under the terms of the prize question, and no doubt also under the influence of Fuchs' paper, Poincaré begins in the wilds of differential equations. After a long hack through this jungle, he finally emerges into the light of geometry.

Fuchs' theory receives a sudden illumination analogous to the illumination of elliptic integrals by the parallelogram tessellation (Figure 1). A differential equation has certain "singular points" analogous to the branch points of an elliptic integral. If a solution of the differential equation is analytically continued round a singular point, it will in general not return to its starting value. A quotient of two solutions, as we saw in Section 2, will in fact undergo a linear fractional transformation. Inversion of the quotient then leads to a function F which is invariant under a certain group G of linear fractional transformations. But is F really a function, i.e., single-valued? To find out one must examine the fundamental polygon R_0 of the group G, which is analogous to the fundamental parallelogram of the group Ω of an elliptic function.

The polygon R_0 is transformed into other polygons R_1, R_2, ... by the elements of G. The function F can be analytically continued from each R_i to the neighboring R_j, hence it will be single—valued provided these regions never overlap. The possibility of local overlapping, as in Figure 8, was ruled out by Fuchs. To do this one only needs to know the polygon angles or their analytic counterpart, the difference between the roots of the indicial equation (called the "determinant equation" by Poincaré). Fuchs himself proceeded analytically, but the geometric significance of the indicial equation was already understood

Figure 8

Figure 9

by Riemann and Schwarz in their papers on the hypergeometric equation. The problem overlooked by Fuchs was the possibility of global overlapping, as in Figure 9. Poincaré's contribution was to recognize the fundamental importance of this problem, and to provide the geometric insight needed to solve it.

The polygons Poincaré is working with are bounded by arcs of circles orthogonal to a fixed circle, like those in the Gauss and Schwarz tessellations (Figures 3 & 4), and hence they are true polygons in the conformal model of the hyperbolic plane. Without mentioning this non—Euclidean interpretation — perhaps he did not notice it yet — Poincaré transforms the curvilinear polygons of the conformal model into rectilinear polygons of the projective model by stereographically projecting the plane onto the sphere, then orthogonally projecting the sphere back onto the plane (as Beltrami [1868b] had done). He then observes that, although angles are not preserved by this process, **the property of being $< \pi$ is preserved**, hence curvilinear polygons with angles $< \pi$ are transformed into rectilinear polygons with angles $< \pi$. The latter polygons are convex and hence Poincaré has proved a new proposition of hyperbolic geometry: a polygon is convex if and only if its interior angles are $< \pi$.

Using this proposition, Poincaré proves a special case of the non—overlapping property: regular quadrilaterals with interior angles of 60° tessellate the hyperbolic plane without overlapping. His method of proof is to construct the tessellation as the union of nested subtessellations $T_0, T_1, T_2, \ldots$

$$T_0 = R_0$$

$$T_{n+1} = T_n \cup \{\text{quadrilaterals adjacent to } T_n\}.$$

A trivial induction shows that all angles on the boundary of T_{n+1} are either 60° or 120°, hence T_{n+1} is convex by the proposition. Thus there is no overlapping in T_{n+1}, for any n, and hence the whole tessellation is without overlapping.

The general non—overlapping result, which has become known as

Poincaré's theorem, is in Section VI of his first Acta paper and says essentially that any hyperbolic polygon with the obvious necessary conditions on its angles and side lengths generates a non-overlapping tessellation. Poincaré was unable to use the convexity argument in the general case, and his proof is rambling and less convincing (see next section). His investigation of the special case was enough, however, to show him the right place to begin a theory of automorphic functions — in non-Euclidean tessellations and their groups. In the three long Acta papers which followed in 1882 and 1883, Poincaré developed the theory from this starting point, and the main lines of his exposition have been followed in textbooks down to the present day, notably Fricke and Klein [1897,1912], Ford [1929] and Lehner [1964].

V. The first Acta paper: Theory of Fuchsian Groups

The theory of regular tessellations is much more complicated in the hyperbolic plane than in the Euclidean. In the Euclidean plane there are only finitely many essentially distinct types of regular tessellation, and those with finite fundamental region ("tile") are all akin to the parallelogram tessellation. The hyperbolic tessellations, on the other hand, come in an infinitude of types which cannot be adequately surveyed until some familiarity with hyperbolic geometry has been established. Poincaré covers the geometry in his Sections I and II, using Beltrami's half plane model in order to have the simplest formulae possible. Viewing the half plane as the upper half of the complex plane, the hyperbolic rigid motions are precisely the linear fractional transformations

$$z \mapsto \frac{az+b}{cz+d} \text{ with } a,b,c,d \in \mathbf{R} \text{ and } ad - bc > 0.$$

At the end of the paper (Section XII), Poincaré carries his theory over to the unit disc by mapping the half plane onto it by $z \mapsto \frac{z+i}{z-i}$. This reduces hyperbolic geometry to a theory of elementary complex functions, and Poincaré can with some justification dispense with the classical view of non–Euclidean geometry as an axiomatic system. Unfortunately he bends over backwards to avoid mentioning non–Euclidean length and area, even when he is using them, and this sometimes hides his motivation.

Section III introduces the group theory needed to discuss regular tessellations. Intuitively, one begins with a tessellation T and considers the group G of motions which map T onto itself. Technically, it is simpler to define G first, as a group of linear fractional transformations acting properly discontinuously on the half plane. This is what Poincaré does, calling such groups **Fuchsian** and motivating proper discontinuity as a necessary condition for the existence of a nonconstant analytic function automorphic with respect to G. But after this one faces a long haul back to the tessellation construction of G. A fundamental region for G is obtained (Section III), cut and pasted into one with hyperbolically straight sides (Section IV), and finally a side pairing of any suitably shaped polygon is shown to define a Fuchsian group G (Section VI). The arguments are not very rigorous, and the result of Section VI in particular ("Poincaré's theorem") received acceptable proofs only recently (see Maskit [1971] and Beardon [1983]).

Once one has a fundamental polygon R_0, interesting combinatorial properties of G are easily derived. The side pairing operations, repeatedly applied, enable R_0 to be transported to any other polygon R_i in the tessellation, hence these finitely many transformations **generate** G. Relations in G correspond to closed chains of polygons, and these in turn reduce to **defining relations** corresponding to circuits around the vertices of the tessellation. Each such circuit corresponds to a **vertex cycle** of R_0, i.e., an equivalence class of vertices identified by the side pairing, of which there are of course only finitely many. Thus G is **finitely presented** – it can be defined abstractly by finitely many generators and relations. Although Poincaré pointed these facts out (Sections III, X), he did not write down any defining relations until his fourth Acta paper. It was left to Klein to obtain the general form of the defining relations of a Fuchsian group in Klein [1882a]. Independently of Poincaré, Klein's student Dyck [1882] developed a theory of finitely presented groups, and in particular the Klein [1878] construction of subgroups of G by uniting polygons in the tessellation for G. Dyck's paper pushes group theory in a direction probably not anticipated by Poincaré. He finds non–trivial results with purely combinatorical content, laying the foundations of the future combinatorial group theory. A substantial part of this theory still draws its inspiration from Fuchsian groups, as may be seen from Magnus [1974].

The difference between Poincaré and Klein is well illustrated by their attitudes to **genus**, which Poincaré introduces in Section VIII. If one identifies the paired sides of a fundamental polygon R_0 for G, one obtains a surface S whose genus p (i.e., number of handles) is called the genus of G. Poincaré uses the number p to classify G, and in his second Acta paper shows how it reappears as the algebraic genus of relations between automorphic functions for G, but he does not use the surface S itself. Klein, on the other hand, uses the topological classification of surfaces to find a way of cutting S down to a canonical fundamental polygon R'_0 from which he derives his canonical presentation for G. He also uses S to obtain automorphic functions for G, whereas Poincaré constructs them directly from G, as we shall now see.

VI. The second Acta paper: On Fuchsian functions

Having established the class of Fuchsian groups as those for which automorphic functions should first be sought, Poincaré asks whether an automorphic function F can be constructed for any Fuchsian group G. Thus F is required to satisfy $F(f_i(z)) = F(z)$ for each $f_i \in G$. If instead one has the Euclidean group

$$\Omega = \{m\omega_1 + n\omega_2 \mid m,n \in \mathbf{Z}\},$$

then it is easy to define an automorphic function for Ω, i.e., an elliptic function, by infinite series. For example, the Weierstrass pe function

$$P(z) = z^{-2} + \sum_{\substack{\omega\in\Omega \\ \omega\neq 0}} [(z-\omega)^{-2} - \omega^{-2}]$$

trivially satisfies $P(z+\omega) = P(z)$ for each $\omega \in \Omega$, and it is convergent except for isolated poles by choice of the exponent -2. No such invariant yet convergent series comes to mind for a Fuchsian group G. However, Poincaré's eagle eye for the right analogy saw another aspect of elliptic functions which could be imitated — their representation as quotients of theta functions (Jacobi [1838]). Briefly, a theta function θ for a group Ω has the property that, for each $\omega \in \Omega$,

$$\theta(z+\omega) = k(z,\omega)\theta(z)$$

for a certain factor k. One can find θ_1 and θ_2 with the samc k, so that θ_1/θ_2 is invariant under the group Ω, and hence an elliptic function.

Poincaré therefore began with the goal of finding "theta Fuchsian" functions Θ with the property

(7) $$\Theta(f_i(z)) = k_i(z)\Theta(z)$$

for each element f_i of a given Fuchsian group G. He constructs such Θ, which are now known as **automorphic forms**, in Section I of this paper. Each Θ is defined by what is now known as a **Poincaré series**

$$\Theta(z) = \sum_{f_i \in G} H(f_i(z)) \left[\frac{df_i(z)}{dz}\right]^m =$$

(8)

$$= \sum_{f_i \in G} H(f_i(z))(\gamma_i z + \delta_i)^{-2m}$$

where $f_i(z) = (\alpha_i z + \beta_i)/(\gamma_i z + \delta_i)$, H is any rational function, and m is a positive integer. The series turns out, for $m \geq 2$, to be absolutely convergent in the unit disc on which G acts, except for isolated poles, essentially as a result of the finite area of this disc. The geometric core of Poincaré's proof is the fact that

(9)
$$\sum \left|\gamma_i z + \delta_i\right|^{-2m} = \sum \left|\frac{df_i(z)}{dx}\right|^m$$

and for $m \geq 2$ the latter series can be compared with the total area of disjoint f_i–images of a certain region. Since these images all lie within the unit disc, their total area is finite and hence the series (8) converges. The convergence of (8), except at poles of the $H(f_i(z))$, follows from this, and hence $\Theta(z)$ is a meromorphic function within the unit disc.

Poincaré also gives a second convergence proof, based on **non–Euclidean** area. Modern versions of both proofs can be found in Lehner [1964].

The characteristic theta property (7) follows from (8) by a simple calculation (Section II):

$$\Theta(z) = \sum_{f_i \in G} H(f_1(z)) \left[\frac{df_i(z)}{dx}\right]^m$$

hence

$$\Theta(f_k(z)) = \sum_{f_i \in G} H(f_i(f_k(z)) \left[\frac{d f_i f_k(z)}{dz}\right]^m =$$

$$= \left[\frac{d f_k(z)}{dz}\right]^{-m} \Theta(z) = (\gamma_k z+\delta_k)^{2m}\Theta(z)$$

using the chain rule and the fact that, as f_i runs over G, so does $f_i f_k$. If Θ_1 and Θ_2 are two Poincaré series for G with the same m value, then $F(z) = \Theta_1(z)/\Theta_2(z)$ is the desired automorphic function. One obtains a large supply of automorphic functions by different choices of H and making suitable rational combinations of Θ's so that the factors $(\gamma_k z + \delta_k)^{2m}$ cancel out.

The basic properties of automorphic functions, which Poincaré derives in Section IV, do not depend on any properties of Poincaré series. He proves them by general function theoretic arguments in the style of Riemann. After a study of the zeroes and poles of Θ functions, Poincaré finds that an automorphic function for G has equal numbers of zeroes and poles in a fundamental polygon for G, and a now-standard sequence of results follows in rapid succession. Any two automorphic functions for G are algebraically related, all automorphic functions for the same G are expressible rationally in terms of two particular ones x,y and the genus of the algebraic relation between x and y is the genus of G. Finally, the inverse of an automorphic function is expressible as the quotient of two solutions of a second order linear differential equation. Singular points of this differential equation correspond to the vertices of a fundamental polygon for G, and the difference between the roots of the indicial equation at a singular point equals the angle sum of the corresponding vertex cycle, divided by 2π. Here we can recognize Poincaré retracing the steps to his starting point, though his proof is in fact extremely condensed and unmotivated. The linear differential equation is derived with bewildering speed from a nonlinear differential equation involving the Schwarzian derivative. A glimpse into the history behind this transformation is provided by Schwarz [1890, p.35], commenting on a special case of it in his own work, Schwarz [1869], and a more special case in Lagrange [1779], where the Schwarzian derivative and the complex analytic theory of conformal mapping has its origin.

Since the properties of automorphic functions can be derived by general arguments, the main virtue of Poincaré series is to prove **existence** of automorphic functions. The rival theory of Klein [1882] was weaker than Poincaré's in this respect, since it was based on the then unproved Dirichlet's principle (though he could in fact have used the rigorous Schwarz–Neumann proof of the Riemann mapping theorem). Klein used this quasi–physical principle to assert the existence of analytic functions F_0 on the surface S obtained by identifying paired sides in a fundamental polygon R_0 for G. An automorphic function F for G is then obtained as the obvious continuation of F_0 to the other polygons R_1, R_2, ... in the tessellation for G.

VII. The third Acta paper: Memoir on Kleinian groups

Since the hyperbolic plane is the disc or half plane in the conformal model, Fuchsian groups are rather special groups of linear fractional transformations, in the sense that they leave the unit circle (respectively, real axis) invariant. This line contains the limit set of the Fuchsian group G in question.
A point p is in the limit set just in case p has a neighborhood containing infinitely many G–images of a particular point. Groups of linear fractional transformations which act discontinuously on parts of the complex plane with more

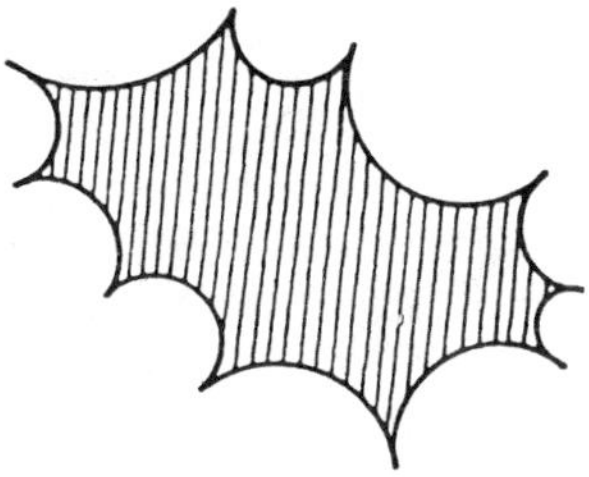

Figure 10

complicated limit sets were given by Schottky [1877], and Poincaré first learned of them from Klein in a letter (Klein [1881]). Klein's example is a group generated by reflection (in the general sense of inversion) in the sides of a zero-angled polygon whose sides are circular arcs (Figure 10).If the sides of this polygon happen to be orthogonal to a circle C, then the vertices will lie on C, the reflections will fill C and the group will be Fuchsian with C as its limit set. If not, the limit set will be a more complicated simple closed curve, though the tessellation will be combinatorically the same and hence define an isomorphic group (called a "deformation" of a Fuchsian group). By starting with more complicated polygons (with holes, for example) one can generate groups not at all like Fuchsian groups, with very complicated limit sets. To bring some sort of order into this class of groups, Poincaré initiated a general theory of groups of linear fractional transformations which act discontinuously on the plane. Those which were not Fuchsian he called **Kleinian**.

The general theory of Kleinian groups still lags far behind that of Fuchsian groups, but the fundamental step was made by Poincaré in this memoir — the leap into the third dimension. A Kleinian group action on the plane turns out to be just the shadow of a discontinuous group of three-dimensional non-Euclidean motions. Since Fuchsian groups are discontinuous groups of two-dimensional non-Euclidean notions, the difference between Fuchsian and Kleinian can be understood as the difference between two and three dimensions. This is of course a vast difference, and the theory of Kleinian groups after Poincaré had to wait for further progress in three-dimensional geometry and topology, much of it coming only recently.

The details of Poincaré's idea are stunningly simple and are presented in Section I of his memoir. He used the Beltrami half-space model H^3 of three-dimensional hyperbolic geometry and views the complex plane as its "sphere at infinity", S^∞. A motion of H^3 is a product of an even number of inversions in spheres S orthogonal to S^∞. Each such inversion is uniquely determined by the inversion of S^∞ in the circle $S \cap S^\infty$, and the product of an even number of inversions in the plane is a linear fractional transformation. Conversely, any linear fractional transformation of the plane is the product of an even number of inversions, and this sequence of inversions in S^∞ uniquely determines a sequence of inversions, and hence a motion, of H^3.

Apart from its value in reasserting the fact that non–Euclidean geometry occurs naturally in complex analysis, Poincaré's three–dimensional interpretation of Kleinian groups is also an important technical tool. To prove an analogue of the "Poincaré theorem" of Section IV of the first Acta paper, Poincaré here (Section V) needs to pass to three–dimensional space, because of the uncontrollable way in which planar tessellations grow from a Kleinian fundamental polygon.

The pathology of Kleinian groups is the subject of a second major result of the memoir (Section VIII). Poincaré investigates the limit set of a group generated by a polygon like Klein's (Figure 10). Although this group is merely a deformation of a Fuchsian group and its limit set is a simple closed curve, Poincaré finds that the curve is nonanalytic ("if one can call it a curve" he says) and fails to have a radius of curvature at a dense set of points. He conjectures that it is in fact nondifferentiable at a dense set of points, and this was proved by Fricke and Klein [1897, pp.405–428]. It is perhaps the first "naturally occurring" nondifferentiable curve in mathematics. Mandelbrot [1983] has reexamined it in the context of his theory of fractals, and criticized Fricke and Klein's attempt to draw the curve.

VIII. The fourth Acta paper: On the groups of linear equations

While large parts of the first three Acta papers have been assimilated and have found their way into textbooks, the fourth remains mysterious and difficult to assess (at least for me). In part this is due to Poincaré's regression to the pre–geometric theory of differential equations in the early paragraphs (Sections I–V) but also, when he finally reduces his problems to topology (Section VIII) and geometry (Section XII) he finds himself facing high–dimensional manifolds too elusive to be properly grasped by the mathematical tools of his time. His arguments become vague and careless but are nevertheless highly stimulating. Since tools are now available for dealing with this type of problem, the fourth Acta paper is the one most likely to be of interest to research mathematicians. Some of its problems have certainly been solved since Poincaré wrote it, others have not, and others again await a precise interpretation.

The basic problem is stated in terms of rather technical questions above differential equations, but in broader terms it is the problem of uniformization. The word "uniform" here is opposed to "multiform", in the context of using single-valued functions in place of multi-valued functions. For example, on the curve

$$x^2 + y^2 = 1 \tag{10}$$

the variable y is a 2-valued function, $\sqrt{1 - x^2}$, of x. One uniformizes the curve by expressing x and y as single-valued functions of a third variable t, for example

$$x = \frac{2t}{1+t^2}, \quad y = \frac{1-t^2}{1+t^2}.$$

Uniformization of curves makes more sense, like everything else, when one allows complex values of the variables, so that an algebraic curve becomes a Riemann surface. The curve (10) then becomes a Riemann surface of genus 0, and it was known that curves of genus 0 could be uniformized by rational functions (as exemplified by the uniformization above). It was also known that curves of genus 1 could be uniformized by elliptic functions, and Poincaré and Klein conjectured that curves of higher genus could be uniformized by automorphic functions. Poincaré also states the analogous uniformization problem, and conjecture, for solutions of linear d.e.'s in this paper (Section V).

The uniformization conjecture was eventually proved by Poincaré [1907] and Koebe [1907]. Their proofs depend in part on giving Klein's approach to automorphic functions a rigorous foundation. As mentioned in Section 6, this foundation was not perceived in the 1880's, and Poincaré first attacked uniformization by a different method, the so-called method of continuity (Section VIII). Ironically, this method runs into worse difficulties, which were only resolved by

Brouwer's topological theorems on the invariance of dimension and domain (Brouwer [1911], [1912]. See also the notes of Freudenthal in Brouwer's [Works 2, p.572]). Leaving these topological difficulties aside, however, Poincaré has the very fruitful idea of considering spaces of Fuchsian groups (now known as Teichmüller spaces) and he makes a bold attempt to investigate their boundaries. Just recently, a better understanding of these boundaries has been gained, from the work of Thurston [1977], Bers [1978] and Kerckhoff [1979], so we finally have some perspective in which to view Poincaré's work. Since this recent research is highly technical — to master it one needs to absorb the books by Laudenbach, Fathi and Poenaru [1979] and Abikoff [1980] — I shall confine myself to illustrating the theory of Teichmüller space in the case of genus 1, concentrating on those aspects which happen to generalize (though the proofs are nontrivial) to higher genus.

Let us return, therefore, to the construction of the torus in Section 2. We take two complex numbers ω_1, ω_2 with $\omega_1/\omega_2 \notin \mathbf{R}$ and identify opposite sides of the parallelogram with vertices 0, ω_1, ω_2, $\omega_1+\omega_2$. The resulting surface can be viewed topologically as in Figure 11, but it is important to endow the torus with the complex

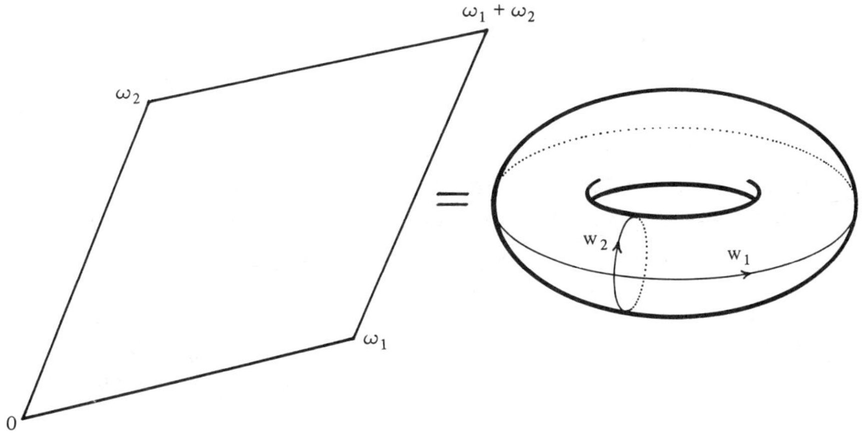

Figure 11

structure lifted from the plane, so that geometrically the torus is "flat" and completely determined by the data ω_1, ω_2. We also wish to retain the specific curves w_1, w_2 on the torus, where the parallelogram sides are joined, as a "marking". The reason for this is that the space of marked tori turns out to be simpler than the space of unmarked tori. One obtains an unmarked torus as the quotient of the plane **C** by the group $\Omega = \{m\omega_1 + n\omega_2 \mid m,n \in \mathbf{Z}\}$. Formally, one has erased the marking by giving the whole group Ω rather than a particular pair of generators ω_1, ω_2.

The Teichmüller space of marked tori is now constructed as follows. We do not distinguish between tori which differ only in scale, so each marked torus can be represented by the single complex number $\tau = \omega_1/\omega_2$. We also choose one of τ, $-\tau$, i.e., an orientation of the torus, so τ can be restricted to lie in the upper half plane H^2. Conversely, each $\tau \in H^2$ determines a marked torus, up to change of scale, with a choice of orientation, and hence the space of marked tori is precisely H^2.

Now let us consider how the different markings of the same torus are related. As we observed in Section 2, two sets of generators $\{\omega_1, \omega_2\}$ and $\{\omega'_1, \omega'_2\}$ of the same group Ω are related by

$$\begin{bmatrix} \omega'_1 \\ \omega'_2 \end{bmatrix} = \begin{bmatrix} a & b \\ c & d \end{bmatrix} \begin{bmatrix} \omega_1 \\ \omega_2 \end{bmatrix}$$

where $a,b,c,d,e \in \mathbf{Z}$ and $ad - bc = \pm 1$. The group of these matrices is called the **extended modular group** $\bar{\Gamma}$; the subgroup of matrices with $ad - bc = 1$ (which give the orientation—preserving changes of marking) is the **modular group** Γ. $\bar{\Gamma}$ acts on H^2 by sending $\tau = \omega_1/\omega_2$ to $\tau' = \frac{a\tau+b}{c\tau+d}$. This is in fact how $\bar{\Gamma}$ is generally known, as the group symmetries of the famous modular tessellation (Figure 12, taken from DuVal [1973]). Each region of this tessellation contains a single marking of each unmarked torus, and hence can be viewed as the space of unmarked tori.

Figure 12

Let us consider the particular fundamental region with vertices 0, i, $e^{i\pi/3}$. Figure 13 shows some of the marked tori associated with points in this region. At 0, the torus degenerates to the single curve w_2. In general, a fundamental region of the tessellation has a cusp at a rational point, say n/m in lowest terms, at which the torus degenerates to the curve $mw_1 + nw_2$.

The changes of marking effected by elements of $\bar{\Gamma}$ can be associated with mappings of the torus in the following way. Up to isotopy ("rubber sheet deformation") a homeomorphism of the torus is determined by the images of the curves w_1, w_2. These curves, in turn, can be deformed into a

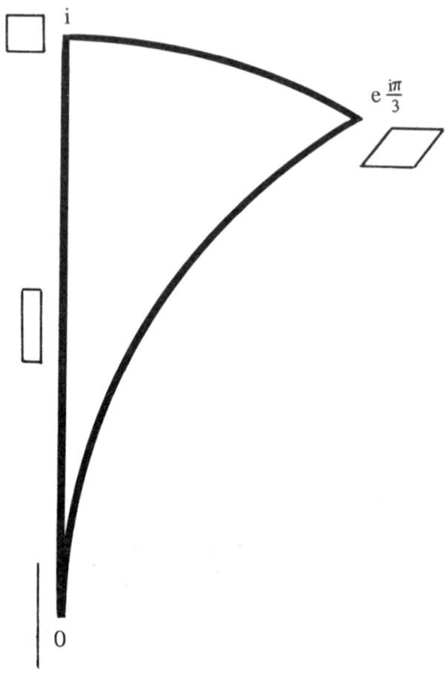

Figure 13

marking. Thus the isotopy class of a homeomorphism h, the so–called **mapping class** [h], corresponds to an element of the extendedmodular group $\bar{\Gamma}$ acting on H^2. But when H^2 is viewed as the hyperbolic plane, these elements are **isometries**, hence we can classify mapping classes of the torus according to the type of isometry they induce on H^2.

If we confine attention to orientation–preserving mapping classes, and hence to motions of H^2, then we have a three part classification:

1. **Rotations**. Each rotation fixes a point τ of H^2, in fact a vertex of the tessellation. The mapping class can therefore be realized by an isometry of the torus τ. (Nontrivial isometries are indeed evident for the square and equilateral tori at the vertices in Figure 13.)

2. **Limit rotations**. Each of these fixes a point on **R**, in fact a rational r. If r = n/m in lowest terms then the mapping class can be realized by a mapping which fixes the curve $mw_1 + nw_2$. It is a

"twist" obtained by cutting along this curve and giving one side of the cut an integral number of twists relative to the other before rejoining.

3. **Translations**. Each of these fixes two points x_1, x_2 on **R**, which are irrational. The mapping class can be realized by a mapping with infinite invariant curves on the torus, corresponding to x_1, x_2. These infinite curves are the limits of a suitable closed curve (corresponding to a rational r between x_1 and x_2) under iteration of the mapping and its inverse (since r moves between the limiting positions x_1 and x_2 under iteration of the translation and its inverse).

The theory of $\bar{\Gamma}$ and its interpretation in hyperbolic geometry was part of the classical theory of elliptic functions. The association with mappings of the torus was at least implicit in this theory, but did not become completely clear until overtaken by Thurston's stunning generalization to surfaces of arbitrary genus (Thurston [1977], Bers [1978]). Thurston's theory is actually the culmination of a long series of investigations, beginning with Poincaré. I shall try to give the gist of the theory by analogy with the torus case, pointing out where Poincaré contributed to it in his fourth Acta paper.

If S is a surface of genus $p > 1$, then S can be constructed by pairing sides of a 4p—gon R_0 in the hyperbolic plane, for example diametrically opposite sides (Section XII). By varying the shape of R_0 one obtains a space M_1 of marked surfaces S (Section XII), i.e., Teichmüller space. There is a "modular" group (the mapping class group) which acts on M_1. Poincaré defined it in terms of transformations of fundamental polygons, i.e., changes of marking. In Section XIII he gives some particular transformations of the 4p—gon for the surface of genus p, and seems to imply that they generate all changes of marking. Unfortunately, Poincaré's set of transformations seems to be incomplete, otherwise one would have a proof that the mapping class group is finitely generated. The latter theorem was claimed by Fricke and Klein [1897, p.291], but their proof is also incomplete. The first correct proof seems to be Dehn [1938].

Fricke and Klein [1897] proved that the dimension of M_1 is 6p—6. The subject was then buried until Teichmüller [1939] rediscovered it and made decisive progress. He found a connection with quadratic differentials and quasiconformal mappings which led to a natural metric on M_1, under which the mapping class group acts as a group of isometries. The metric also enabled Teichmüller to show that M_1 is homeomorphic to $\mathbf{R}^{6p-6}$. Unfortunately, M_1 is **not** a hyperbolic space under the Teichmüller metric. The surprise is that the mapping class motions are qualitatively just like the three types in H^2, and their fixed points have analogous interpretations.

This comes from Thurston's contribution, the construction of boundary "sphere at infinity" of Teichmüller space to which the action of the mapping class group extends continuously. It contains a dense set of "rational" points corresponding to systems of simple closed curves on S, and the irrational points are infinite curve systems to which closed curves tend under iteration of the mapping or its inverse. The set of rational points, and the action of the mapping class group on it, actually appears in Dehn [1938], but as a completely self-contained representation of the mapping class group. It was left to Thurston to identify simple curve systems as belonging to the boundary of Teichmüller space, and to understand the nature of the other boundary points.

Poincaré's investigation of the boundary of Teichmüller space was intended to serve as his first approach to the uniformization theorem. As we have said, this approach became obsolete in 1907. Now that the boundary of Teichmüller space is important again, it might be worth reexamining the "limit polygons" (Section XII) in Poincaré's boundary and relating them to the Thurston boundary.

It appears that Poincaré is considering the degeneration of unmarked polygons. Bill Abikoff has pointed out to me that, if so, the limit polygons correspond to "pinched" Riemann surfaces, which form a boundary of moduli space first described by Mumford and Mayer (about 1965, unpublished). Moduli space is the quotient Teichmüller space/mapping class group, which results from "forgetting the marking". If, instead, marked polygons degenerate, one can get the regions considered by Nielsen [1929], which turn out to correspond to Thurston's boundary points.

REFERENCES

Abikoff, W., **The Real Analytic Theory of Teichmüller Space**, Springer–Verlag, New York, 1980.

Beardon, A., **The Geometry of Discrete Groups**, Springer–Verlag, New York, 1983.

Beltrami, E., Saggio di interpretazione delle geometria non–Euclidea, **Giornale di Math.**, **VI** (1868), 284–312.

Beltrami, E., Teoria fondamentale degli spazii di curvatura costante, **Annali di Math. Ser. II**, (1868), 232–255.

Bers, L., An extremal problem for quasiconformal mappings and a theorem of Thurston, **Acta Math.**, **141** (1978), 73–98.

Brouwer, L.E.J., Beweis der Invarianz der Dimensionzahl, **Math. Ann.**, **71** (1911), 161–165.

Brouwer, L.E.J., Beweis der Invarianz des n–dimensionales Gebiets, **Math. Ann.**, **71** (1912), 305–313.

Brouwer, L.E.J., **Collected Works, Vols. 1,2**, North Holland, 1976.

Dehn, M., Die Gruppe der Abbildungsklassen, **Acta Math.**, **69** (1938), 135–206.

DuVal, P., **Elliptic Functions and Elliptic Curves**, London Math. Society Lecture Notes, (1973).

Dyck, W., Gruppentheoretischen Studien, **Math. Ann.**, **20** (1882), 1–44.

Euler, L., **Institutionum Calculi Integralis**, Petropoli, 1768.

Ford, L., **Automorphic Functions**, McGraw–Hill, New York, 1929.

Fricke, R. and Klein, F., **Vorlesungen über die Theorie der automorphen Functionen I, II**, Teubner, Leipzig, 1897, 1912.

Fuchs, L., Uber eine Klasse von Funktionen mehrerer Variablen welche durch Umkehrung der Integrale von Lösungen der linearen Differentialgleichungen mit rationalen Coeffizienten entstehen, **J. f. reine u. angew. Math.**, **89** (1880), 150–169.

Gauss, C.F., **Werke**, Königliche Gesellschaft der Wissenschaften, Göttingen, 1900.

Hermite, C., Sur la résolution de l'équation du cinquième degré, **C.R. de l'Acad. des Sciences, XLVI** (1858), 508–513.

Jacobi, C.G.J., Theorie der elliptischen Functionen aus den Eigenschaften der Thetareihen abgeleitet, **Gesammelte Werke, I** (1838), 497–538.

Kerckhoff, S., The asymptotic geometry of Teichmüller space, **Topology, 19** (1979), 23–41.

Klein, F., Über die Transformation siebenter Ordnung der elliptischen Funktionen, **Math. Ann., 14** (1878), 428–471.

Klein, F., Letter to Poincaré, 2 July 1881, **Ges. Math. Abh., Vol.III** (1881), 596–599.

Klein, F., Neue Beiträge zur Riemannschen Funktionentheorie, **Math. Ann., 20** (1882), 141–218.

Koebe, P., Zur Uniformisierung der beliebiger analytischer Kurven, **Gött. Nach.**, (1907), 191–210.

Lagrange, J.L., Sur la construction des cartes geographiques, **Nouveaux Memoires de l'Academie Royale des Sciences et Belles–Lettres**, (1779), 161–185.

Laudenbach, F., Fathi, A. and Poenaru, V., Travaux de Thurston sur les Surfaces, **Astérisque**, (1979), 66–67.

Lehner, J., Discontinuous groups and automorphic functions, **Mathematical Surveys, Number VIII**, American Math. Soc., Providence, 1964.

Liouville, J., Note IV to Monge's **Application de l'analyse à la geometrie**, 5th ed., Bachelier, Paris, 1850.

Magnus, W., **Non–Euclidean Tesselations and Their Groups**, Academic Press, New York, 1974.

Mandelbrot, B., Self–inverse fractals osculated by sigma–discs and the limit sets of inversion groups, **Math. Intelligencer, 5** (1983), 9–17.

Maskit, B., On Poincaré's theorem for fundamental polygons, **Adv. in Math., 7** (1971), 219–230.

Milnor, J., Hyperbolic geometry: the first 150 years, **Bull. Amer. Math. Soc. 6**(1982), 9–24.

Nielsen, J., Untersuchungen zur Topologie der geschlossenen zweiseitigen Flächen II, **Acta Math., 53**(1929), 1–76.

Poincaré, H., Extrait d'un memoire inedit de Henri Poincaré sur les fonctions Fuchsiennes, **Acta Math., 39** (1923), 58–93.

Poincaré, H., Theorie des groupes Fuchsiens, **Acta Math., 1** (1882), 1–62.

Poincaré, H., Memoire sur les fonctions Fuchsiennes, **Acta Math., 1** (1882), 193–294.

Poincaré, H., Memoire sur les groupes Kleinéens, **Acta Math., 3** (1883), 49–92.

Poincaré, H., Sur les groupes des equations lineaires, **Acta Math., 4** (1884), 201–312.

Poincaré, H., Sur l'uniformisation des functions analytiques, **Acta Math., 31** (1907), 1–63.

Poincaré, H., **Science et Methode**, Flammarion, Paris (1918).

Riemann, G.F.B., Grundlagen für eine allgemeine Theorie der **Funktionen einer veränderlichen komplexen Grösse, [1851], Collected Works**, Dover, New York, (1953), 3–43.

Riemann, G.F.B., Uber die Hypothesen, welche der Geometrie zur Grunde liegen, [1854], **Collected Works**, Dover, New York (1953), 272–287.

Riemann, G.F.B., **Collected Works**, Dover, New York, 1953.

Schottky, F., Uber die conforme Abbildung mehrfach zusammenhängender ebener Flächen, **J. f. reine u. angew. Math., 83** (1877), 300–351.

Schwarz, H.A., Uber einige Abbildungsaufgaben, **J. f. reine u. angew. Math., 70** (1869), 105–120.

Schwarz, H.A., Uber diejenigen Fälle, in welchem die Gaussische hypergeometrische Reihe eine algebraische Function ihres vierten Elementes darstellt, **J. f. reine u. angew. Math., 75** (1872), 292–335.

Schwarz, H.A., **Gesamellte Math. Abhandlungen**, Julius Springer, Berlin (1890).

Teichmüller, O., Extremal quasikonforme Abbildungen und quadratische Differentiale, **Abh. Preuss. Akad. Wiss., 22** (1939), 1–197.

Thurston, W., On the geometry of dynamics of diffeomorphisms of surfaces, I, (preprint, 1977).

ANNOTATED TABLE OF CONTENTS

I. Real transformations

Linear fractional transformations $z \mapsto t = \frac{az+b}{cz+d}$ are conformal, map circles onto circles, preserve cross ratios. Can be written

$$\frac{t-\alpha}{t-\beta} = K \frac{z-\alpha}{z-\beta}$$

if α, β are the fixed points, $\alpha \neq \beta$, otherwise

$$\frac{1}{t-\alpha} = \frac{1}{z-\alpha} + C$$

Real transformations ($a, b, c, d \in \mathbf{R}$) preserve upper half plane and map circles orthogonal to real axis onto other such circles. They are elliptic, parabolic or hyperbolic according as $(a+d)^2 <,=,> 4$.

II. Congruent figures.

Are transforms of each other by real transformations. This is congruence in the metric $ds = \frac{|dz|}{y}$ and gives a model of hyperbolic geometry.

III. Discontinuous groups.

Discontinuity of a group G a necessary condition for the existence of a nonconstant meromorphic function automorphic with respect to G. Fundamental region and tessellation for G. Side pairing of fundamental region, vertex cycles. Side pairing transformations generate G, vertex cycles correspond to defining relations.

IV. Generating polygons

Cutting and pasting fundamental regions. Can always obtain one which is a (hyperbolic) convex polygon.

V. Classification into families

According to the types of vertex cycles. Examples III, IV are standard ones for a surface of genus 2. Angle sum of a vertex cycle is an aliquot part of 2π.

VI. Existence of Fuchsian groups.

Polygon determines a Fuchsian group if its paired sides are congruent and vertex cycle angle sums are aliquot parts of 2π (Poincaré's theorem).

VII. Main examples.

First is triangle group, and its specialization to the modular group.

VIII. Classification by genus.

Genus p is that of the surface obtained by identifying paired sides of fundamental region. Computed from Euler characteristic $\chi = 2 - 2p$.

IX. Simplification of the generating polygon.

By cutting and pasting.

X. Isomorphism.

Occurs when two groups have the same presentation. When no relations are present it occurs when the numbers of generators are the same. In general, it suffices to have fundamental regions with isomorphic side pairing for which corresponding vertex cycles have the same angle sums.

XI. Effective formation of Fuchsian groups.

By computing coefficients of generators. Bounds, between which these parameters can very arbitrarily, are computed for certain examples.

XII. Generalization.

Groups G with an invariant circle — **Fuchsian groups**. All concepts carried over from the half—plane in the obvious way.

XIII. Historical.

References to Hermite, Dedekind, Klein on modular group, Schwarz on triangle functions, Fuchs on differential equations, and Dyck for representation of groups by tessellations.

On Fuchsian Functions (Poincaré [1882b]) 128

I. Thetafuchsian series.

Poincaré series for automorphic forms Θ for group G. Their absolute convergence by Euclidean area argument. Sum of series is continuous in the parameters of G, by uniform continuity argument based on non—Euclidean area.

II. Classification and general properties.

Functional equation

$$\Theta\left[\frac{\alpha z+\beta}{\gamma z+\delta}\right] = \Theta(z)(\gamma z+\delta)^{2m}$$

Classification of Θ's according to the classification of groups G (<u>Theory of Fuchsian Groups</u>, Section V). Θ can be identically zero. Behavior of Θ at vertices of fundamental region.

III. Zeroes and infinities.

By integrating Θ around the boundary of a fundamental region, one finds the difference between the number of zeroes and the number of poles to be proportional to m.

IV. Fuchsian functions.

Constructed as Θ_1/Θ_2 where Θ_1, Θ_2 have the same m value. Thus a Fuchsian function has equal number of zeroes and poles, and hence two Fuchsian functions for the same G are algebraically related. All Fuchsian functions for the same G are expressible rationally in terms of two particular ones x,y. The genus of the relation between x and y is the genus of G (Theory of Fuchsian Groups, Section VIII). The inverse of a Fuchsian function is the quotient of two solutions of a second order differential equation, singular points of this differential equation correspond to vertices of a fundamental region, and the difference between the roots of the indicial equation equals the (vertex cycle angle sum)/2π.

V. First family; genus zero.

The problem of adjusting parameters to obtain a Fuchsian equation arises. (To be treated in On Groups of Linear Equations, Section Vff.)

VI. First family; arbitrary genus.

Group of genus n depends on 6n − 6 parameters. Dimension of the space of automorphic forms.

VII. Second family; genus zero.

VIII. Third family.

IX. Fifth family; genus zero.

X. Resumé.

XI. Historical.

Modular functions and triangle functions. Observation that Schwarz mapping of curvilinear polygons implies the existence of automorphic functions when the polygon is symmetric.

I. Complex transformations.

Classification into parabolic, hyperbolic, elliptic and loxodromic. Representation as product of an even number of inversions used to define action on half—space. The nature of elliptic, hyperbolic, parabolic, loxodromic space transformations. Their interpretation as three—dimensional non—Euclidean motions.

II. Discontinuous groups.

Proper and improper discontinuity. The modular group is properly discontinuous on the upper half—plane, but improperly on the real axis. A discontinuous group action on the half—space is properly discontinuous.

III. Generating polygons.

Definition of a Kleinian group as one which acts properly discontinuously on the plane. Fundamental regions in the plane; piecewise equivalence. Vertex cycles.

IV. Generating polyhedra.

Fundamental regions in the half space; face pairing, edge cycles (angle sum = aliquot part of 2π). Obtain region bounded by spheres by cut and paste. Faces of fundamental polyhedra on the sphere at infinity are fundamental polygons.

V. Existence of Kleinian groups.

From fundamental polyhedra, in the same way that Fuchsian groups come from fundamental polygons in the Theory of Fuchsian Groups, Section VI. It follows that a Kleinian group is also obtained from a fundamental polygon.

VI. Classification.

Some Kleinian groups are deformations of Fuchsian groups, others are not.

VII. Third family.

Fundamental polygons bounded by disjoint circles. Discontinuity is evident without passing to half space.

VIII. Second family.

Fundamental polygon has all angles zero. Discontinuity is evident and limit set decomposes plane into two regions. However, the limit set is a nonanalytic curve.

IX. Symmetrical groups.

Includes an example with non—simply connected domain of discontinuity, for which discontinuity seems impossible to prove without passing to half space.

X. First family.

Construction of a Kleinian group by deformation of a Fuchsian group. Limit set is closure of the set of hyperbolic and loxodromic fixed points.

XI. Kleinian functions.

Construction, and convergence proof, are analogous to Fuchsian case, as are algebraic relations between functions and the associated second order differential equation.

XII. Historical.

References to Schottky and Klein.

On the Groups of Linear Equations (Poincaré [1884]) 357

I. Fundamental invariants.

The group of a linear differential equation, and its invariants. The first problem is to find the group.

II. Numerical calculation of the fundamental invariants.

Methods of Fuchs and Hamburger. These give generators of

the group.

III. Properties of the fundamental invariants.

As functions of the coefficients of the equation.

IV. Inverse functions.

Automorphic functions obtained by inversion. Counting parameters shows that "in general" one can find a second order differential equation admitting a given group. Roots of the indicial equation determine angles of the fundamental polygon. (Cf. On Fuchsian Functions, Section IV).

V. Enunciation of the second problem.

To dispose the parameters of a second order differential equation so as to obtain a Fuchsian function by inverting the ratio of two integrals. Reduces to the problem of finding a Fuchsian equation within a given "type". A solution will lead to uniformization of algebraic curves and integrals of linear differential equations.

VI. Subordination of types.

VII. Fundamental lemma.

A Fuchsian type contains at most one Fuchsian equation. This result was also found by Klein [1882].

VIII. First glance at the method of continuity.

One– and two–dimensional examples where a homeomorphism of one manifold into another is onto. Difficulties in higher dimensions. The plan is to define spaces of equation types and spaces of Fuchsian groups, a continuous map from groups to types (which is injective by the fundamental lemma), and to try to show that this map is onto. This requires investigation of the space of groups and its boundary.

IX. Second lemma.

Parameters of an equation type are continuous functions of the group parameters, by On Fuchsian Functions, Section I. Limit polygons and corresponding limit equations.

X. Symmetric types.

Contain Fuchsian equations whose G has a symmetric fundamental polygon. This is proved by the method of continuity when there are ≤ 5 singular points, and the method is said to generalize.

XI. Generalization of the preceding theorem.

Leads to an instance of uniformization.

XII. Limit polygons.

The (Teichmüller) space M_1 of groups of a given class, defined as a space of polygons. Limit polygons occur at the boundary of M_1. An example for which M_1 is two-dimensional. The fundamental group of the surface of genus p:

$$\langle S_1, S_2, \ldots, S_{2p}; S_1S_2 \ldots S_{2p} = S_{2p} \ldots S_2S_1\rangle.$$

XIII. Reduced polygons.

The mapping class group action on M_1. Said to be discontinuous. Vague construction of a particular fundamental region D_0 of "reduced" polygons, analogous to standard fundamental region for the modular group. Claim that D_0 has a finite number of faces. Limit polygons in D_0. Cut and paste operations which are said to reduce fundamental polygons for the surface of genus p.

XIV. The method of continuity.

The manifold M of equation types; without boundary by a dimension argument. D_0 is mapped continuously into M, and the map is claimed to be onto by dimension arguments and the method of continuity.

VI. A particular application.

Using the two-dimensional space of groups from Section XII.

XVI. Theory of subgroups.

Index and normality, which arise in the question of algebraic integrability of differential equations.

XVII. Third problem. Symmetric types.

Finding the Fuchsian equation in a given symmetric Fuchsian type.

XVIII. Third problem; general case.

XIX. Reflections on the convergence of the preceding series.

XX. Resume.

ON FUCHSIAN FUNCTIONS

Comptes rendus de l'Academie des Sciences 92 (14 February 1881)

The goal which I propose, in the work which I have the honor of presenting to the Academy, is to investigate whether there exist analytic functions analogous to elliptic functions which permit the integration of various linear differential equations with algebraic coefficients. I shall show that there is a very extensive class of functions which satisfy these conditions and which I give the name **Fuchsian functions** in honor of Fuchs, whose works have been very useful to me in these researches.

Here is the notation I shall use. Let z be a complex variable represented by a point in the plane. If K_1 is the operation which changes z to $f_1(z)$, and K_2 is that which changes z_2 to $f_2(z)$, then I shall habitually write

$$zK_1 = f_1(z) \ , \ zK_2 = f_2(z) \ , \ zK_1K_2 = f_2[f_1(z)] \ .$$

When zK_1 ranges over a certain region S as z_1 ranges over a region R I shall write

$$S = RK_1$$

I call the circle with center at the origin and radius 1 the **fundamental circle**, the **hyperbolic group** is the group of operations which change z to $\frac{az+b}{cz+d}$ (where a,b,c,d are constants) and which leave the fundamental circle fixed; a **discontinuous group** is any group which does not contain an infinitesimal operation, i.e., an operation which changes z to a quantity infinitely close to z; a **Fuchsian group** is any discontinuous group contained in the hyperbolic group.

I call any uniform function of z which is unchanged under the operations of a Fuchsian group a **Fuchsian function.**

It is necessary first to construct all Fuchsian groups; this I have done with the aid of non–Euclidean geometry, which will not be discussed here. I have shown that the disc bounded by the fundamental circle can be decomposed (in infinitely many ways) into an infinity of regions R_0, R_1, R_2, ..., R_i, ... satisfying the following conditions:

1. The regions are curvilinear polygons whose sides are arcs of circles which cut the fundamental circle orthogonally.

2. For any index i,

$$R_i = R_0 K_i \ ,$$

where K_i is an operation of the hyperbolic group.

It is clear that the different operations K_i form a discontinuous group contained in the hyperbolic group, i.e., a Fuchsian group.

PROBLEM I. **Is it possible to effect this decomposition in such a way that the first of the regions, R_0, is a given curvilinear polygon?**

We take a particular example; imagine two curvilinear triangles ABC, BCD whose sides are arcs of circles which cut the fundamental circle orthogonally. Suppose that the curvilinear angles of these triangles are equal, respectively,

$$\text{BAC and BDC to } \pi/\alpha \ ,$$

$$\text{CBA and CBD to } \pi/\beta \ ,$$

$$\text{BCA and BCD to } \pi/\gamma \ ,$$

where α, β, γ are positive integers (finite or infinite) such that

$$1/\alpha + 1/\beta + 1/\gamma < 1 \ .$$

One can decompose the interior of the fundamental circle into an infinity of regions R_0, R_1, ..., R_i, ... satisfying conditions I and II and such that R_0 is precisely the quadrilateral ABCD. This mode of decomposition corresponds to a Fuchsian group which I call the group (α,β,γ).

I shall later solve problem I in the general case and show how one can construct all Fuchsian groups and classify them rationally from two different points of view.

Among the Fuchsian groups there are some which merit particular attention:

1. The group $(2,3,\div)$, which is isomorphic to the group of operations which change z to $\frac{az+b}{cz+d}$, where a,b,c,d are integers such that $ad - bc = 1$.

2. Certain groups isomorphic to groups of linear transformations with integer coefficients which leave invariant an indefinite ternary quadratic form with integer coefficients.

The existence of these groups emphasizes the intimate connection between the theory of numbers and the analytic question we are concerned with.

I give the name **theta Fuchsian function** to any function $\Theta(z)$ uniform in z which (for each K_i in a Fuchsian group) identically satisfies

$$\Theta(zK_i) = \Theta(z)\left[\frac{dzK_i}{dz}\right]^{-m}$$

for some positive integer m.

In other words, for an infinity of values a,b,c,d such that

$$ad - bc = 1$$

we have identically

$$\Theta\left[\frac{az+b}{cz+d}\right] = \Theta(z)(cz+d)^{2m}.$$

I shall show that there is an infinity of theta Fuchsian functions defined by the convergent series

$$\sum_{i=1}^{i=\div} H(zK_i)\left[\frac{dzK_i}{dz}\right]^m \quad ;$$

m is an integer greater than 1; K_i is any operation of a Fuchsian group G; H(z) is a rational function of z.

Two cases can occur: (1) All points of the fundamental circle are essential singular points of $\Theta(Z)$; then there are really two distinct functions: the first exists inside the fundamental circle, and the second outside it, since we cannot pass continuously from one to the other; (2) $\Theta(Z)$ has an infinity of essential singular points on the fundamental circle, but these singular points are isolated, so that the function exists throughout the plane.

This function is always meromorphic, except on the fundamental circle; I shall indicate how to calculate the number of its distinct zeroes and distinct infinities.

ON FUCHSIAN FUNCTIONS

Comptes rendus de l'Academie des Sciences 92 (21 February 1881)

The quotient of two theta Fuchsian functions corresponding to the same Fuchsian group and the same value of the integer m is a uniform function F(z) of z, and such that

$$F(zK_i) = F(z)$$

It is therefore a Fuchsian function, according to the definition given in the preceding note. In other words, we have identically, for infinitely many values of the constants a, b, c, d

$$F\left(\frac{az+b}{cz+d}\right) = F(z)$$

I prove two theorems:

1. **Between two Fuchsian functions with the same group and without any essential singular point which is not a consequence of their definition, there is an algebraic relation.**

2. **Any Fuchsian function** F(z) **permits the integration of a linear equation with algebraic coefficients in the following way. If we put**

$$x = F(z), \; y_1 = \sqrt{\frac{dF}{dz}}, \; y_2 = z\sqrt{\frac{dF}{dz}}$$

then y_1 **and** y_2 **satisfy the differential equation**

$$\frac{d^2y}{dx^2} = y\phi(x)$$

where $\phi(x)$ **is algebraic in** x.

In particular, let the equation be

$$(1) \qquad \frac{d^2y}{dx^2} = y\left[\frac{\frac{1}{\alpha^2}-1}{4x^2} + \frac{\frac{1}{\beta^2}-1}{4(x-1)^2} + \frac{1+\frac{1}{\gamma^2}-\frac{1}{\alpha^2}-\frac{1}{\beta^2}}{4x(x-1)}\right]$$

where α, β, γ are finite or infinite positive integers such that

$$\frac{1}{\alpha}+\frac{1}{\beta}+\frac{1}{\gamma} < 1 .$$

If z is the ratio of the integrals then we have

$$x = f(z)$$

where f(z) is a Fuchsian function for the group (α,β,γ).

It exists only inside the fundamental circle, and it can be regarded as the quotient of two theta–Fuchsian functions

$$\frac{\left(\frac{df}{dz}\right)^m}{[f(z)]^p[f(z)-1]^q} \qquad \frac{\left(\frac{df}{dz}\right)^m f(z)}{[f(z)]^p[f(z)-1]^q}$$

m, p and q are integers which satisfy the (always compatible) inequalities

$$1-\frac{p}{m} \geq \frac{1}{\alpha}, \quad 1-\frac{q}{m} \geq \frac{1}{\beta}, \quad \frac{p+q-1}{m}-1 \geq \frac{1}{\gamma} .$$

These two functions, which exist only inside the fundamental circle, are holomorphic there.

If $\alpha = \beta = \gamma = \infty$, equation (1) reduces to the equation which determines the periods of sin am x as a function of the square of the modulus.

Next I return to theta Fuchsian functions of the arithmetic invariants which were defined in a note I had the honor of presenting to the Academy in November 1879.

Let F(z) be any Fuchsian function; we set $x = F(z)$.

By a **system of zeta Fuchsian functions** I mean a system of functions $\theta_1(z), \theta_2(z), \ldots, \theta_n(z)$ which are uniform in z and such that the determinant

$$\begin{vmatrix} \frac{d^{\alpha_1}\theta_1}{dx^{\alpha_1}} & \frac{d^{\alpha_2}\theta_1}{dx^{\alpha_2}} & \cdots & \frac{d^{\alpha_n}\theta_1}{dx^{\alpha_n}} \\ \frac{d^{\alpha_1}\theta_2}{dx^{\alpha_1}} & \frac{d^{\alpha_2}\theta_2}{dx^{\alpha_2}} & \cdots & \frac{d^{\alpha_n}\theta_2}{dx^{\alpha_n}} \\ \cdots & \cdots & \cdots & \cdots \\ \frac{d^{\alpha_1}\theta_n}{dx^{\alpha_1}} & \frac{d^{\alpha_2}\theta_n}{dx^{\alpha_2}} & \cdots & \frac{d^{\alpha_n}\theta_n}{dx^{\alpha_n}} \end{vmatrix}$$

is a Fuchsian function of z for any integers $\alpha_1, \alpha_2, \ldots, \alpha_n$.

It is clear that $\theta_1, \theta_2, \ldots, \theta_n$ satisfy a linear differential equation whose coefficients are algebraic in x.

I show that one can construct an infinity of zeta Fuchsian functions, which I define by various series expressions, and which permit the integration of an infinity of differential equations, which include among others **all linear differential equations with rational coefficients whose only singular points are two at a finite distance and one at infinity**.

We give a particular application.

Let K and K′ be the periods of an elliptic function, and ω

the square of its modulus.

Let ϕ be an algorithm such that

$$\omega = \phi\left[\frac{K + \sqrt{-1}\,K'}{K - \sqrt{-1}\,K'}\right].$$

Suppose we have a linear differential equation with rational coefficients whose singular points are

$$x = 0\ ,\ x = 1\ ,\ x = \infty\ .$$

We put $x = \phi(z)$ and let $\theta_1(z)$, $\theta_2(z)$, ..., $\theta_n(z)$ be the integrals of the given equation:

$\phi(z)$ will be a Fuchsian function, θ_1, θ_2, ..., θ_n the zeta Fuchsian functions.

These functions exist only within the fundamental circle.

They are holomorphic inside this circle and consequently can be represented by power series whose coefficients are easy to calculate.

In summary, there exists a very extensive class of functions of which the elliptic functions are a special case. They permit the integration of a great number of differential equations. Various properties emphasize their analogy with the elliptic transcendants, and those of the theta Fuchsian and zeta Fuchsian functions with the Θ and Z functions.

THEORY OF FUCHSIAN GROUPS

Acta Mathematica I (1882)

In a series of memoirs presented to the Academy of Sciences, I have defined certain new functions which I call Fuchsian, Kleinian, theta–Fuchsian and zeta–Fuchsian. Just as the elliptic and abelian functions permit the integration of algebraic differentials, the new transcendants permit the integration of linear differential equations with algebraic coefficients. I have succinctly summarized the results obtained in a note in the **Mathematische Annalen**. Intending now to give a detailed exposition, I begin, in the present work, by studying the properties of Fuchsian groups, reserving until later their consequences from the point of view of function theory. [4]

I. Real transformations*

(*Translator's Note: Poincaré calls them **substitutions**)

Let z be a complex variable defined by the position of a point in the plane; let t be a complex function of this variable defined by the relation

$$t = \frac{az+b}{cz+d} \tag{1}$$

I assume, without loss of generality, that we have

$$ad - bc = 1 .$$

If the point z describes two arcs which cut at a certain angle α, then the point t for its part will describe two arcs which cut at the same angle α, i.e., the transformation $\left(z, \frac{az+b}{cz+d}\right)$ preserves angles.* (*In what follows I use the notation of Jordan. The transformation [z,f(z)] or [x,y;f(x,y),ϕ(x,y)] is the operation which changes z to f(z), or that which changes x to f(x,y) and y to ϕ(x,y). The inverse transformation to [z,f(z)] is [f(z),z]; the product of two transformations is the operation which consists in making the two transformations successively. A system of transformations forms a **group** if the inverse of each transformation of the system, and the product of any two transformations of the system, is likewise in the system.) (Translator's note: Poincaré then defines **isomorphismholoèdrique** and **isomorphism mériédrique**, which are our **isomorphism** and **homomorphism**, and I shall translate them as such below.)

The function $\frac{az+b}{cz+d}$ is in fact monogenic.

If z describes a circle, then t also describes a circle; i.e., the transformation $\left[z,\frac{az+b}{cz+d}\right]$ changes circles into circles.

Finally, if z_1, z_2, z_3, z_4 are four values of z and if t_1, t_2, t_3, t_4 are the corresponding values of t, then

$$\frac{t_1-t_2}{t_1-t_3}\,\frac{t_4-t_3}{t_4-t_2} = \frac{z_1-z_2}{z_1-z_3}\,\frac{z_4-z_3}{z_4-z_2} \tag{2}$$

There are in general two values of z which equal the corresponding values of t; we call them the **fixed points*** of the transformation (1). (*Translator's Note: Poincaré calls them **double points**.)

If $(a+d)^2 \gtrless 4$, the fixed points are distinct; and if we call them α and β the relation (1) can be written

$$\frac{t-\alpha}{t-\beta} = K\,\frac{z-\alpha}{z-\beta} \tag{3}$$

where K is a constant I call the **multiplier**.

If on the other hand

$$a + d = \pm 2$$

The fixed points coincide and we have $\alpha = \beta$.

Relation (1) can then be written

(4) $$\frac{1}{t-\alpha} = \frac{1}{z-\alpha} \pm C$$

These are the main properties of the linear transformations $\left[z, \frac{az+b}{cz+d}\right]$. However, we have to make one more assumption; we assume that the coefficients a, b, c, d, are real. I then say that the transformation (1) is a **real transformation**.

It follows that the imaginary part of t is positive, zero or negative according as the imaginary part of z is likewise positive, zero or negative; i.e., the transformation (1) preserves the real axis, which I shall henceforth call X, and it likewise maps the half plane* above the axis onto itself. (*Translator's Note: Poincaré calls it the **part of the plane**.)

If z describes a circle with its center on X, t will also describe a circle with its center on X. If z_1 and z_2 are complex conjugates, then the corresponding values t_1 and t_2 of t will also be complex conjugates.

If α and β are two values of z, if γ and δ are the corresponding values of t, and if α', β', γ', δ' are their conjugates, then we have, by virtue of relation (2), that

$$\frac{\alpha-\alpha'}{\alpha-\beta'}\,\frac{\beta-\beta'}{\beta-\alpha'} = \frac{\gamma-\gamma'}{\gamma-\delta'}\,\frac{\delta-\delta'}{\delta-\gamma'}$$

If we make the abbreviation

$$\frac{\alpha-\alpha'}{\alpha-\beta'}\,\frac{\beta-\beta'}{\beta-\alpha'} = (\alpha,\beta)$$

then the latter relation can be written

(5) $$(\alpha,\beta) = (\gamma,\delta)\ .$$

The real transformations have been studied by different geometers, in particular by Klein in his researches on modular functions; they are divided into elliptic, parabolic and hyperbolic transformations.

The elliptic transformations are those for which

$$(a+d)^2 < 4 .$$

The fixed points α and β are conjugate imaginaries; hence one of them is above X and the other below; the relation (1) can be put in the form (3) and the constant K is an imaginary or negative quantity whose absolute value is unity. If z describes a circle through α and β, then t likewise describes a circle through α and β, which cuts the first circle at an angle equal to the argument of K.

The transformation (1) leaves invariant each circle whose center is on the prolongation of $\alpha\beta$ and which cuts the latter segment harmonically.

The parabolic transformations are those for which

$$(a+d)^2 = 4 .$$

The fixed points coincide in a single one, which is situated on X. The relation (1) comes under the form (4), so that α must be real. If z describes a circle through α, then t likewise describes a circle through α, tangent to the first. The transformation (1) leaves invariant the circles which are tangent to X at α. Let C be such a circle, and m_0 a point on it. The transformation (1) changes m_0 to another point m_1 on the same circle, m_1 to another point m_2 of C, m_2 to another point m_3, etc. As κ tends to infinity, the point m_κ tends to α. Let m_{-1} be the point which the transformation (1) changes to m_0, m_{-2} the point which it changes to m_{-1}, etc. As κ tends to infinity, the point $m_{-\kappa}$ also tends to α.

I let C_κ denote the circle with center on X which passes through α and m_κ, and which consequently cuts the circle C orthogonally at α and m_κ. It is clear that the transformation (1) changes C_{-1} into C_0, C_0 into C_1, C_1 into C_2, etc., and in general C_κ into $C_{\kappa+1}$. Moreover, if κ is positive or negative infinity, C_κ reduces to a circle of infinitely small radius. I.e., if we apply the transformation (1), or its inverse, infinitely many times to a circle through α with its center on X, the radius of this circle becomes infinitely small.

It follows that an arc of finite length which does not meet X cannot meet infinitely many of the circles C_κ, i.e., the successive transforms of a circle C_0 through α with its center on X.

The hyperbolic transformations are those for which

$$(a+d)^2 > 4 .$$

The fixed points α and β are distinct and situated on X. The relation (1) takes the form (3) with K real and positive. Also I can always assume that $K > 1$. If z describes a circle through α and β, then t likewise describes a circle through α and β, and tangent to the first. The transformation (1) leaves invariant the circles which pass through α and β.

As above, I let C denote a circle through α and β, and let

$$\dots\ m_{-2}\ ,\ m_{-1}\ ,\ m_0\ ,\ m_1\ ,\ m_2\ ,\ \dots$$

be a series of points such that the transformation (1) changes m_κ into $m_{\kappa+1}$. It is clear that the point m_κ tends to β as κ tends to $+\infty$, and to α as κ tends to $-\infty$.

I also let C_κ denote the circle, with center on X, which passes through α and m_κ. As κ tends to $+\infty$, C_κ tends to the circle with diameter $\alpha\beta$; as κ tends to $-\infty$ the radius of C_κ decreases indefinitely. i.e., if we apply the transformation (1) an infinite number of times to a circle C_0 passing through α and with its center on X, we obtain in the limit a circle with diameter $\alpha\beta$. If we apply the inverse transformation to C_0 an infinite number of times, the limit radius will be zero.

If, on the contrary, we apply the transformation (1) infinitely often to a circle through β and with its center on X, the limit radius will be zero, while if we apply the inverse transformation to it infinitely often we obtain in the limit the circle with diameter $\alpha\beta$.

It follows that an arc of finite length which does not cut X will meet an infinite number of the circles C_κ, i.e., the successive transforms of the circle C_0, or else a finite number of these transforms, according as it does or does not meet the circle with diameter $\alpha\beta$.

If we have $(\alpha,\beta) = (\gamma,\delta)$ then there is a real transformation* which changes α into γ and β into δ. (*It is necessary that α and γ (and consequently β and δ) be on the same side of X; if not, the determinant ad − bc will be negative. N.E.N.) This transformation is defined by the relation

$$\frac{t-\gamma}{t-\delta}\,\frac{\gamma'-\delta}{\gamma'-\gamma} = \frac{z-\alpha}{z-\beta}\,\frac{\alpha'-\beta}{\alpha'-\alpha}$$

There is another property of real transformations which I want to point out; in differentiating the relation (1) one finds

$$\frac{dt}{dz} = \frac{1}{(cz+d)^2}$$

Moreover, if I let y be the imaginary part of z, and let Y be the imaginary part of t, then I find

$$\left|\frac{dt}{dz}\right| = \frac{Y}{y}\,.$$

II. Congruent figures

I say that two figures are **congruent** if one is the transform of the other by a **real** transformation. The real transformations form a group; it is clear that two figures congruent to a third are congruent to each other.

I can begin by stating the following theorems:

In two congruent figures, the corresponding angles are equal.

If, in two congruent figures, the point γ corresponds to α and the point δ corresponds to β, then

$$(1) \qquad (\alpha,\beta) = (\gamma,\delta)$$

This relation can take another form.

Consider the quantities α, β and their conjugates α', β', so that

$$(\alpha, \beta) = \frac{\alpha-\alpha'}{\alpha-\beta'} \frac{\beta-\beta'}{\beta-\alpha'} .$$

The four points α, β, α', β' are on the same circle with center on X. Suppose in addition that α and β are both above X. The circle cuts X at two points I call h and k; h will be the one on the arc $\beta\beta'$, and k the one on the arc $\alpha\alpha'$. I put

$$[\alpha, \beta] = \frac{\alpha-h}{\alpha-k} \frac{\beta-k}{\beta-h} .$$

$[\alpha, \beta]$ is essentially real, positive and greater than 1. Moreover,

$$(\alpha, \beta) = \frac{4[\alpha, \beta]}{([\alpha, \beta]+1)^2}$$

If γ is a point on the arc $\alpha\beta$ of the circle, we have

$$[\alpha, \gamma][\gamma, \beta] = [\alpha, \beta] .$$

It is now clear that, by employing this new notation, we can put the relation (1) in the form

$$[\alpha, \beta] = [\gamma, \delta] .$$

We look at what happens when α and β become infinitely close.

Let

$$z = x + y\sqrt{-1}$$

$$dz = dx + dy\sqrt{-1}$$

$$|dz| = \sqrt{dx^2 + dy^2} .$$

Neglecting infinitely small terms of higher order, we have

$$[z,z+dz] = 1 + \frac{|dz|}{y} .$$

or

$$\log[z,z+dz] = \frac{|dz|}{y} .$$

Thus we see that the napierian logarithm of [z,z+dz] is proportional to the absolute value of dz and independent of its argument.

The integral

$$\int \frac{|dz|}{y}$$

along an arbitrary curve will be called the L of that curve.* (*Translator's Note: Poincaré has used L for log a few lines above.)

The double integral

$$\iint \frac{dxdy}{y^2}$$

taken over the interior of any plane region, will be called the S of that region.

It follows from the preceding that two congruent arcs have the same L and two congruent areas have the same S. The L of the arc $\alpha\beta$ of a circle with center on X will be the napierian logarithm of $[\alpha,\beta]$.

I cannot pass in silence the connection between the preceding notions and the non–Euclidean geometry of Lobachevsky.

Suppose that we agree to strip the words **line**, **length**, **distance**, **area**, of their usual meaning, and let a line be any circle with the center on X, let the length of a curve be what we have called its L, let the distance between two points be the L of the arc of the circle which joins the two points and has its center on X, and finally let the area of a plane region be what we have called its S.

Suppose moreover that we let the words **angle** and **circle** retain their usual meaning, but agree to call the **center of a circle** the point which is at a constant distance (in the new sense of the word) from all points of the circle, and let the radius of the circle be this constant distance.

If we adopt this terminology, **the theorems of Lobachevsky are true**, i.e., the new quantities satisfy all the theorems of ordinary geometry, except those which depend on Euclid's postulate.

This terminology has been of great service in my researches, but to avoid any confusion, I do not employ it here.

III. Discontinuous groups

We envisage an infinity of transformations of the form

$$\left(z , \frac{a_i z+b_i}{c_i z+d_i}\right) , \tag{1}$$

which, if we use the abbreviation

$$f_i(z) = \frac{a_i z+b_i}{c_i z+d_i}$$

can be written in the form

$$[z,f_i(z)] .$$

To abbreviate further, I speak simply of the transformation

$$f_i(z) .$$

I suppose that i is an index which varies from zero to infinity.

Moreover, I suppose that

$$a_0 = d_0 = 1 , b_0 = c_0 = 0 ,$$

whence

$$f_0(z) = z .$$

Finally, I assume that the transformations form a group.

I am going to define certain symbols to be used below. I put

$$f_i^2(z) = f_i[f_i(z)],\ f_i^3(z) = f_i[f_i^2(z)],\ \dots\ f_i^m(z) = f_i[f_i^{m-1}(z)]\ \dots$$

Likewise, I define the symbol $f_i^m(z)$ when m is zero or negative by a natural extension. I put

$$f_i^0(z) = z\ ,\ z = f_i^{-1}[f_i(z)]\ ,\ f_i^{-m}(z) = f_i^{-1}[f_i^{-m+1}(z)]\ .$$

If $f_1(z)$ is one of the transformations in the group considered, then all transformations comprised by the general formula

$$(2) \qquad f_1^m(z)$$

make up a subgroup.

If $f_2(z)$ is a transformation in the group not covered by the formula (2), then all transformations of the form

$$(3) \qquad f_1^{\alpha}(f_2^{\beta}(f_1^{\gamma}(f_2^{\delta}(\ \dots\ (f_1^{\lambda}(f_2^{\mu}(z)))\ \dots))$$

make up a group.

Now if $f_3(z)$ is a transformation in the group not covered by the formula (3), i.e., not a combination of f_1 and f_2, if f_4 is a transformation in the group which is not a combination of f_1, f_2 and f_3, etc., and finally if f_p is a transformation which is not a combinations of f_1, f_2, ..., f_{p-1}, then the transformations comprised by the general formula

(4) $$f_{\alpha_1}^{\beta_1}(f_{\alpha_2}^{\beta_2}(\dots (f_{\alpha_n}^{\beta_n}(z)) \dots)$$

(where $\alpha_1, \alpha_2, \dots, \alpha_n$ are indices which can be 1, 2, 3, ..., p−1 or p, and where $\beta_1, \beta_2, \dots, \beta_n$ are positive or negative integers, likewise make up a subgroup.

It may be that we have now exhausted all transformations in the group considered, so that every transformation in the group is a combination of $f_1, f_2, \dots, f_p$. We then say that the group is **generated** by $f_1, f_2, \dots, f_p$, or that $f_1, f_2, \dots, f_p$ is a system of **generators*** for the group. (*Translator's Note: Poincaré calls them **fundamental transformations**.) It is evident that a group can have infinitely many systems of generators, but a single one of these systems suffices to determine the group completely.

One can also conceive groups which cannot be defined by finite sets of generators. However, we systematically leave these aside.

Suppose then that G is a group generated by p generators $f_1, f_2, \dots, f_p$, so that all its transformations are of the form (4).

I call the sum of absolute values of the integers $\beta_1, \beta_2, \dots, \beta_n$ in a transformation its **exponent**.

It may be that not all the transformations contained in the formula (4) are distinct, and that we have identically

(5) $$f_{\alpha_1}^{\beta_1}(f_{\alpha_2}^{\beta_2}(\dots f_{\alpha_n}^{\beta_n}(z)) \dots) = f_{\gamma_1}^{\delta_1}(f_{\gamma_2}^{\delta_2}(\dots f_{\gamma_n}^{\delta_n}(z) \dots))$$

The relation (5) can always be put in the form

(6) $$f_{\alpha_1}^{\beta_1}(f_{\alpha_2}^{\beta_2}(\dots f_{\alpha_n}^{\beta_n}(z) \dots)) = z .$$

The α and β here have the same meaning as in the formula (4).

The relations of the form (6) can, in general, all be regarded as consequences of a certain number among them which we call **defining relations*** and which are the only really distinct ones. (*Translator's Note: Poincaré calls them **fundamental relations**.)

A group H will be a homomorph of G if it is generated by the same number of generators, between which the same relations hold. If the transformations in H satisfy no other relations (6) than those of G, then the homomorphism is invertible and consequently an **isomorphism***. (*Translator's Note: Poincaré uses the term **isomorphisme mériédrique** and **isomorphisme holoèdrique** for **isomorphism** and **homomorphism** respectively. When **mériédrique** and **holoèdrique** are used on their own I shall translate them as **many—to—one** and **one—to—one**.)

As a result of the relations (6), the same transformation can be put in the form (4) in infinitely many ways; thus the above definition of the **exponent of a transformation** has a certain ambiguity. We therefore let the exponent of a transformation be the **smallest** value which can be taken by the sum of the absolute values of $\beta_1, \beta_2, \ldots, \beta_n$ when the transformation is written in the form (4).

This is the place to make an important distinction between the different kinds of groups consisting of real transformations. We must first put aside the groups which contain only a finite number of transformations and which have previously been studied in depth by several geometers. But if the group contains infinitely many transformations one can make two different hypotheses:

One can assume that it is possible to choose a transformation

$$(z, f_i(z))$$

in the group such that $f_i(z)$ differs infinitesimally from z (whatever z may be), i.e., that the group contains an infinitesimal transformation.

One can also assume that the group contains no such transformation.

The groups of the first kind are **continuous**, those of the second kind are **discontinuous**.

There can be no uniform analytic function of z which remains unchanged under the transformations in a continuous group, since this function would have the same value at infinitely close points and consequently it would be constant. Accordingly, the continuous groups have no interest for us, and we reserve the name **Fuchsian groups** for the discontinuous groups of real transformations.

If the group G is discontinuous, then it is clear that we can divide the plane, or a part of the plane, into an infinity of regions which enjoy the following properties:

Each of them corresponds to one of the transformations in the group G. The one which corresponds to the transformation

$$(z,f_i(z))$$

will be called the region R_i, and consequently the one which corresponds to the transformation

$$(z,f_0(z)) \quad \text{or} \quad (z,z)$$

will be called R_0.

When z **is in the interior of the region** R_0, $f_i(z)$ **is in the interior of** R_i. In other words, R_i is the transform of R_0 under the transformation

$$(z,f_i(z))$$

If z is now in the interior of the region R_k, corresponding to

$$(z,f_k(z)) ,$$

then $f_k^{-1}(z)$ will be in the interior of R_0, and consequently $f_i(f_k^{-1}(z))$ will be in the interior of R_i.

To say that R_i is the transform of R_0 under a real transformation is to say that these regions are congruent, and consequently all the regions R are congruent to each other.

To use an expression which is common on the other side of the Rhine, I say that the division of a plane into an infinity of regions is **regular** when the new mode of subdivision can be made to coincide with the old by continuous deformation, under which each region of the new subdivision comes to coincide with a region of the old, and such that any given region of the new subdivision can be made to coincide with any region, **equally arbitrary**, of the old. In the mode of subdivision which concerns us, it is clear that if we apply the transformation

$$[z, f_i(z)]$$

to different regions R, then each of these regions R changes into another region R, and R_0 changes into R_i. This says that the division of the plane is regular.

The problem of investigating Fuchsian groups therefore reduces to the following: **to subdivide the plane, or a part of the plane, in a regular fashion, into an infinity of mutually congruent regions.**

Two regions will be called **adjacent** when they meet along an arc which serves as their common frontier.

Let R_p be a region adjacent to R_0 along an arc AB of its perimeter; this arc AB will be called **one of the sides** of R_0. But we can assume that not the whole plane, but only a part of it has been divided into an infinity of regions R, all congruent to each other. In that case, R_0 could meet the perimeter of the subdivided portion of the plane along an arc CD. The arc CD will again be **one of the sides** of R_0. Sides such as AB are called the **first kind**, sides such as CD are the **second kind**.

I always assume that the sides of the second kind are segments of the X axis.

I shall say a few words in justification of this hypothesis. If we divide a part of the plane into an infinity of regions R, and if the region R_0 is contiguous to the unsubdivided part of the plane, we can always extend the subdivision to a larger part of the plane. I extend until R_0 ceases to be contiguous to the unsubdivided part, or until R_0 reaches to X axis. This hint will suffice, I think, to make my reasons for adopting this hypothesis comprehensible.

I give the name vertices of R_0 to the extremities of its sides and I am led to consider: 1° the vertices above X; 2° the vertices on

X which separate two sides of the first kind; 3° the vertices on X which separate a side of the first kind from one of the second kind. I call these different vertices, vertices of the first, second and third category.

The points

$$z,\ f_1(z),\ f_2(z),\ \ldots,\ f_i(z)\ \ldots.$$

will be called **corresponding**.

When the point z is in the interior of R_0, all the points $f_i(z)$ will be in regions R different from R_0. Thus there cannot be two corresponding points in the interior of R_0, and an interior point of R_0 cannot correspond to more than one point on the perimeter of that region.

On the contrary, a point on one of the sides of R_0, λ_p for example, belongs to two regions R_0 and R_p. If the point z is on λ_p, i.e., on the perimeter of R_p, the point $f_p^{-1}(z)$ will be on the perimeter of R_0 on a side I call λ'_p and which separates R_0 from a region R'_p corresponding to the transformation

$$(z, f_p^{-1}(z))\ .$$

The sides λ_p and λ'_p will be called **equivalent.*** (*Translator's Note: Poincaré calls them **conjugate**.)

One can draw the following conclusions from this:

1° the sides of the first kind are even in number and equivalent in pairs;

2° two equivalent sides are congruent;

3° the points on the perimeter of R_0 above the X axis (not counting the vertices, which will concern us later) correspond in pairs.

It is necessary to make a special convention in order to let the general case cover a special case which will sometimes occur.

It may happen that we have identically

$$f_p(z) = f_p^{-1}(z) \quad \text{or} \quad z = f_p[f_p(z)] .$$

Then R_p coincides with R'_p and λ_p with λ'_p. But in this case the transformation

$$(z, f_p(z))$$

is elliptic and the relation (1) (Section I) can be put in the form (3) (Section I) in such a way that

$$K = -1 .$$

The transformation $(z, f_p(z))$ does not change the side λ_p, which must pass through the point α (fixed point of the transformation) and be divided into two congruent halves.

Under these circumstances, we can regard the point α as a vertex and the two halves of the side λ_p as distinct sides, equivalent to each other; we are then reduced to the general case.

Each vertex belongs to two or more different regions; it follows that several vertices of R_0 may correspond. I say that corresponding vertices belong to the same cycle.

It follows from the preceding that the number of sides of the first kind is even; let the number be 2n. These 2n sides are equivalent in pairs. Let $\lambda_1, \lambda_2, \ldots, \lambda_n$ be n of the sides, and let $\lambda_{n+1}, \lambda_{n+2}, \ldots, \lambda_{2n}$ be the other n sides equivalent to them, so that λ_p, λ_{n+p} are equivalent. Let R_p be the region adjacent to R_0 along the side λ_p, and let

$$(z, f_p(z))$$

be the corresponding transformation. In this way we have

$$f_{n+p}(z) = f_p^{-1}(z) .$$

It is easy to see that if we leave a region R_i corresponding to

$$(z, f_i(z))$$

by the side corresponding to λ_p, then we enter a region corresponding to the transformation

$$(z, f_i[f_p(z)]) \ .$$

Now nothing is easier than to find the transformation corresponding to a given region R_{β_ν}. Let A be an interior point of R_0, B an interior point of R_{β_ν}, and join these two points by any arc AMB. If I suppose that this arc AMB leaves the region R_0 by the side λ_{α_1}, to enter the region R_{α_1}, then leaves R_{α_1} by the side corresponding to λ_{α_2} to enter the region R_{β_2}, then leaves R_{β_2} by the side corresponding to λ_{α_3} to enter R_{β_3}, etc., and finally leaves $R_{\beta_{\nu-1}}$ by the side corresponding to λ_{α_ν} to enter R_{β_ν}, then the transformation corresponding to the region R_{β_ν} will be

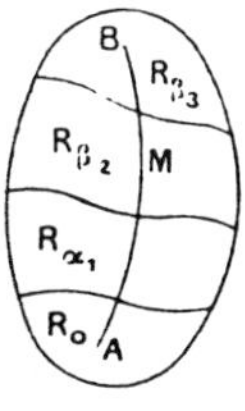

Figure 1

$$[z, f_{\alpha_1}(f_{\alpha_2}(f_{\alpha_3}(\dots (f_{\alpha_{\nu-1}}(f_{\alpha_\nu}(z) \dots)] \ .$$

In this expression, α_1, α_2, ..., α_ν denote indices which can be 1, 2, ..., or 2n.

Thus every transformation which corresponds to a region which

can be reached from R_0 by passing over a finite number of edges is a combination of (z,f_1), (z,f_2), ..., (z,f_n). We leave aside all groups for which one cannot pass from one region to another by crossing a finite number of sides. Then each transformation will be a combination of

$$(z,f_1),\ (z,f_2),\ \ldots,\ (z,f_n)\ . \tag{7}$$

How can we find the relations of the form (6) which hold between the functions f_1, f_2, ..., f_n?

It evidently will suffice to show how one can express the transformation

$$(z,z) \quad \text{or} \quad (z,f_0(z))\ ,$$

which corresponds to R_0, as a combination of the transformations (7). We apply the rule explained above, i.e., we describe any closed curve AMA issuing from an interior point A of R_0. If this curve traverses the regions

$$R_0,\ R_{\alpha_1},\ R_{\beta_2},\ \ldots,\ R_{\beta_{\nu-1}},\ R_{\beta_\nu} = R_0$$

in succession, leaving by the sides corresponding to

$$\lambda_{\alpha_1},\ \lambda_{\alpha_2},\ \ldots,\ \lambda_{\alpha_\nu}$$

respectively, then the transformation corresponding to R_0 will be

$$[z,f_{\alpha_1}(f_{\alpha_2}(\ \ldots\ (f_{\alpha_\nu}(z)\ \ldots)]\ ,$$

which may be written

$$z = f_{\alpha_1}(f_{\alpha_2}(\ldots(f_{\alpha_\nu}(z)\ \ldots)\ .$$

The relations (6) obtained in this way may not all be necessary to be definitive. To obtain the defining relations, it suffices to describe infinitesimal closed curves around each of the vertices of R_0.

Since we have thus found all relations of the form (6), the transformations (7) are generally independent and consequently form a system of generating transformations for the group considered.

Moreover, we see that the exponent of a transformation f_i is the minimum number of edges one can cross in passing from R_0 to the region which corresponds to the transformation f_i.

Apart from the Fuchsian groups, or discontinuous groups of real transformations, I have been led to consider discontinuous groups of arbitrary linear transformations, which I have called **Kleinian groups**. I shall not discuss them in the present work, reserving my exposition of the results I have obtained about them to a special memoir.

IV. Generating polygons

The regions R_i are defined by the condition that each of them contains **no more than one** point corresponding to a given point z. But this condition does not suffice to define them completely. In fact, if we are given a Fuchsian group G, we have infinitely many ways to subdivide the plane so that each region has no more than one point corresponding to a given point z.

On the other hand, given the decomposition of the plane into an infinity of regions such as R, the group G will be completely determined.

Suppose we are given a group G, and consider the different decompositions of the plane into regions R_i which correspond under this group. These decompositions are infinite in number. How can one of them enable us to derive all the others? Let S_0 be a portion of the region R_0, S_i the corresponding portion of the region R_i. We choose an arbitrary transformation

$$(z, f_p(z))$$

among those in the group G. Let i_p be the index of the transformation

$$[z, f_p(f_i(z))],$$

so that

$$f_p[f_i(z)] = f_{i_p}(z) .$$

From the different regions R_i I subtract the portion S_i and add the region S_{i_p} to what remains. In this way I obtain an infinity of regions

$$R'_i = R_i + S_{i_p} - S_i .$$

These regions R'_i cover the same part of the plane as the regions R_i, and do not cover it more than once. Each of them contains no more than a single point corresponding the a given point z. The same group G therefore corresponds to the decomposition of the plane into regions R'_i as corresponds to the decomposition into regions R_i. If S_0 is contiguous to the boundary of R_0 from the inside and if S_p is contiguous to the boundary of $R_0 - S_0$ from the outside, then the region R'_0 will be a **single piece without holes**. Most of the time, for greater convenience, we shall choose the region R'_0 so that it is a single piece without holes, but this is not essential.

Operating on the regions R'_i as we have operated on the regions R_i, we obtain a new decomposition of the plane into regions R''_i which still correspond to the group G considered. Continuing indefinitely in this way, we obtain all decompositions which correspond to the group G.

One is led naturally to the following generalization:

Until now I have assumed the region S_0 to be entirely inside the region R_i, and hence S_i to be inside the region R_i, and in fact without this hypothesis one would not know what is meant by the region

$$R'_i = R_i + S_{i_p} - S_i ,$$

at least without making a special convention.

One is led in this way to imagine regions divided into two parts, of which one is considered positive and the other negative.

Thus if S_i is not part of R_i we consider the region R'_i to consist of a positive part $R_i + S_{i_p}$ and a negative part S_i. This is not an absolutely new idea.

If we consider a quadrilateral ABCD whose opposite sides AB and CD cut at a point M, then this quadrilateral, formed by two triangles ADM and BCM meeting at a vertex, will be called a **concave** quadrilateral.

The formula which gives the area of a convex quadrilateral applies to such a figure when we regard its area as the difference and not the sum of the areas of the two triangles ADM and BCM. In a concave quadrilateral one of these two triangles must be regarded as positive and the other as negative. Here we have an analogous situation. A **concave region** R'_i consists of a positive portion $R_i + S_{i_p}$ and a negative portion S_i. The number of points in the positive part of R'_i which corresponds to a given point z is always finite and it exceeds by one the number of points in the negative part corresponding to z.

We shall make little use of this subdivision of the plane into concave regions R_i, because the subdivision into convex regions is infinitely more convenient, but it serves as an intermediary in passing from one convex subdivision to another which is likewise convex.

Since there are infinitely many ways of subdividing the plane into convex regions R_i, we need to choose the most simple and most convenient from among them.

Here is how we can proceed. We determine a region S_0, and the region S_p which results from S_0 by the transformation $f_p(z)$. We adjoin S_p to R_0 and remove S_0. We thereby obtain a new region R'_0 which can serve as the basis of a new subdivision of the plane into regions R'_i corresponding to the group G. To obtain the region R'_i it suffices to remove from R_i the transform S_i of S_0 by $f_i(z)$ and to adjoin the transform S_{i_p} of S_p by $F_i(z)$. What now is a suitable choice of the regions S_0 and S_p?

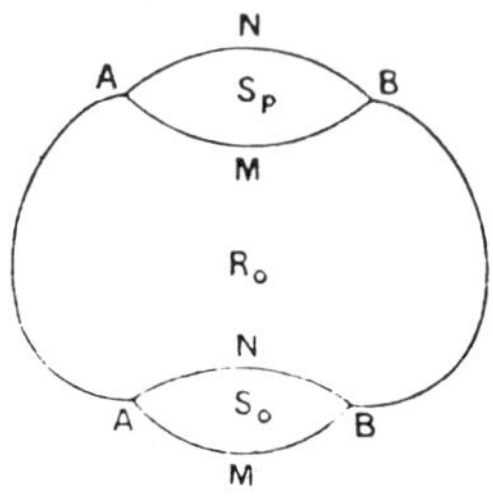

Figure 2

Let AMB be a side of R_0, and let A′M′B′ be the equivalent side. These two sides are congruent and the real transformation which changes AMB to A′M′B′ is a transformation $f_p(z)$ in the group G. We join the points A and B by an arc ANB of a circle with its center on X. The region S_0 will be the region AMBNA between the arc AMB and the arc ANB of the circle. The transform of ANB by $f_p(z)$ will be an arc A′N′B′ of a circle with its center on X, and the transform S_p of S_0 will be the region A′M′B′N′A′. If we adjoin S_p to R_0 and remove S_0, we obtain a region $R_0{}'$ analogous to R_0, but in which the equivalent sides AMB, A′M′B′ have become arcs ANB, A′N′B′ of circles with their centers on X.

Operating in the same way on each pair of equivalent sides, we reduce all sides to arcs of circles with their centers on X. Thus we can always assume that the region R_0 is a polygon whose sides are of two kinds: arcs of circles with centers on X, and segments of the axis X itself. Such a polygon will be called a **normal polygon.**

However, a special difficulty can occur in certain cases.

We have to deal with the possibility that the region

$$S_0 = \text{AMBNA}$$

does not lie entirely inside R_0. It would then happen that the normal polygon to which we reduce the region R_0 is concave.

Here is how we get out of trouble in such a case. It is clear that the positive and negative parts of R_0 must both be normal polygons. Suppose, to fix ideas, that the positive part contains at most two points corresponding to a given point z. Among the points of R_0 there will be some in the positive part with no correspondents in either the positive or negative part. They form a certain normal polygon P.

The negative part will form a certain normal polygon P', and each point of this negative part will correspond to two points in the positive part, the sets of which form two normal polygons P'', P''', both congruent to P'. The positive part of R_0 will be formed by the three polygons P, P'' and P'''. We then remove P'' from R_0 and adjoin the congruent polygon P'. In this way we obtain a new region R'_0 which can serve, as R_0 does, to generate the Fuchsian group G. This region no longer has the negative part and it is formed by two polygons P and P'''.

It will then be a **normal convex polygon**.

We see in this way that, in all possible cases, we can assume that R_0 is a normal convex polygon.

If we know the normal polygon R_0 and the partition of its sides into equivalent pairs, then the group G is completely determined*. (*If the two extremities of a side AB of the first kind are on the X axis, then it is the same for the extremities of the equivalent side $A'B'$. There then exist an infinity of real transformations which change AB into $A'B'$. In that case, we have a polygon R_0 which generates an infinity of Fuchsian groups. N.E.N.)

In fact, we refer back to what we saw in the preceding section.

The generating transformations of G are those which correspond to the adjacent regions of R_0; we obtain them by considering the sides of the first kind on R_0 and finding the transformations which change each side into its equivalent. Now suppose AB and $A'B'$ are two equivalent sides. By Section III, these two sides must be congruent, i.e., we have

$$(A,B) = (A',B') .$$

But in this case, by Section I, there is a real transformation which changes AB into $A'B'$ and, in general, this transformation is uniquely determined. The generating transformations of the group G, and hence the group G itself, are therefore uniquely determined.

It is for this reason that I call the polygon R_0 a **generating polygon*** for the group G. (*Translator's Note: Nowadays we speak of a **fundamental polygon** or **fundamental region**, however I have retained Poincaré's term because his meaning is clear.)

V. Classification into families

We have seen above that two, three or more of the vertices of the polygon R_0 can be corresponding points, and we have called a complete set of mutually corresponding vertices of the polygon a **cycle**.

The vertices of R_0 are partitioned in this way into a certain number of cycles. We see how one can find the distribution of vertices into cycles from the pairing of sides of the first kind. To do this, it will suffice to apply the following rule, which is proved without difficulty:

Suppose that we traverse the perimeter of the polygon R_0 in a certain sense, encountering successively a side, then a vertex, then another side, then another vertex, and so on. Having agreed on this ordering, we depart from any vertex; we consider the next side, then its equivalent if the side is of the first kind; then the next vertex, then the next side, then its equivalent if the side is of the first kind, and so on. We can continue in this way until we return to the starting vertex, or else until we arrive at a side of the second kind.

In the first case, all the vertices encountered form a cycle. In the second case it will be necessary, in order to complete the cycle, to return to the starting vertex and consider in succession the preceding side, then its equivalent if this side is of the first kind, then the preceding vertex, then the preceding side, and so on until we finish by arriving at a side of the second kind.

It is easy to see that there are three categories of cycles: each point corresponding to a point z above X is itself above X, hence each vertex corresponding to a vertex of the first category (Section III) is itself of the first category. We then have cycles consisting entirely of vertices of the first category; they are the **cycles of the first category**. On leaving a vertex of the first category and successively taking the sides and vertices given by the above rule, we encounter only vertices of the first category, and hence only sides of the first kind also. Thus we are not stopped until we return to the starting vertex, i.e., the cycle is **closed**.

Each vertex corresponding the a vertex of the second or third category must itself belong to one of these two categories. This leads us to consider two new classes of cycles.

The **cycles of the second category** contain only vertices of the second category; in applying the above rule to them, one does not encounter sides of the second kind, i.e., the cycle will be **closed**.

The **cycles of the third category** contain vertices of the third category and may also contain some of the second; in applying the above rule to them, we encounter sides of the second kind, i.e., the cycle will be **open**. This permits us to classify the polygons R_0, and consequently the groups G, into families.

The polygon R_0 will be:

of the 1st family if its cycles are of the 1st category only;
of the 2nd family if its cycles are of the 2nd category only;
of the 3rd family if its cycles are of the 3rd category only;
of the 4th family if its cycles are of the 2nd and 3rd categories only;
of the 5th family if its cycles are of the 1st and 3rd categories only;
of the 6th family if its cycles are of the 1st and 2nd categories only;
of the 7th family if its cycles are of all three categories.

Some examples will make my idea more comprehensible.

Example 1: The polygon R_0 is a hexagon ABCDEF; all sides are the first kind; the equivalent sides are

AB and AF

BC and FE

CD and ED

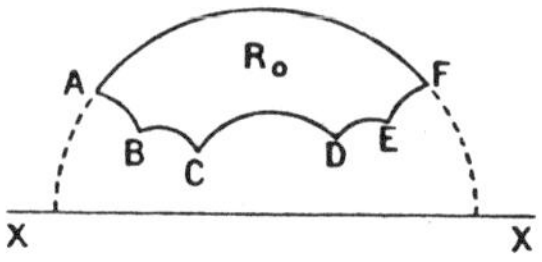

Figure 3

Applying the rule, we successively encounter:

1° Departing from vertex A: vertex A, side AB, side AF and vertex A;

2° Departing from B: vertex B, then BC, then EF, then F, then FA, then AB, then B;

3° Departing from C: vertex C, then CD, then ED, then E, then EF, then BC, then C;

4° Departing from D: vertex D, then DE, then CD, then D.

Thus there are four cycles, consisting respectively of:

1° The vertex A;

2° The vertices B and F;

3° The vertices C and E;

4° The vertex D.

Example II. The polygon R_0 is again a hexagon ABCDEF, all sides of which are of the first kind; but the opposite sides

AB and ED

BC and FE

CD and AF

are equivalent. Applying the rule, we successively encounter:

1° Departing from A: the vertex A, then AB, then DE, then E, then EF, then BC, then C, then CD, then FA, then A;

2^0 Departing from B: the vertex B, then BC, then EF, then F. then FA, then CD, then D, then DE, then AB, then B.

Thus there are two cycles, consisting respectively of:

1^0 The vertices A, C and E;

2^0 The vertices B, D and F.

Example III. The polygon R_0 is an octagon ABCDEFGH, all sides of which are the first kind; the opposite sides

AB and FE

BC and GF

CD and HG

DE and AH

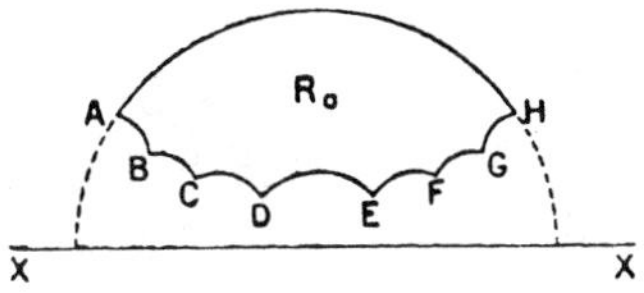

Figure 4

are equivalent. Departing from the vertex A, we successively encounter: A, AB, EF, F, FG, BC, C, CD, GH, H, HA, DE, E, EF, AB, B, BC, FG, G, GH, CD, D, DE, HA, A.

Thus all vertices belong to the same cycle.

Example IV. Suppose now that the equivalent sides are

AB and DC

BC and ED

EF and HG

FG and AH

Again departing from A, we encounter: A, AB, CD, C, DE, BC, C, CD, AB, B, BC, DE, E, EF, GH, H, HA, FG, G, GH, EF, F, FG, HA, A.

Thus all vertices again belong to the same cycle.

In the four preceding examples, all sides are of the first kind, hence all cycles are of the first or second category.

Example V. The polygon R_0 is again an octagon ABCDEFGH. The sides AB, CD, EF, GH are of the first kind; the sides BC, DE, FG, HA are of the second kind. The sides

AB and CD

EF and GH

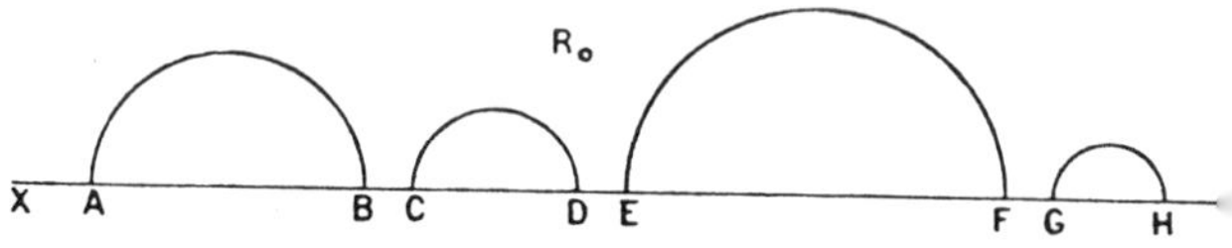

Figure 5

are equivalent. We successively encounter:

1° Departing from A: the side AB, then CD, then D, then DE, which is of the second kind;

2° Departing from C: the side CD, then AB, then B, then BC, which is of the second kind;

3° Departing from E: the side EF, then GH, then H, then HA, which is of the second kind;

4° Departing from G: the side GH, then EF, then F, then FG, which is of the second kind.

Thus the vertices are partitioned into four cycles, consisting respectively of:

1° A and D;

2° C and B;

3° E and H;

4° G and F.

These cycles are open, because each of them leads to an edge of the second kind. They are therefore of the third category.

The classification of polygons R_0 and groups G can be carried further. Here are some considerations which can be used to that effect.

First consider a cycle of the first category and a vertex A_0 of R_0 which is in that cycle. Let B_1 be the next side, C_1 its equivalent, A_1 the next vertex, B_2 the next side, C_2 its equivalent, A_2 the next vertex and so on; we continue in this way until we arrive at a vertex A_n coincident with A_0.

We can continue indefinitely in this way without ever being stopped, because we never encounter a side of the second kind. But vertex A_n is the same as A_0, A_{n+1} is the same as A_1 and, in general, the vertex A_h is A_k if we have

$$h \equiv k \pmod{n}$$

The cycle then consists of n distinct vertices, namely $A_0, A_1, \ldots, A_{n-1}$.

Now suppose that we describe an infinitely small circuit around A_0; we leave R_0 by the side B_1, to enter the polygon R_1. Considered as a side of the latter polygon, B_1 will be homologous to C_1; consequently the point A_0 will be homologous to A_1 and the next side will be homologous to B_2. The infinitely small circuit around A_0 therefore leaves the polygon B_1 by crossing the side homologous to B_2 and entering a certain polygon which I call R_2. Considered as a side of the latter polygon, the side which in R_1 is homologous to B_2 will now be homologous to C_2; the point A_0 will be homologous to A_2 and the next side will be homologous to B_3.

Continuing in this way, the rule is manifest. One passes successively through the polygons R_0, R_1, R_2, ..., R_{q-1} and finally into R_q which coincides with R_0.

One leaves the polygon R_k to enter the polygon R_{k+1} by crossing the side which, as a side of R_k, is homologous to B_{k+1} and which, as a side of R_{k+1} is homologous to C_{k+1}. The vertex A_0, reviewed as belonging to R_k, is homologous to A_k.

But the polygon R_q must coincide with R_0; then the vertex A_q cannot differ from A_0, and we have

$$q \equiv 0 \pmod{n} .$$

whence

$$q = pn .$$

The number of polygons which have a vertex at A_0 is therefore divisible by n. The figure below may clarify this. Suppose we have

$$n = 3 \ , \ p = 2 \ .$$

We have drawn only the parts of the polygons in an infinitesimal neighborhood of A_0; thus each polygon is represented by just an angle. In the neighborhood of each vertex A (or of each side B or C), in the interior of each polygon R_i, we have marked a letter indicating to which vertex (or side) or R_0 this vertex A (or side B or C) is homologous as a part of the polygon R_i.

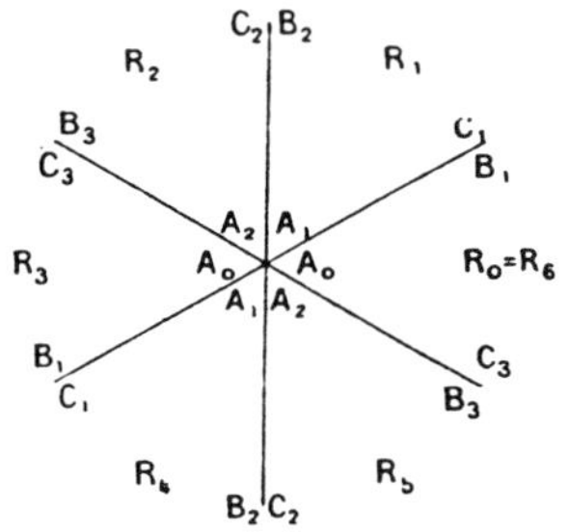

Figure 6

Around the point A_0 we then have pn angles whose sum equals 2π, and since homologous angles in congruent polygons are equal we have the following relation between the angles A:

$$A_0 + A_1 + A_2 + \dots + A_{q-1} = 2\pi .$$

But we have

$$A_{kn} + A_{kn+1} + A_{kn+2} + \dots + A_{kn+n-1} = A_0 + A_1 + \dots + A_{n-1} \ .$$

Hence

$$A_0 + A_1 + A_2 + \dots + A_{n-1} = \frac{2\pi}{p} \ ,$$

which permits us to state the following theorem:

The sum of the angles of R_0 which correspond to vertices in

the same cycle of the first category is an aliquot part of 2π.

Two cases can occur here: either the sum of angles will equal 2π, or else it will be 2π divided by an integer greater than 1.

In the first case I say that the cycle belongs to the first subcategory; in the second case it belongs to the second subcategory.

The properties of the cycles of the two subcategories are not entirely the same; and this compels us to make a cut in the first family, separating, under the name of the first order, the polygons R_0 whose cycles are all of the first category, from the other polygons R_0 which have a cycle of the second subcategory and which form the second order.

We return to the polygons R_0, R_1, ..., R_{q-1} which have a vertex at A_0.

Consider the real transformation which changes R_0 into R_n, and which evidently belongs to the group G. This transformation does not change the point A_0, and consequently it must be a fixed point. The transformation is therefore elliptic; it can be put in the form (3) (Section I) with

$$K = e^{\frac{2i\pi}{p}} .$$

Suppose now that the point A_0, instead of belonging to a cycle of the first category, belongs to a cycle of the second category. All the preceding reasoning applies, because we have assumed only that the cycle is closed, and this is as true for cycles of the second category as for those of the first. It is just that the number q, and consequently the number p, will be infinite. The sum of the angles corresponding to the different vertices in the cycle will therefore be zero, which is easy to foresee, because each angle of a normal polygon at a vertex of second category is evidently zero. The real transformation which changes R_0 into R_n does not change A_0. This vertex will therefore be a fixed point and, since it is on X, the transformation will be parabolic or hyperbolic.

Let S be this transformation; let C_k be the side which must be crossed in passing from R_k to R_{k+1}. C_{k+n} will be the transform of C_k by S, so that the different sides C_k will be successive transforms of C_0, C_1, ..., C_{n-1} under the transformation S and its inverse.

Suppose first that the transformation S is parabolic. We have seen (Section 1) that a finite arc which does not cut X cannot meet more than a finite number of successive transforms of a circle such as C_0 by a transformation such as S and its inverse. Such an arc therefore passes through no more than a finite number of the polygons which have a vertex at A_0.

Suppose, on the other hand, that the transformation S is hyperbolic. Let A_0 and A'_0 be the two fixed points of this transformation, and let C be the circle with diameter $A_0A'_0$. We have seen (Section I) that a finite arc which does not cut x meets an infinite or finite number of successive transforms of a circle such as C_0 according as it does or does not cut the circle C. Such an arc therefore passes through an infinite number of the polygons R_0, R_1, ..., R_q, ... if it cuts C, and through only a finite number if it does not cut C. If the transformation S is parabolic, we say that the cycle to which A_0 belongs is of the third subcategory; if it is hyperbolic we say that the cycle is of the fourth subcategory.

This allows us to subdivide the second, fourth, sixth and seventh families.

A group in one of these families belongs to the first order of that family if it does not contain a cycle of the fourth subcategory, and to the second order if it does not contain cycles of the fourth subcategory*. (*We can omit the latter subdivision. Poincaré shows later that we can avoid cycles of the fourth subcategory by suitably modifying the polygon R_0. However, this proof invites criticism; in reality, the fundamental polygon does not admit cycles of the fourth subcategory. N.E.N.)

VI. Existence of Fuchsian groups

Until now, we have reasoned on the assumption that Fuchsian groups exist, and we have not proved their existence. We have seen that each Fuchsian group can be considered given by a normal polygon whose sides are of two kinds; those of the first kind are arcs of circles with their centers on X, those of the second kind are segments of the axis X itself. The sides of the first kind are 2n in number and partitioned into equivalent pairs. When we know the polygon R_0 and the partition of its sides into pairs the corresponding group G is, in general, completely determined. As for the vertices, they are partitioned into a certain number of cycles.

In order to give rise to a Fuchsian group G, the polygon R_0 must satisfy two conditions:

1° **Two equivalent sides must be congruent**;

2° **The sum of the angles in a cycle of the first category must be an aliquot part of** 2π.

These conditions are necessary; are they sufficient? That is what I am going to prove.

I suppose that the polygon R_0 is given, along with the partition of its sides of the first kind. This information determines the generating transformations of the corresponding group and consequently the possibility of constructing the polygons R adjacent to R_0*. (*It is necessary to except the case raised in the note on p.77. N.E.N.) Operating on them as we have operated on R_0, we construct the polygons adjacent to the polygons previously obtained and, by continuing the operation indefinitely, we construct an infinity of polygons congruent to R_0. However, several hypotheses can be made here:

1° The polygons so constructed cover a part of the plane but not more than once, and without encroaching on each other. They then form a sort of checkerboard; the potions of the plane in question is partitioned into a certain number of mutually congruent regions. **The existence of the Fuchsian group corresponding to R_0 is then**

demonstrated.

2° Otherwise, the polygons encroach on each other in such a way as to cover a part of the plane several times or even infinitely often. In that case the group to which R_0 leads (i.e., the group whose generating transformations change each side of R_0 of the first kind into its equivalent) is continuous; it is not a Fuchsian group.

Now how can we tell which of the two cases we are dealing with? Consider an arbitrary point A, inside R_0, and another arbitrary point B. We join A to B by any arc AMB. This arc leaves R_0 by a certain side C_0; we then construct the polygon R_1 adjacent to R_0 along C_0; if AMB leaves R_1 by a side C_1, we construct the polygon R_2 adjacent to R_1 along C_1, and so on.

The arcs AMB are of two kinds:

1° After a finite number of operations of the above type one arrives at a polygon R_n which the arc AMB does not leave, so that the point B is inside R_n; the point B is then in the part of the plane covered by the polygons R and the arc AMB does not leave this part of the plane: we then say the arc AMB is **of the first kind**.

2° Otherwise, no matter how many operations are carried out, we never find a polygon R_n which the arc AMB does not leave: the arc AMB is then **of the second kind**.

As far as the point B is concerned, we can make three hypotheses:

1° All arcs AMB which can be drawn from A to B are of the second kind; the point B is then outside the part of the plane covered by the polygons R.

2° The arcs AMB from A to B include some of the first kind; each of these arcs leads, as above, to a polygon R_n containing B. Moreover, this polygon R_n is the **same** for each arc AMB of the first kind joining the points A and B. In this case the point B is in the part of the plane covered by the polygons R; moreover, these polygons do not cover this part of the plane more than once, and the existence of the Fuchsian group corresponding to R_0 is demonstrated.

3° The arcs AMB from A to B include some of the first kind, each of which leads to a polygon R_n containing B; but this polygon changes when the arc AMB is changed. In this case the point B is in the part of the plane covered by the polygons R, but these polygons cover this part of the plane more than once, so that there is no Fuchsian group corresponding to R.

This leads us to the following rule:

Join two points A and B by two arcs AMB, ANB of the first kind and investigate whether these two arcs lead to the same polygon R_n.

This can be modified as follows: trace a closed curve AMA whose initial and final point A is inside R_0; if the arc AMA leaves R_0 by a side C_0 we construct the polygon R_1 adjacent to R_0 along C_0; if it leaves R_1 by a side C_1 we construct the polygon R_2 adjacent to R_1 along C_1, and so on. I also assume that the arc AMA is of the first kind, i.e., that after a finite number of such operations we arrive at a polygon R_n which the arc AMA does not leave. **For the Fuchsian group to exist, it is necessary and sufficient that the polygon R_n be precisely R_0 for each curve AMA of the first kind.**

Suppose first that the polygon R_0 in question does not have any cycle of the fourth subcategory.

I then enunciate the following theorems:

I. **Each arc AMB which cuts X is of the second kind.** In fact, the polygon R_0 is entirely above X and consequently the same is true of all the polygons R congruent to it. Suppose that we have constructed the polygons R_1, R_2, ..., R_n as above. They will all be in the half plane above X and hence, since the arc AMB must leave this part of the plane, it must leave R_n, no matter how large n is.

Q.E.D.

II. **Each arc AMB which does not cut X is of the first kind.**. In fact, I construct the polygons

$$R_1, R_2, \ldots, R_p, \ldots$$

as above, and I let λ_p denote the portion of the arc AMB inside the polygon R_p. The arc AMB will be decomposed into a series of arcs λ_0, λ_1, λ_2, ... corresponding to the series of polygons R_0, R_1, R_2, ... The number of these arcs (which is the number of corresponding polygons) will be finite if AMB is of the first kind (which is what we propose to show) and infinite in the contrary case.

It is evident that λ_p joins two sides of the first kind in the polygon R_p, since it does not cut X and consequently cannot leave by a side of the second kind, since these sides are segments of X. Moreover, we can make two hypotheses:

1° We can assume that the two sides connected by the arc λ_p are consecutive sides of the polygon R_p, separated only by a vertex A_p of this polygon. I then say

that λ_p is of the first species and that it subtends the vertex A_p. Examples: the arc CD and FG of Figure 7.

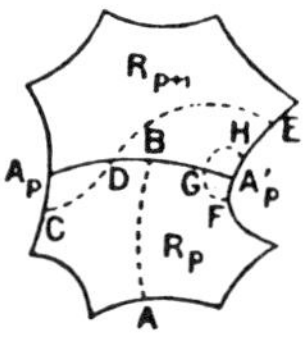

Figure 7

2° We can assume that the two sides connected by the arc λ_p are not consecutive, and in that case I say that λ_p is of the second species. Example: the arc AB of the figure. Since R_p is congruent to R_0, the arc λ_p will be congruent to a certain arc λ joining two points on the perimeter of R_0 and belonging to two non—consecutive sides. It is clear that the distance between these two points cannot be infinitely small and, consequently, that the L (see Section II) of this arc will also not be infinitely small, but will be greater than a certain given quantity K.

Let L_0 be the L of the arc AMB; it is finite because this arc does not cut X. The number of arcs of the second species will then be less than L_0/K, and consequently **bounded**.

We can then take q so large that the arcs λ_{q+1}, λ_{q+2}, ... , and in general the arcs λ_p, where $p > q$, are all of the first species; it remains to show that the number of these arcs λ_p $(p > q)$ is finite.

I envision two consecutive arcs λ_p and λ_{p+1} and the arc $\mu_p = \lambda_p + \lambda_{p+1}$ formed by their union; I say that the arc μ_p is of the first category if the arcs λ_p and λ_{p+1} subtend the same vertex, and of the second category in the contrary case. Thus the arc CE of the figure will be of the second category and the arc FH will be of the first, since the arc CD subtends the vertex A_p while the arcs FG, GH and DE subtend the vertex A'_p.

We envision an arc μ_p of the second category: it ends at two sides of the first kind belonging respectively to R_p and R_{p+1}, and it crosses the side C_p common to the boundaries of the two polygons; moreover, the three sides have no common point.

Since the polygon R_p is congruent to R_0, the side C_p will be congruent to one of the sides H of R_0. Let R' be the polygon among the R which is adjacent to R_0 along H; the arc μ_p will then be congruent to a certain arc μ with one extremity on a side B of R_0, and the other on a side B' of R'; this arc crosses the side H. Moreover, the three sides B, H and B' have no point in common.

Under these conditions, it is clear that the L of the arc μ, and consequently that of μ_p, cannot be infinitely small, and that it is greater than a fixed quantity K. It follows from this, as above, that the number of arcs μ_p of the second category is **bounded**, and that we can take q so large that all the arcs μ_{q+1}, μ_{q+2}, ... and in general μ_p $(p > q)$ will be of the first category; in other words, all the arcs λ_{q+1}, λ_{q+2}, ... and in general all λ_p $(p > q)$ will subtend the same vertex D.

It remains to show that the number of these arcs λ_p subtending the vertex D is finite. We can make two hypotheses:

1° Either D belongs to a cycle of the first or third category, and then this vertex does not belong to more than a finite number of polygons R, and consequently the number of arcs λ_p is necessarily finite;

2° Or else D belongs to a cycle of the second category, i.e., to the third subcategory, because we have assumed that the polygon R_0 does not admit cycles of the fourth subcategory. We have seen (Section V) that an arc of a curve which does not cut X cannot traverse more than a finite number of polygons to which the vertex D belongs. It follows that the number of arcs λ_p is finite.

Thus, in all cases, the number of arcs λ_p is finite; then the arc AMB is of the first kind.

Q.E.D.

Then the part of the plane covered by the polygons R **is the half plane above** X.

Now consider a closed curve AMA of the first kind, i.e., not cutting X; we apply to it the rule described above and suppose that in following it we successively encounter the polygons R_0, R_1, ... , arriving at a polygon R_n as we return to the point A. If the polygon R_n is precisely R_0, I say that AMA is of the first species and, in the contrary case, that it is of the second species. For the Fuchsian group to exist, it is necessary and sufficient that all contours AMA of the first kind be of the first species. I prove the following series of theorems on this subject:

III. **If the contour** AMA **is composed: first, of an arc** AMB, **second, of an infinitely small closed contour** BNB, **and third, of the arc** AMB **traversed in the opposite sense, then it is of the first species.**

In fact, we successively construct, by the above rule, the polygons R_0, R_1, R_2, ..., R_m encountered in traversing the arc AMB. We next traverse the infinitely small contour BNB. We can make three hypotheses concerning the latter contour;

1° Either, BNB lies entirely in the interior of a polygon R_m, so that one does not leave this polygon in traversing the contour;

2° Or, BNB crosses one of the sides C_m of R_m, but without enveloping one of the vertices of this polygon. Then, in following the contour, we initially leave R_m to cross the side C_m and enter a new polygon R_{m+1}, but we leave R_{m+1} by crossing the side C_m in the contrary sense and consequently reenter R_m;

3° Or else, BNB envelopes one of the vertices of R_m. This vertex is necessarily of the first category, because it cannot be on X. This vertex belongs to a cycle comprising, say, n vertices, and the sum of the corresponding angles will be $\frac{2\pi}{p}$. Then, in traversing BNB, we successively encounter pn polygons R_{m+1}, R_{m+2}, ..., R_{m+pn}, since the sum of the angles at D of all these polygons is 2π, R_{m+pn} will coincide with R_m.

Then, in all possible cases, we fall back on the polygon R_m after traversing the contour BNB. We now traverse the arc AMB in the opposite sense; we successively encounter the polygons R_m, R_{m+1}, ..., R_1, R_0, i.e., we meet those encountered the first time on the arc AMB, but in the opposite order. When we return to the point A, we fall back on the polygon R_0. Then the contour AMA will be of the first species.

Q.E.D.

IV. **If two contours** AMBPA, APBNA **are of the first kind and of the first species, then the same is true of the contour** AMBNA **formed by their union.**

In fact, if AMBPA, APBNA are of the first kind, then neither AMB nor ANB leaves the region covered by the polygons R and consequently AMBNA is also of the first kind.

If AMBPA is of the first species, we arrive at the same polygon R_n by following the arc AMB or the arc ABP; if APBNA is of the first species we arrive at the same polygon R_n in following the arc APB or the arc ANB. Then we arrive at the same polygon R_n in following the arc AMB or the arc ANB. Thus AMBNA is of the first species.

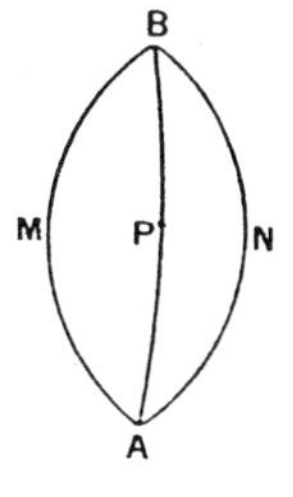

Figure 8

Q.E.D.

V. **If a contour** AMA **is of the first kind, then each contour**

ANA **inside it is also of the first kind**.

In fact, if AMA does not cut X, then the same is true of the interior contour ANA.

Q.E.D.

VI. **Any contour** AMA **of the first kind is of the first species.**

In fact, this contour can be decomposed into an infinity of contours like those considered in Theorem III. By virtue of Theorem V, these contours are all of the first kind; then by virtue of Theorem III each of them is of the first species, hence by virtue of Theorem IV the total contour AMA is itself of the first species.

Q.E.D.

Thus the existence of the Fuchsian group corresponding to R_0 **is shown**.

The polygons R resulting from R_0 by the different transformations in the group cover the half plane above X, and they cover it just once.

I now suppose that R_0 **admits cycles of the fourth subcategory***. (*See the note on p.86.)

Theorems I, III and IV are evidently still true. But this is not the case for Theorem II. Let us see how we have to proceed under these circumstances.

We retain the notation of Theorem II. We show as above that we can take q so large that each of the arcs λ_p, for $p > q$, will be of the first species and will subtend the same vertex D. There is nothing to add in the case where D is of the first or third category, or of the third subcategory. Suppose now that D is of the fourth subcategory. Is the number of arcs λ_p $(p > q)$ finite or infinite? We have seen (Section V) that an arc which does not cut X traverses an infinite number of polygons R which have D as a vertex, or else a finite number, according as it does or does not cut a certain circle C passing through D and with its center on X.

It follows that an arc AMB which does not cut X cannot be of the second kind.

We consider an arc AMB of the second kind which ends at a circle C, and suppose that it is deformed in a continuous manner; it cannot cease to be of the second kind without ceasing to cut C, as is easy to comprehend.

Theorem V is again true, but a new proof is lacking. We envisage a closed contour AMA of the first kind, and a second contour ANA inside the first; I claim that this second contour cannot be of the second kind. In fact, if it were so, it would have to cut a certain circle C; we could pass from the contour ANA to the contour AMA in a continuous fashion, and while its kind changed, the contour ANA would not cease to cut C: it therefore cannot be of the second kind unless AMA is, contrary to what we have supposed.

Now that Theorem V is proved, Theorem VI must be equally true, and consequently **the existence of the Fuchsian group corresponding to R_0 is proved**. The polygons R resulting from R_0 by the transformations in the group cover a part S of the plane, and do not cover it more than once. This region S is entirely above X and bounded by segments of this axis and an infinity of circles such as C, with their centers on X.* (*If R_0 is chosen suitably, the polygons R will always cover the whole part of the plane above X. N.E.N.)

Thus we have shown the existence of the Fuchsian group corresponding to R_0 in all possible cases; but the proof can be considerably simplified in certain particular cases, for example when we have no vertex of the first category. In this case, in fact, the polygon R_0 has 2n sides of the first kind and it divides the half plane above X into 2n+1 regions, namely the polygon R_0 itself and each region between X and a side of the first kind. We consider a side C_1, the region S_1 between C_1 and X, and the polygon R_1 adjacent to R_0 along C_1. The polygon R_1 subdivides the region S_1 into 2n smaller regions, namely the polygon R_1 itself and each region between the axis X and one of the remaining free sides of the first kind. When we have constructed several polygons R_0, R_1, ..., R_p , the half plane situated above X is found to be divided into a certain number of regions, namely the polygons R_0, R_1, ..., R_p and the regions S_1, S_2, ..., S_q between X and each of the remaining free sides of the first kind. If we look for a new polygon adjacent to R_0, R_1, ..., R_p we find it entirely inside one of the regions S_1, S_2, ..., S_q. It is then impossible for any of the polygons so constructed to overlap. The existence of the Fuchsian group corresponding to R_0 is therefore proved.

VII. Main examples

Before giving some examples of those Fuchsian groups whose existence I have shown generally, I want to explain a notation which will serve to define certain properties of the polygon R_0 in what follows. I commence by designating each of the sides of R_0 by a number, in such a way that, when the perimeter of this polygon is traversed in a suitable sense, we successively encounter side 1, then side 2, etc. Once this is done, I write the number of one of the sides of the first kind, then that of its equivalent, then a comma, then another side of the first kind, then its equivalent, then a comma, etc. Finally I write the sequence of numbers of sides of the second kind. For example, if R_0 is an octagon whose sides of odd rank are of the first kind, while those of even rank are of the second kind, and whose opposite sides of the first kind are equivalent, then I say that the polygon R_0 is of the system

$$(15,37;2468)$$

I have thus defined the number of sides of each kind and the distribution of sides in pairs. From the distribution of sides in pairs we deduce the distribution of vertices into cycles.

I now want to study some examples, using the preceding notation.

Example I. (14,23)

The polygon R_0 is a quadrilateral whose vertices I denote by A,B,C,D; side 1 will be the side AB; side 2 the side BC; side 3 the side CD; side 4 the side DA. All sides are of the first kind, and hence arcs of circles with their centers on X; the pairs of equivalent, and hence congruent, sides are AB, AD and CB, CD. Finally, the angle A is equal to $\frac{2\pi}{\alpha}$, angle B+D is $\frac{2\pi}{\beta}$, and angle C is $\frac{2\pi}{\gamma}$, where the numbers α, β, γ are integers.

We can join the two opposite vertices A and C by an arc of a circle with its center on X, thus decomposing the quadrilateral R_0 into two curvilinear triangles ACD and ABC.

I now want to define an expression which will be called on frequently in what follows. We consider a transformation of the plane by reciprocal radii; let 0 be the center of the transformation and let K^2 be its parameter; the circle C with center 0 and radius K remains unaltered by the transformation. I say that this transformation is a **reflection in the circle** C, and that two figures which are thereby transformed into each other are **symmetric with respect to this circle**.

It is then clear that the two triangles ACD, ABC are symmetric with respect to the circle AC. It follows that the curvilinear angles CAD and CAB, ACD and ACB, ABC and ADC are equal, and consequently that the triangle ABC has its three angles equal to $\frac{\pi}{\alpha}$, $\frac{\pi}{\beta}$, $\frac{\pi}{\gamma}$ respectively.

Now let R_1 be a quadrilateral adjacent to R_0 along one of its sides, for example AB; these two polygons are symmetric with respect to the circle AB.

This leads us, in the case under consideration, to a very simple construction of the system of polygons R.

We first construct a triangle ABC whose three angles are $\frac{\pi}{\alpha}$, $\frac{\pi}{\beta}$, $\frac{\pi}{\gamma}$ respectively; we can construct an infinity of such triangles, but they are all congruent; we then construct the triangles symmetric to ABC along its sides, and so on along the remaining free sides. In this way we partition the plane into an infinity of triangles congruent to ABC or ACD, and by reuniting these in pairs we have partitioned the plane into an infinity of quadrilaterals congruent to R_0 = ABCD.

To see what happens if one of the integers α, β, γ

becomes infinite, suppose for example that γ becomes infinite; the angle C which equals $\frac{2\pi}{\gamma}$ becomes zero, the cycle of the vertex C ceases to be of the first category and becomes of the third subcategory. R_0 again consists of two triangles ABC, ADC, symmetric with respect to the circle AC. The two sides AC and BC are arcs of circles which are tangential at the point C on X; the side AB is also an arc of the circle cutting AC and BC at the angles $\frac{\pi}{\alpha}$ and $\frac{\pi}{\beta}$.

Suppose in particular that

$$\alpha = 3 , \beta = 2 .$$

Suppose also, in order to determine the triangle ABC, that the sides AC and BC reduce to two lines perpendicular to X, that the point C is thrown to infinity and the prolongation of the line BC passes through the origin 0, and that the distance between the two parallels AC and BC equals 1/2. The circle AB then has center 0 and radius 1. Thus we fall back on the Fundamentaldreieck of Klein (**Mathematische Annalen, Vol. XIV**). The corresponding Fuchsian group is extremely remarkable; it consists of all the transformations $\left(z, \frac{az+b}{cz+d}\right)$ such that a,b,c,d are integers with ad − bc = 1. Consideration of this group and its subgroups is the basis of Klein's researches on the modular functions.

Example II. (16,25,34):

This is the example I of paragraph V. The polygon R_0 is a hexagon ABCDEF and the vertices form four distinct cycles: A; B and F; C and E; D; with the curvilinear angles A, B + F, C + E, D respectively equal to $\frac{2\pi}{\alpha}$, $\frac{2\pi}{\beta}$, $\frac{2\pi}{\gamma}$, $\frac{2\pi}{\delta}$, where α, β, γ, δ are positive integral or infinite numbers.

A case of particular interest is that where the hexagon decomposes into two quadrilaterals ABCD, ADEF, symmetric with respect to the circle AD. In this case we have the following values for the curvilinear angles of the two quadrilaterals:

$$BAD = DAF = \frac{\pi}{\alpha} ; \qquad CDA = EDA = \frac{\pi}{\delta}$$

$$B = F = \frac{\pi}{\beta} ; \qquad C = E = \frac{\pi}{\gamma} .$$

We can then construct all the hexagons R by operating as in the preceding example; we construct the quadrilateral ABCD, then the quadrilaterals symmetric to it in each of its sides, then the quadrilaterals symmetric to them in each of their sides, and so on.

We now consider a more general example. Suppose that R_0 is a polygon of 2n sides of the first kind whose vertices, in order, are

$$A_1,\ A_2,\ \ldots,\ A_n,\ A_{n+1},\ B_n,\ B_{n-1},\ \ldots,\ B_2\ .$$

Suppose that the sides A_1A_2 and A_1B_2, A_2A_3 and B_2B_3, ..., $A_{n-1}A_n$ and $B_{n-1}B_n$, A_nA_{n+1} and B_nA_{n+1} are equivalent. The vertices are distributed in n+1 cycles:

$$A_1;\ A_2B_2;\ A_3B_3;\ \ldots;\ A_{n-1}B_{n-1};\ A_nB_n;\ A_{n+1}$$

and the polygon R_0 is written

$$(1 \cdot 2n,\ 2 \cdot 2n-1,\ 3 \cdot 2n-2,\ \ldots\ ,n-1 \cdot n+2,\ n \cdot n+1)$$

in the notation we have adopted. In a certain particular case, this polygon R_0 divides into two halves $A_1A_2A_3 \ldots A_{n+1}$ and $A_1B_2B_3 \ldots B_nA_{n+1}$ symmetric with respect to the circle A_1A_{n+1}. Each of these halves is a curvilinear polygon whose sides have their centers on X and whose angles are aliquot parts of π. To construct all the polygons R it suffices to start with $A_1A_2A_3 \ldots A_{n+1}$ and construct the polygons symmetric to it in each of its sides, then the polygons symmetric to them in each of their sides, and so on.

We could cite a great number of other examples of generating polygons of Fuchsian groups. We mention only the examples previously cited in Section V. In the notation we have adopted, the polygons R_0 of Examples II, III, IV and V are written:

(14,25,36)

(15,26,37,48)

(13,24,57,68)

(13,57;2468)

What are the conditions the polygon R_0 must satisfy in these examples in order to give rise to a Fuchsian group?

In the first three examples the equivalent sides must be congruent. Moreover, in Example II, the sum of the angles of even rank, and likewise that of the angles of odd rank, must be an aliquot part of 2π; in Examples III and IV it is the sum of all the angles which must divide 2π.

In Example V, the polygon R_0 is not subject to **any** condition.

VIII. Classification by genus

In the preceding, I have shown how one can partition the half plane above X into an infinity of congruent curvilinear polygons R. Now consider the half plane below X, and in it the curvilinear polygons R'_0, R'_0, ... symmetric to R_0, R_1, ... respectively with respect to X. It is clear that if R_i is transformed into R_0 by a certain transformation $f_i(z)$ in the group G, then R'_i will be transformed into R'_0 by the same transformation. The transformations in the group G which transform R_0 into R_1, R_2, ... will then transform R'_0 into R'_1, R'_2, ... However, there are several cases to consider: if R_0 has no vertex of the fourth subcategory, the set of polygons R and R′ will cover the whole plane; if, on the contrary, we have vertices of this subcategory, then the polygons R and R′ will not meet the regions inside an infinity of circles with their centers on X.* (*See the note on p.95. N.E.N.)

Here there is a second distinction which is very important in what follows. If we suppose that the group G is of the first, second or sixth family, then we have neither a vertex of the third category, nor an edge of the second kind. Since the polygon R_0 has no segment of X as a side, it will be completely separated from its mirror image R'_0 or else they will meet at a vertex only (if there are vertices of the second category) and not along a whole edge. If, on the contrary, the group G is of the third, fourth, fifth or seventh family, the polygon R_0 will have one or more sides of the second kind and it will meet its mirror image R'_0 all along these sides. Suppose that we suppress the sides of the second kind which serve as the common boundary of R_0 and R'_0: we can consider the union $R_0 + R'_0$ of the two figures as a single region bounded only by arcs of circles with their centers on X, i.e., by the sides of the first kind on R_0 and R'_0. The set of these circular arcs in general forms several separate closed curves.

We now have to speak of a new classification of Fuchsian groups. First consider a group of the first, second or sixth family; the polygon R_0 has no side of the second kind; consequently the points on the perimeter of R_0 will correspond in pairs, because each point on a side of the first kind corresponds to a point on its equivalent; the interior points of R_0 correspond to no others inside the polygon or on its perimeter; finally, all vertices in the same cycle correspond. Suppose that we cut out the polygon R_0, then deform it continuously so that corresponding points of its perimeter are pasted together; after this deformation R_0 will have become a closed surface. If, for example, we take the quadrilateral ABCD in Example I of the preceding paragraph, cut it out and deform it so that AB is pasted to AD, and BC to CD, then R_0 will take on the appearance of a closed convex surface.

Now consider a quadrilateral ABCD with system (13,24), so that AB is equivalent to DC and BC to AD. If we fold up the quadrilateral and paste AB to DC, it takes the form of an open cylinder, with the ends, which are the sides BC and AD, remaining free but now closed curves; if we than paste AC to BD, the quadrilateral takes the form of a sort of closed ring.

We know that the closed surfaces can be classified by genus in the following way: on a sphere, for example, one cannot trace a closed curve without dividing the surface of the sphere into two distinct regions; it is not the same on a torus, where a meridian circle, for example, does not subdivide the surface into two distinct regions; however we cannot trace two disjoint closed curves on the torus without obtaining such a subdivision.

The genus of a closed surface is then the maximum number of disjoint closed curves which can be traced on the surface without dividing it into two distinct regions. Thus a closed surface in which we have made p holes will be of genus p.

The genus of a Fuchsian group is the genus of the closed surface obtained by the deformation of R_0 described above.

How do we obtain an expression for this genus?

Suppose that we have subdivided a surface of genus p into F curvilinear polygons, with a total of A sides and S vertices. Then we have the relation

$$F + S - A = -2p + 2 .$$

Returning to the polygon R_0; let 2n be the number of its sides of the first kind and let q be the number of closed cycles. It follows from this information; since the sides are pasted together in pairs, that the number of distinct sides remaining will be n; likewise, the number of distinct vertices remaining will be q.

We then have

$$F = 1 , A = n , S = q ,$$

whence

$$q + 1 - n = 2 - 2p$$

or

$$p = \frac{n+1-q}{2}$$

Returning to Examples I and II of the preceding section, we have

$$q = n + 1 ,$$

whence

$$p = 0 .$$

I want to give another application of the preceding rule.

Suppose that the polygon R_0 has 2n sides of the first kind and that opposite sides are equivalent. To find the genus of the corresponding Fuchsian group it is necessary to find the number of cycles into which the 2n vertices are partitioned. Applying the rule of Section V, we find that all vertices belong to the same cycle if n is even, and if n is odd they form two cycles, one consisting of the vertices of even rank, and the other of the vertices of odd rank. We than have

$$q = 1 \text{ or } 2$$

according as n is even or odd, and consequently

$$p = \frac{n}{2}$$

if n is even, and

$$p = \frac{n-1}{2}$$

if n is odd.

Now suppose that the polygon R_0 admits sides of the second kind; it will be contiguous to R'_0 along these sides, so that the region $R_0 + R'_0$ will be a single piece; the points in the interior of this region cannot correspond to any other point of the region, the points of the perimeter, which all belong to sides of the first kind of R_0 or R'_0, will correspond in pairs; finally, vertices of the same cycle will be corresponding points. We now cut out the region $R_0 + R'_0$ and fold and deform it so that corresponding points on the perimeter are pasted together. The genus of the Fuchsian group will by definition be the genus of the surface so obtained.

Let 2n again be the number of sides of R_0 of the first kind, and let q be the number of closed cycles; R'_0 will also have 2n sides of the first kind and q closed cycles. Suppose that we have h sides of the second kind and consequently h open cycles common to R_0 and R'_0.

We return to the formula

$$F + S - A = 2 - 2p .$$

We have $F = 2$, since there are two polygons R_0 and R'_0; we have $A = 2n + h$, since there are n pairs of sides of the first kind belonging to R_0 and n pairs of sides of the first kind belonging to R'_0, and h sides of the second kind; finally we have $S = 2q + h$ because there are 2q closed cycles and h open cycles.

It then follows that

$$p = n - q .$$

Suppose for example that R_0 is of the third family; i.e., that all vertices are of the third category. We then have

$$q = 0 , \quad p = n .$$

Thus, in Example V of Section V, the genus is equal to 2.

As a second example, I take a polygon R_0 whose distribution of sides is given by the following formula:

$$(15,24;3)$$

or, more generally

$$(1 \cdot 2n+1,\ 2 \cdot 2n,\ 3 \cdot 2n-1,\ \ldots,\ n \cdot n+2;\ n+1),$$

so that the $(n+1)^{th}$ side is the only one of the second kind, and the sides m and $2n + 2 - m$ are equivalent. In this case we have

$$q = n\ ,$$

whence

$$p = 0\ .$$

IX. Simplification of the generating polygon

We have seen at the beginning of Section IV that the regions R_0 are not completely defined by the condition that each of them contain a single point corresponding to a given point z. We have subsequently imposed another condition on these regions, that they be curvilinear polygons whose sides are circular arcs and segments of X. However, the regions R_0 (or **generating polygons**) are still not completely determined. In fact, we have seen that we can adjoin to R_0 any region S_0, provided we remove a region S_p resulting from S_0 by one of the transformations in the Fuchsian group, and the region obtained, $R_0 + S_0 - S_p$, can serve to generate the group in the same way as R_0. If the regions R_0, S_0 and S_p are contiguous and if all three are normal polygons, i.e., if their sides are segments of X or arcs of circles with their centers on X, then the resulting region $R_0 + S_0 - S_p$ is equally a normal polygon. It follows that the same Fuchsian group can be generated from infinitely many polygons R_0, and that we can exploit this indetermination to simplify the polygon.

Here is how we can effect such a simplification. We join two points A and B on the perimeter of R_0 by an arc of a circle with its center on X, so as to divide the polygon into two others, S_0 and T_0. We consider two equivalent sides CD and EF of R_0 and suppose that CD belongs entirely to the perimeter of S_0, and EF to that of T_0. Let R_1 be the polygon adjacent to R_0 along EF and suppose that it decomposes into two polygons S_1 and T_1, congruent to S_0 and T_0 respectively. The polygon $R'_0 = S_1 + T_0$ can serve as well as R_0 to generate the Fuchsian group G. It may be that R'_0 is more advantageous to consider than R_0, and by applying to R'_0 an analogous operation to that on R_0 we may arrive at a polygon R''_0 which is simpler than the other two.

In each particular case we let the circumstances of the problem guide us, and I do not want to dwell long on this point. I confine myself to a few examples, returning to the notation of Section VII.

1° A polygon R_0 of the system (16,23,45) can always be reduced to the system (16,25,34).

2° A polygon of the system (12,34,56,78) can always be reduced to the system (18,27,36,45).

3° An octagon of the system (13,24,57,68) can always be reduced to an octagon of the system (15,26,37,48).

Under this transformation of the curvilinear polygon R_0, what are the properties which remain invariant?

1° The family of the polygon R_0 does not change, with one exception I shall mention.

2° Its genus does not change either.

3° The number of closed cycles of the third category does not vary.

4° The number of cycles of the fourth subcategory also does not vary.

5° The number of closed cycles of the second subcategory does not vary either. We have seen that the sum of the angles corresponding to different vertices of such a cycle must equal $\frac{2\pi}{p}$, where p is an integer greater than one. The value of this number p is likewise invariant.

6° On the contrary, we can in general arbitrarily increase or diminish the number of closed cycles of the first subcategory, which are those for which the sum of the corresponding vertex angles is 2π.

X. Isomorphism

We have seen at the beginning of Section III that a Fuchsian group H is isomorphic to another Fuchsian group G if the number of generators is the same and the relations of the form (6) (Section III) which hold between the generators of G are the same as those which hold between the generators of H. To recognize isomorphism of two given groups, we must therefore be able to find the relations of the form (6) (Section III) which exist between the transformations in a given group, and in particular the **defining relations**, of which all the others are combinations.

In Section III we have seen that each relation of the form (6) is obtained by describing a closed curve AMA which leaves from a point A inside R_0 and does not go outside the half plane above X. Suppose that this curve traverses successively the regions $R_0, R_{\alpha_1}, R_{\beta_2}, \ldots, R_{\beta_{\nu-1}}$ and a final region R_{β_ν} coincident with R_0, and that in passing from $R_{\beta_{\kappa-1}}$ to R_{β_κ} it crosses a side of $R_{\beta_{\kappa-1}}$ equivalent to the side of R_0 which the latter has in common with R_{α_κ}. We can then write the identity

$$z = f_{\alpha_1}[f_{\alpha_2}(f_{\alpha_3} \ldots (f_{\alpha_\nu}(z)) \ldots]$$

which is of the form (6), and all relations of this form are obtained in this way. However, if we are able to write the defining relations it will not be necessary to describe all possible curves AMA; we can confine ourselves to infinitely small curves around the vertices of R_0. Imagine then the successive vertices of R_0, and describe an infinitesimal curve around each. Since each curve AMA must not leave the part of the plane above X, we cannot describe such a curve around the vertices of R_0 which are on X itself. Thus the only vertices of the polygon R_0 which we have to consider are of the first category.

I suppose that the different sides of R_0 are numbered so that, if the perimeter of the polygon is traversed in a suitable sense, one encounters the sides C_1, C_2, C_3, ..., C_q in succession; with q as the total number of sides, A_1 will be the vertex between C_q and C_1; A_2 the vertex between C_1 and C_2, etc. In general, A_i will be the vertex between C_{i-1} and C_i. If C_i is of the first kind, then the region adjacent to R_0 along C_i will be called R_i, and the corresponding generator will be

$$[z, f_i(z)].$$

Consider any vertex of the first category, which I call A_{α_1}. This vertex makes up part of a certain cycle of the first category; we find the other vertices of this cycle by applying the rule of Section V. Suppose that the rule gives the vertices

$$A_{\alpha_1},\ A_{\alpha_2},\ \ldots,\ A_{\alpha_p}$$

in precisely that order. Suppose that the sum of the angles corresponding to these vertices is $\frac{2\pi}{\lambda}$, where λ is an integer. We make the abbreviations

$$F(z) = f_{\alpha_1}[f_{\alpha_2}[f_{\alpha_3} \cdots [f_{\alpha_p}(z)] \ldots],$$

$$F[F(z)] = F^2(z) \qquad F[F^2(z)] = F^3(z), \ \ldots$$

$$F[F^{\lambda-1}(z)] = F^{\lambda}(z).$$

The defining relation to which we are led by the above rule, using an infinitesimal curve around A_{α_1} will be

$$z = F^{\lambda}(z).$$

The vertices A_{α_2}, A_{α_3}, ..., A_{α_p} which belong to the same cycle as A_{α_1} lead to the same relation. Thus:

The number of defining relations which exist between the generators of a Fuchsian group G is precisely the number of cycles of the first category in the corresponding polygon R_0.

The polygons of the second, third and fourth families do not admit any cycles of this category. Thus there is no defining relation between the transformations in any group from these families, and all relations of the form (6) which we can find between their transformations reduce to identities.

Here is the first consequence we can derive from this fact: **each group H with n generators is homomorphic to a Fuchsian group G of the second, third or fourth family, provided that the latter group likewise has n generators.** In fact, the first condition of isomorphism, which demands that the number of generators be the same, is satisfied, and the second, which requires that each relation between transformations in G hold between those of H is satisfied automatically, because there are no such relations. There is only one difference. If there are relations of the form (6) between the transformations of H, then G is not isomorphic to H and consequently the homomorphism is many—to—one. If, on the contrary, there is no such relation, then the homomorphism is invertible and consequently one—to—one. It follows that **two Fuchsian groups of the second, third or fourth family are isomorphic provided that the number of generators is the same, i.e., provided that the corresponding polygons R_0 have the same number of sides of the first kind.**

Now consider any two polygons R_0 and R'_0 whose sides of the first and second kind are distributed in the same manner, so that they are designated by the same formula, if we use the notation explained at the beginning of Section III. The two polygons are evidently of the same genus and same family, and the number of cycles formed by the vertices is the same for both polygons, so that each cycle of R_0 corresponds to a cycle of R'_0 and conversely. Moreover, a closed cycle of the first polygon corresponds to a cycle in the second polygon which is likewise closed. If, in addition, the sum of the angles of each closed cycle of R_0 is the same as the sum of the angles of the corresponding cycle of R'_0, then the Fuchsian groups generated by the two polygons will be isomorphic.

XI. Effective formation of Fuchsian groups

A Fuchsian group is completely determined when we know its generators, since by combining them in all possible ways we obtain all transformations in the group. Thus to form a Fuchsian group is to calculate the coefficients of its generators. This problem does not present any difficulty. Suppose in fact that we have constructed the polygon R_0 corresponding to the group to be formed, and that we have calculated the coordinates, and consequently the affixes, of all its vertices. The generators sought are the transformations which send each side of the first kind to its equivalent. Let α, β and γ, δ be the affixes of two equivalent sides of R_0. Suppose first of all that the quantities α, β, γ, δ are all imaginary, so that none of these vertices is on X. If α', β', γ', δ' denote the complex conjugates of α, β, γ, δ we must have

$$(\alpha, \beta) = (\gamma, \delta),$$

i.e.,

$$\text{(1)} \qquad \frac{\alpha-\alpha'}{\alpha-\beta'} \frac{\beta-\beta'}{\beta-\alpha'} = \frac{\gamma-\gamma'}{\gamma-\delta'} \frac{\delta-\delta'}{\delta-\gamma'}.$$

The real transformation which changes $\alpha\beta$ into $\gamma\delta$ is then perfectly determinate; to write it in a way which shows the reality of the coefficients I put

$$\alpha = \alpha_1 + i\alpha_2 \quad \beta = \beta_1 + i\beta_2 \quad \gamma = \gamma_1 + i\gamma_2 \quad \delta = \delta_1 + i\delta_2$$

$$\alpha' = \alpha_1 - i\alpha_2 \quad \beta' = \beta_1 - i\beta_2 \quad \gamma' = \gamma_1 - i\gamma_2 \quad \delta' = \delta_1 - i\delta_2.$$

The desired transformation is then written

$$\left(z, \frac{Az + B}{Cz + D}\right)$$

where

$$A = \begin{vmatrix} \alpha_1\gamma_1 - \alpha_2\gamma_2 & \gamma_1 & 1 \\ \beta_1\delta_1 - \beta_2\delta_2 & \delta_1 & 1 \\ \beta_1\delta_2 + \beta_2\delta_1 & \delta_2 & 0 \end{vmatrix} \qquad B = \begin{vmatrix} \alpha_1\gamma_1 - \alpha_2\gamma_2 & \alpha_1 & \gamma_1 \\ \beta_1\delta_1 - \beta_2\delta_2 & \beta_1 & \delta_1 \\ \beta_1\delta_2 + \beta_2\delta_1 & \beta_2 & \delta_2 \end{vmatrix}$$

$$C = \begin{vmatrix} \alpha_1 & \gamma_1 & 1 \\ \beta_1 & \delta_1 & 1 \\ \beta_2 & \delta_2 & 0 \end{vmatrix} \qquad D = \begin{vmatrix} \alpha_1\gamma_1 - \alpha_2\gamma_2 & \alpha_1 & 1 \\ \beta_1\delta_1 - \beta_2\delta_2 & \beta_1 & 1 \\ \beta_1\delta_2 + \beta_2\delta_1 & \beta_2 & 0 \end{vmatrix}$$

But we can again make two hypotheses. Suppose that α and γ are real, while β and δ remain imaginary; condition (1) is then automatic; the real transformation which changes $\alpha\beta$ into $\gamma\delta$ is again perfectly determinate and when it is written

$$\left(z, \frac{Az + B}{Cz + D}\right)$$

then the expressions for the coefficients A,B,C,D are

$$A = \begin{vmatrix} \alpha\gamma & \gamma & 1 \\ \beta\delta & \delta & 1 \\ h & 0 & 0 \end{vmatrix} \qquad B = \begin{vmatrix} \alpha\gamma & \alpha & \gamma \\ \beta\delta & \beta & \delta \\ h & -1 & 0 \end{vmatrix}$$

$$C = \begin{vmatrix} \alpha & \gamma & 1 \\ \beta & \delta & 1 \\ -1 & 0 & 0 \end{vmatrix} \qquad D = \begin{vmatrix} \alpha\gamma & \alpha & 1 \\ \beta\delta & \beta & 1 \\ h & -1 & 0 \end{vmatrix}$$

The real quantities α, β, γ, δ which can be the vertices of a polygon R_0 are subject to the inequality

$$\left[\frac{1}{\delta-\alpha} - \frac{1}{\gamma-\alpha}\right]\left[\frac{1}{\gamma-\alpha} - \frac{1}{\beta-\alpha}\right] < 0$$

As for h, it is an arbitrary parameter which is assumed real and subject to the inequality

$$(\gamma+h)(\delta+h)(\delta-\gamma)(\beta-\alpha) > 0$$

which says that the determinant AD − BC is positive.

Thus, when we know the affixes of the vertices of R_0, we can calculate the coefficients of the generators of G. It is therefore a matter of choosing the affixes in such a way that the polygon R_0 satisfies the conditions of Section VI, which are necessary if it is to give rise to a Fuchsian group. This is a purely algebraic problem which does not present any difficulty. The great variety of possible cases does not permit me to examine all of them in detail, as this would involve repetitions and be unprofitably long. I therefore confine myself to looking at a few examples.

FIRST EXAMPLE

We first seek to construct Fuchsian groups of the third family, with n generators. The corresponding polygon R_0 has 2n sides of the first kind and 2n of the second. To determine such a polygon completely it is necessary to impose 4n conditions arbitrarily. However, we have seen in Section IX that the same group can be generated by an infinity of polygons R_0. If we take this circumstance into account, then we easily recognize that it is necessary to impose 3n conditions arbitrarily in order to determine completely an n generator group of the third family. But this is precisely the total number of coefficients of n arbitrary real transformations. It follows that it is sufficient to take n real transformations at random in order to obtain a Fuchsian group of the third family; but **the coefficients of these transformations are not subject to any conditions of equality, they must only satisfy certain inequalities**.

What are these inequalities? This is the problem which remains to be resolved. First we find without difficulty that all the transformations must be hyperbolic, and consequently their fixed points must be real and situated on X. We can enunciate a general result concerning the distribution of these fixed points on X. In fact, we consider two equivalent sides ab and cd of R_0 and suppose that they are of the first kind and belong to the same pair. Suppose also, to fix ideas, that the point ∞ is in R_0, so that this region is a part of the plane above X and exterior to the different circles which are its sides of the first kind. If this is not the case, we can bring it about by a suitable change of variable. The generator which sends ab to cd then has **one of its fixed points on the segment** ab **of** X, **and the other on the segment** cd.

To push further the study of the inequalities which govern the generators of a group in the third family, I want to take a particular example, namely the octagon in Example V of Section V. The method I employ extends to the more general case.

The two generators change AB to DC and EF to HG. First consider the transformation S which changes AB to DC; its two fixed points M and N are situated on the segments AB and CD, respectively, of the axis X. We describe a circle on diameter AN, and consider the curvilinear triangle ABN; we similarly consider the circle on diameter DN and the curvilinear triangle DCN; the transformation S changes the triangle ABN into DCN. Then, by virtue of the principles of Section IX, we can replace the octagon R_0 = ABCDEFGH by the heptagon R'_0 = ANDEFGH. We similarly consider the transformation S_1 which changes HG into EF, and its two fixed points M_1 and N_1, situated on

EF and GH respectively. By an argument like that above, we see that we can replace the heptagon R'_0 by the hexagon $R''_0 = ANDEN_1H$. We remark that R''_0 is of the fourth family and of the second order of that family, while R_0 is of the third family. This is the exception to the principles of Section IX which I announced earlier; we can simplify the generating polygon R_0 so as to reduce a polygon of the third family to a polygon of the second order in the fourth family, or else of the second family, but never to a polygon of the first order in the fourth family, or to the second family. We now consider the operation S_2 which consists in first making the transformation S, then applying the transformation inverse to S_1. The transformation S changes AN to DN, and the inverse of S_1 changes DN to a certain circle QP inside N_1H. The transformation S_2 then changes AN to QP,

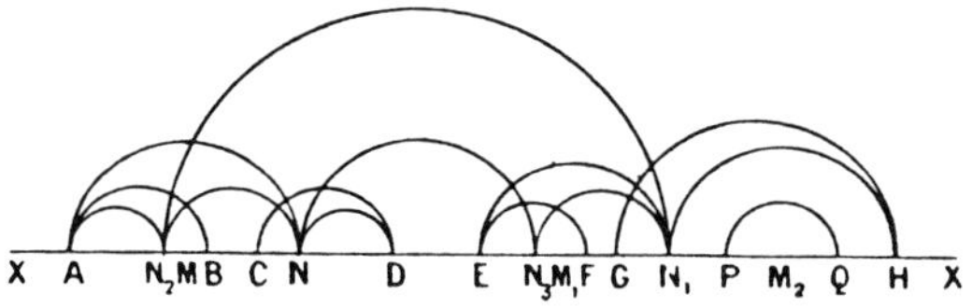

Figure 9

and its two fixed points N_2 and M_2 are situated on segments AN, PQ respectively. The transformation S changes N_2 to a certain point N_3; the transformation S_1 must also change N_2 and N_3, so N_3 is situated on the segment M_1E. We describe the circles on diameters N_2N, N_2N_1, N_1N_3, NN_3; the transformation S changes N_2N to N_3N, the transformation S_1 changes N_2N_1 to N_3N_1, and this permits us, by virtue of the principles of Section IX, to replace the octagon R_0 by the quadrilateral $R'''_0 = N_1N_2NN_3$. This quadrilateral is of the second family, and of the second order of this family.

We can now find the inequalities which govern the coefficients of S and S_1; we can always assume that the fixed points of S_1 are 0 and ∞, so that

$$N_1 = 0 \qquad M_1 = \infty$$

because if this condition is not satisfied we can bring it about by a very simple change of variable.

The two transformations are then written:

$$S = \left(z, \frac{az+b}{cz+d}\right) \qquad S_1 = (z, Kz) .$$

K is positive and greater than 1. Moreover, we can always assume that c is positive by changing the signs of all four quantities a,b,c,d. The points M and N are the fixed points of S and consequently roots of the equation

(1) $$cz^2 + (d - a)z - b = 0 .$$

The points M_2 and N_2 are the fixed points of the transformation

$$\left(z, \frac{az+b}{K(cz+d)}\right)$$

or the roots of the equation

(2) $$cz^2 + (d - \frac{a}{K})z - \frac{b}{K} = 0 .$$

The point N_3 and another point M_3 are the fixed points of the transformation

$$\left(z, \frac{az+bK}{cz+dK}\right)$$

and hence the roots of the equation

(3) $$cz^2 + (dK-a)z - bK = 0 .$$

Inspection of the figure shows that the quantities M, N, M_2, N_2, M_3, N_3 are all real and of the same sign, positive for example; whence the relations

$$b < 0 \qquad (d-a)^2 + 4bc > 0 \qquad a > d$$

$$\frac{b}{K} < 0 \qquad (dK-a)^2 + 4Kbc > 0 \qquad a > dK$$

$$bK < 0 \qquad (dK-a)^2 + 4Kbc > 0 \qquad a > dK$$

which reduce to

$$b < 0 \qquad a > dK \qquad (a+d)^2 > 4 \qquad a > d$$

$$\left(\frac{a}{\sqrt{K}} + d\sqrt{K}\right)^2 > 4$$

by taking into consideration that

$$ad - bc = 1 \ .$$

We now express the fact that M_2 and N_2 are smaller than M and N, and that M_3 and N_3 are greater. In this way we find the conditions

$$b < 0 \qquad a^2 + bc > 1 > d^2 + bc$$

$$b < 0 \qquad \frac{a^2}{K} + bc < 1 < d^2K + bc \qquad a - dK > a - d > \frac{a}{K} - d \ .$$

These conditions reduce to the following:

$$a > 0 \qquad b < 0 \qquad d < 0 \qquad a+d > 2$$

$$\frac{a}{\sqrt{K}} + d\sqrt{K} < -2 \ .$$

This method applies to all Fuchsian groups of the third family. We can always reduce the polygon R_0 as above, to a polygon of the second order in the third family. The vertices of the new polygon are known, since they are the fixed points of generators or certain of their combinations. To find the desired inequalities, it suffices to express the fact that these vertices are ordered in a certain way.

SECOND EXAMPLE

As a second example we take groups of the first family and genus 0 and, to be specific again, we choose the hexagon ABCDEF of Example II of Section VII.

Here we have three generators:

S_1 which changes AB to AF

S_2 which changes BC to FE

S_3 which changes CD to ED

In addition we consider

S_4 , the product of S_2 with S_1 inverse

S_5 , the product of S_3 with S_2 inverse

We consider the group as generated by S_1, S_4 and S_5.

Let a, b, c, d be the affixes of the points A, B, C, D and a′, b′, c′, d′ their complex conjugates. Let α, β, γ, δ be the angles A, B+F, C+E, D, which are aliquot parts of 2π. The transformations S_1, S_4, S_5 and S_3^{-1} are written:

$$\left(\frac{z-a}{z-a'},\ e^{i\alpha}\frac{z-a}{z-a'}\right) \qquad \left(\frac{z-b}{z-b'},\ e^{i\beta}\frac{z-b}{z-b'}\right)$$

$$\left(\frac{z-c}{z-c'},\ e^{i\gamma}\frac{z-c}{z-c'}\right) \qquad \left(\frac{z-d}{z-d'},\ e^{i\delta}\frac{z-d}{z-d'}\right)$$

By expressing the fact that the product of the four transformations S_3^{-1}, S_5, S_4, S_1 in succession is the identity transformation (z,z); we arrive at three relations between a, b, c, d, α, β, γ, δ.

Suppose that these relations are satisfied. Is this sufficient for the group generated by the transformations S_1, S_5 S_4 to be discontinuous? No, it is also necessary that the points A, B, C, D be able to form four vertices of a hexagon ABCDEF with angles

$$A = \alpha \qquad B+F = \beta \qquad C+E = \gamma \qquad D = \delta\ .$$

Consider the curvilinear angles of the quadrilateral ABCD obtained by joining the four points A, B, C, D by arcs of circles with their centers on X. It is necessary that the four angles A, B, C, D of this quadrilateral be respectively smaller than α, β, γ, δ; whence the inequalities

$$(2) \qquad \begin{aligned} 0 &< \arg \frac{d-a}{d-a'}\,\frac{b-a'}{b-a} < \alpha \\ 0 &< \arg \frac{a-b}{a-b'}\,\frac{c-b'}{c-b} < \beta \\ 0 &< \arg \frac{b-c}{b-c'}\,\frac{d-c'}{d-c} < \gamma \\ 0 &< \arg \frac{c-d}{c-d'}\,\frac{a-d'}{a-c} < \delta \end{aligned}$$

We can now recognize whether three given transformations S_1, S_4, S_5 are generators of a Fuchsian group analogous to the one we are concerned with. We form the product S_3^{-1} of S_1 inverse, S_4 inverse and S_5 inverse, in that order, and calculate the fixed points a, a'; b, b'; c, c'; d, d' of these four transformations, and their multipliers

$e^{i\alpha}$, $e^{i\beta}$, $e^{i\gamma}$, $e^{i\delta}$. Two conditions must then be satisfied:

1° α, β, γ, δ must be aliquot parts of 2π;

2° a, a′, b, b′, c, c′, d, d′, α, β, γ, δ must satisfy (2).

THIRD EXAMPLE

As third example we take the groups of second family, first order and genus 0; among these we again choose the one generated by the hexagon R_0 = ABCDEF considered in Example II of Section VII. But now this hexagon must be of the second family, with all its vertices on X. As in the preceding example, we consider the transformations

S_1 which changes AB to AF
S_2 which changes BC to FE
S_3 which changes DC to DE
S_4 the product of S_2 with S_1 inverse
S_5 the product of S_3 with S_2 inverse.

Let a, b, c, d, e, f be the affixes of the vertices of R_0; these will be real quantities. The transformations S_1, S_4, S_5 and S_3 must be parabolic, because the group is assumed to be of the first order, and they are written

$$S_1 = \left(\frac{1}{z-a}, \frac{1}{z-a} + \frac{1}{f-a} - \frac{1}{b-a}\right)$$

$$S_4 = \left(\frac{1}{z-b}, \frac{1}{z-b} + \frac{1}{h-b} - \frac{1}{c-b}\right)$$

$$S_5 = \left(\frac{1}{z-c}, \frac{1}{z-c} + \frac{1}{b-c} - \frac{1}{k-c}\right)$$

$$S_3 = \left(\frac{1}{z-d}, \frac{1}{z-d} + \frac{1}{e-d} - \frac{1}{c-d}\right)$$

where h and k are two quantities defined by the equations

$$\frac{1}{e-a} = \frac{1}{h-a} + \frac{1}{f-a} - \frac{1}{b-a}$$

$$\frac{1}{f-d} = \frac{1}{k-d} + \frac{1}{e-d} - \frac{1}{c-d}$$

Now what are the necessary conditions for the three transformations S_1, S_4, S_5 to give rise to a Fuchsian group? It is necessary that the six quantities a, b, c, d, e, f occur in precisely that circular order, or else in the inverse order, so that we have

$$(3) \qquad \frac{1}{b-a} < \frac{1}{c-a} < \frac{1}{d-a} < \frac{1}{e-a} < \frac{1}{f-a}$$

or else

$$\frac{1}{b-a} > \frac{1}{c-a} > \frac{1}{d-a} > \frac{1}{e-a} > \frac{1}{f-a}$$

Then we express the fact that the product of the transformations S_3^{-1}, S_5, S_4, S_1 is the identity, the result is

$$(4) \qquad (b-a)(d-c)(f-e) = (c-b)(e-d)(f-a)$$

REMARK

In the two preceding examples it is possibly advantageous to consider, not the hexagon ABCDEF of Example II in Section VII, which is denoted (16,25,34), but the hexagon denoted (12,34,56), which is equivalent to it, as we have seen in Section IX.

Let ABCDEF be the latter hexagon, with AB and BC, CD and DE, EF and FA as equivalent sides. We consider the transformations

S_1 which changes BA to BC
S_2 which changes DC to DE
S_3 which changes FE to FA
S_4 the product of S_1, S_2 and S_3

These four transformations are elliptic and their multipliers are $e^{i\alpha}$, $e^{i\beta}$, $e^{i\gamma}$, $e^{i\delta}$, where α, β, γ, δ are aliquot parts of 2π.

B is one of the fixed points of S_1, D is one of S_2, F is one of S_3, A is one of S_4; the other fixed points of these transformations are the respective complex conjugates of B, D, F, A; C is the transform of A by S_1, and E is that of C by S_2. For the group generated by S_1, S_2, S_3 to be discontinuous, it is sufficient that the normal curvilinear hexagon ABCDEF be convex; whence the conditions

$$0 < \arg \frac{d-a}{d-v'} \frac{f-b'}{f-b} < \alpha$$

(5) $$0 < \arg \frac{f-d}{f-d'} \frac{b-d'}{b-d} < \beta$$

$$0 < \arg \frac{b-f}{b-f'} \frac{d-f'}{d-f} < \gamma$$

To recognize whether the group generated by S_1, S_2, S_3 is discontinuous, we calculate the fixed points b, b′, c, d′, f, f′, a, a′ and the multipliers $e^{i\alpha}$, $e^{i\beta}$, $e^{i\gamma}$, $e^{i\delta}$ of S_1, S_2, S_3 and their product S_4, and investigate whether these quantities satisfy the inequalities (5) and whether α, β, γ, δ are aliquot parts of 2π.

Now suppose that the hexagon abcdef is no longer of the first family, but of the first order in the second family, so that the quantities a, b, c, d, e, f are real. These quantities must satisfy the inequalities (3) for the hexagon abcdef to be convex. The transformations S_1, S_2 S_3 are written:

$$S_1 = \left[\frac{1}{z-b}, \frac{1}{z-b} + \frac{1}{c-b} - \frac{1}{a-b}\right]$$

$$S_2 = \left[\frac{1}{z-d}, \frac{1}{z-d} + \frac{1}{e-d} - \frac{1}{c-d}\right]$$

$$S_3 = \left[\frac{1}{z-f}, \frac{1}{z-f} + \frac{1}{a-f} - \frac{1}{e-f}\right]$$

If we express the fact that the product S_4 of S_1, S_2, S_3 is a parabolic transformation, we find that the quantities a, b, c, d, e, f satisfy equation (4). This equation, together with the inequalities (3), is the condition for the group generated by S_1, S_2, S_3 to be discontinuous.

FOURTH EXAMPLE

As the fourth example we take a quadrilateral abcd whose opposite sides ab, dc and bc, ad are equivalent; the four vertices form a single cycle, and the sum Σ of the angles is an aliquot part of 2π. This quadrilateral generates a group of the first family and genus 1. We join the two opposite vertices bd by an arc of a circle with its center on X. We thus obtain a triangle abd, and on its sides we mark three points; α on bd, β on da, γ on ab, so that

$$(b,\alpha) = (\alpha,d) \qquad (d,\beta) = (\beta,a) \qquad (a,\gamma) = (\gamma,b).$$

Then there is a transformation S_1 which changes αb into αd, another S_2, which changes βd into βa, and another, S_3, which changes γa into γb; all three have multiplier -1. The group generated by S_1, S_2, S_3 will be discontinuous and as generating polygon it has the hexagon bαdβaγ, whose sides bα and αd, dβ and βa, aγ and γb are equivalent and colinear. It is therefore a particular case of a group generated by a hexagon abcdef, studied in the remark concerning the preceding example.

The generating transformations of the group generated by the quadrilateral abcd are S_5, which changes ab to dc, and S_6, which changes bc to ad, and it is easy to see that S_5 is the product of S_3 and S_1, and S_6 is the product of S_2 and S_1. This gives the following rule for the formation of all Fuchsian groups derived from a quadrilateral such as abcd:

We take three transformations S_1, S_2, S_3 with multiplier -1 which satisfy the conditions stated in the remark concerning the preceding example; then the products of S_3 and S_1, and S_2 and S_1, are the generators of the group.

XII. Generalization

Until now we have assumed that all transformations are real; but a first generalization can be made immediately.

Let

$$(1) \qquad \left(z, \frac{az+b}{cz+d}\right)$$

be any real transformation, and corresponding to it we take the transformation

$$(2) \qquad \left(\frac{\alpha z+\beta}{\gamma z+\delta}, \frac{a\dfrac{\alpha z+b}{\gamma z+\delta} + b}{c\dfrac{\alpha z+\beta}{\gamma z+\delta} + d}\right)$$

where α, β, γ, δ are any complex constants. Suppose that we fix the values of α, β, γ, δ and let the four quantities a, b, c, d run through all real values such that $ad - bc = 1$. It is clear that the transformations (2) form a group, and this group enjoys the following properties:

1° The transformations leave invariant the circle whose equation is

$$\text{imaginary part of } \frac{\alpha z+\beta}{\gamma z+\delta} = 0$$

and which I call the **fundamental circle**.

2° The transform of a point z by one of the transformations (2) will be inside or outside the fundamental circle according as the point z itself is inside or outside this circle.

We will suppose for example that, if z is inside the fundamental circle, the point $\frac{\alpha z+\beta}{\gamma z+\delta}$ is above X.

Now suppose that we successively equate the coefficients of (1) with those of the transformations in a Fuchsian group G. We then obtain an infinity of transformations (2) which form a group G′, and the latter group is evidently discontinuous. Corresponding to the group G there is a decomposition of the half plane above X into an infinity of normal polygons R_0, R_1, ..., R_i,... all congruent to each other. Suppose that $\frac{\alpha z+\beta}{\gamma z+\delta}$ runs around one of the polygons R_i whose sides are, as we have seen, either segments of X or arcs of circles with their centers on X; the point z will then run around the sides of a certain polygon S_i whose sides are either arcs of the fundamental circle, or arcs of circles which cut the fundamental circle orthogonally.

Thus, just as the group G corresponds to a division of the half plane above X into an infinity of congruent normal polygons, the group G′ corresponds to division of the interior of the fundamental circle into an infinity of congruent normal polygons, if the following terminology is adopted:

A normal polygon is a curvilinear polygon whose sides are arcs of the fundamental circle or arcs of circles which cut the fundamental circle orthogonally.

Two figures are congruent if we can pass from one to the other by a transformation such as (2), i.e., by a transformation which leaves the fundamental circle invariant. All that we have said about the distribution of sides in pairs and the vertices in cycles, and the classification of normal polygons into families, orders, and genera, remains true in the particular case to which we have confined outselves here.

Now what are the conditions under which a normal polygon gives rise to a discontinuous group G′? They are the same as the conditions we have found in the particular case of groups of real transformations, but it is convenient to modify the statement of these conditions:

1° The sum of the angles corresponding to the different vartices in a cycle must be an aliquot part of 2π;

2° If ab and cd are two equivalent sides then we must have, as in the case of real transformations,

$$\frac{a-a'}{a-b'} \frac{b-b'}{b-a'} = \frac{c-c'}{c-d'} \frac{d-d'}{d-c'} ;$$

however, a′, b′, c′, d′ no longer denote the complex conjugates of a, b, c, d but the points symmetric to a, b, c, d in the fundamental circle (see the definition in Section VII). In other words, if α is the center of the fundamental circle and ρ is its radius,

$$a' = \alpha + \text{complex conjugate of } \frac{\rho^2}{a-\alpha} ,$$

and similarly for b′, c′, d′.

We give the name **Fuchsian groups** to discontinuous groups such as G′, since they do not differ essentially from the groups of real transformations, and we reserve the name **Kleinian groups** for those groups which do not leave a fundamental circle invariant. We shall make these groups the subject of a special memoir.

The particulars which can be presented are the same as those for the corresponding groups of real transformations.

If the generating polygon R_0 is of the second order in the second, fourth, sixth or seventh family, then the set of polygons R_i does not cover the whole interior of the fundamental circle, but only the interior of a certain domain bounded by an infinity of circles which cut the fundamental circle orthogonally.

If the polygon R_0 is of the third, fourth, fifth or seventh family, then it has sides of the second kind. We have seen at the beginning of Section VIII that in this case it is advantageous to adjoin to each polygon R_i the polygon R'_i symmetric to it in X; this divides the whole plane into an infinity of regions $R_i + R'_i$ bounded by one or more separate perimeters. In the present case, we adjoin to each polygon R_i the polygon R'_i symmetric to it in the fundamental circle, so that the whole plane is again divided into an infinity of regions $R_i + R'_i$.

For the moment we leave aside the regions we have called R'_i and concern ourselves with the polygons R_i themselves. The area of the surfaces of these polygons, all interior to the fundamental circle, will be **finite**, which is very important from the point of view of later applications. There is an exception when the fundamental circle reduces to a line, as in the particular case of groups of real transformations; however, as we have seen at the beginning of this section, a linear change of variable suffices to return us to the general case.

In my later work I shall assume, to fix ideas, that the fundamental circle has center at the origin and radius 1. If this is not the case, we can easily bring it about by a very simple change of variable.

XIII. Historical

The first example of a discontinuous group consisting of linear transformations is the one encountered in studying the modulus k of an elliptic function (usual notation) or the modulus J (notation of Klein) as a function of the ratio of the periods. Hermite has made a profound study of this sort of transcendant and, in showing that it is uniform, has shown at the same time that the corresponding group is discontinuous.

The functions k and J, and hence the corresponding discontinuous group, have been studied by Dedekind, Fuchs and Klein, and more recently by Hurwitz. We mention in particular the important works of Klein found in the **Mathematische Annalen**, and a remarkable memoir of Fuchs in **Volume 83** of **Crelle's Journal**.

It is evident that any group G contains infinitely many others which are all discontinuous if G is; thus knowledge of a single discontinuous group easily permits the formation of infinitely many others. It is this remark which is the point of departure of the beautiful researches of Klein on the transformation of elliptic functions, and on modular functions in general.

Apart from the groups contained in the modular group, whose discontinuity is evident, there is another group whose discontinuity has been noted by Schwarz in a memoir in **Volume 75** of **Crelle's Journal**; this is Example I of Section VII. This is the first time we arrive at such a result without starting from the theory of elliptic functions. Finally, Fuchs takes up an analogous question in the work in **Volume 89** of **Crelle's Journal** and in the proceedings of the Göttingen Society. Although the groups studied in this latter work all reduce to previously known groups, it was a reading of this remarkable memoir which guided me in my first researches and which enabled me to find the law for the generation of Fuchsian groups, and to give a rigorous proof of it.

I first published them in a memoir I had the honour of submitting to the judgement of the Academie des Sciences in the concourse for the Grand Prix of mathematical sciences on 1st July 1880, and I have pursued the study of the groups in a series of works in **Comptes rendus** for the year 1881.

The mode of representation which I employ, i.e., the division of a portion of the plane into an infinity of curvilinear polygons, can be very useful for the study of general properties of a group; this is what I have sought to show. To geometers who wish to pursue this idea in the study of a Fuchsian group, or any other group, I recommend reading the **Habilitationsschrift** of Walter Dyck of the University of Leipzig, as it employs an analogous mode of representation and brings out its numerous advantages.

ON FUCHSIAN FUNCTIONS

Acta Mathematica I (1882), 193–294

I. Thetafuchsian series

In a previous memoir (Theory of Fuchsian groups), I have shown how it is possible to form discontinuous groups from transformations of the form

$$(1) \qquad \left[z, \frac{\alpha_i z+\beta_i}{\gamma_i z+\delta_i}\right]$$

by choosing the coefficients α_i, β_i, γ_i, δ_i in such a way that the different transformations in the group leave invariant a certain circle, called the **fundamental circle**. In what follows, I assume that this fundamental circle has center at the origin and radius one, so that its equation is

$$|z| = 1 .$$

I consider one of these discontinuous groups, called **Fuchsian groups**, which I name G. This group corresponds to a decomposition of [the disc within] the fundamental circle into an infinity of normal polygons R, all congruent to each other.

I propose to show that there is always a system of uniform functions of z which remain unaltered by the different transformations in the group G, and which I call **Fuchsian functions**.

For this purpose, I consider the different transformations in G to be comprised by the formula (1), and I set

$$\frac{\alpha_i z+\beta_i}{\gamma_i z+\delta_i} = f_i(z)$$

for brevity, as I have done in Section III of the memoir cited. I then form the series

$$(2) \qquad \sum_{i=0}^{\infty} \left|\frac{df_i(z)}{dz}\right|^m ,$$

where m is a positive integer greater than 1.

I shall show that this series is convergent under a series of different hypotheses:

1° Suppose first that z is inside the fundamental circle: it will then be in one of the polygons R, for example in the polygon R_h corresponding to the transformation in G with index h, which is written

$$[z, f_h(z)].$$

Let

$$f_k(z) = f_i[f_h(z)].$$

The transformation

$$[z, f_k(z)]$$

will be in the group G and will correspond to a certain polygon R_k. Since z is assumed to be in R_h, $f_i(z)$ will be in R_k.

Suppose that we describe a very small contour C_0 around z situated entirely inside R_h; the transform of C_0 by $[z, f_i(z)]$ will be a certain very small contour C_i around the point $f_i(z)$ and entirely inside R_k.

Then, to establish the convergence of the series (2), we shall have to prove in succession a certain number of lemmas.

LEMMA I. **The sum of the areas enclosed by all the contours C_i is a finite quantity** C.

In fact, these infinitely many contours C_i are all inside the fundamental circle; moreover, they have no parts in common, since each of them is entirely inside one of the polygons R. The sum of their areas is therefore smaller than the area of the fundamental circle, and hence finite.

Q.E.D.

LEMMA II. **The ratio of the greatest absolute value taken by** $\frac{df_i}{dz}$ **inside** C_0 **to the smallest is smaller than a certain quantity** K **independent of** i.

In fact, we have

$$\frac{df_i}{dz} = \frac{1}{(\gamma_i z+\delta_i)^2}.$$

The absolute value of $\frac{df_i}{dz}$ is therefore equal to $\frac{1}{|\gamma_i^2|}$ divided by the square of the distance from the point z to the point $-\frac{\delta_i}{\gamma_i}$. The points $-\frac{\delta_i}{\gamma_i}$ are the different transforms of the point ∞; they are therefore all outside the fundamental circle, since the point ∞ is itself outside, and they cannot be infinitely close to each other except in the neighborhood of this circle. Let M_i and m_i be the greatest and smallest absolute value of $\frac{df_i}{dx}$ when z is inside C_0; let a and b be the greatest and smallest distance of the point $-\frac{\delta_i}{\gamma_i}$ from the contour C_0; we evidently have

$$\frac{M_i}{m_i} = \frac{a^2}{b^2}.$$

But all the points $-\frac{\delta_i}{\gamma_i}$ are outside the fundamental circle. Let A and B be the greatest and smallest distances from the origin, the center of the fundamental circle, to the contour C_0. These two distances are smaller than 1, because C_0 is entirely inside the fundamental circle. We then have

$$a < 1 + A^{*} \qquad b > 1 - B ,$$

(*This inequality is not satisfied. But if we suppose, as is not inconvenient, that C_0 is a circle, we find that
$b > 1-A$ and

$$\frac{a}{b} = 1 + \frac{a-b}{b} = 1 + \frac{A-B}{b} < 1 + \frac{A-B}{1-A} = \frac{1-B}{1-A}$$

whence

$$\frac{M_i}{m_i} < \left(\frac{1-B}{1-A}\right)^2 = K .$$

If C_0 is just any very small contour, then we can reason as follows. The different transforms of the point ∞ are at a bounded distance from the origin. Suppose that they are all inside a circle with radius R and center at the origin. We then have

$$a < R + A \qquad b > 1 - A$$

whence

$$\frac{M_i}{m_i} < \left(\frac{R+A}{1-A}\right)^2 = K$$

so K is independent of i. N.E.N.)

whence

$$\frac{M_i}{m_i} < \left(\frac{1+A}{1-B}\right)^2 .$$

Moreover, $\left(\frac{1+A}{1-B}\right)^2 = K$ is independent of i. Q.E.D.

LEMMA III. **For any** i

$$M_i^2 < K^2 \frac{c_i}{c_0}$$

In fact, let $z = x + iy$; the area bounded by the contour C_0 is then written

$$c_0 = \iint dx\, dy$$

and the area bounded by the contour C_i is

$$c_i = \iint \left|\frac{df_i}{dz}\right|^2 dx\, dy$$

the two integrals being taken over the interior of the contour C_0. We then have

$$c_i > \iint m_i^2\, dx\, dy = m_i^2 c_0$$

or

$$c_i > \frac{M_i^2}{K^2} c_0$$

or finally

$$M_i^2 < K^2 \frac{c_i}{c_0}$$

Q.E.D.

Nothing could now be easier than establishing the convergence of the series (2). Suppose first that m = 2; we have to consider the series

$$\text{(2)} \qquad \sum \left|\frac{df_i}{dx}\right|^2 = \sum |\gamma_i z+\delta_i|^{-4}$$

But we have

$$\left|\frac{df_i}{dz}\right|^2 \leqslant M_i^2 < \frac{K^2}{c_0} c_i$$

The terms of the series (2) are therefore smaller than $\frac{K^2}{c_0}$ multiplied by the corresponding terms of the series ΣC_i, whose sum is a finite number c, by Lemma 1.

Then the series (2) also has a finite sum S. **A fortiori**, for any i

$$\left|\frac{df_i}{dz}\right| < \sqrt{S}$$

We therefore have, provided m > 2, that

$$\left|\frac{df_i}{dz}\right|^m < \left|\frac{df_i}{dz}\right|^2 S^{\frac{m-2}{2}}$$

i.e., each term of the series $\sum \left|\frac{df_i}{dz}\right|^m$ is smaller than the constant $S^{\frac{m-2}{2}}$ multiplied by the corresponding term of the series $\sum \left|\frac{df_i}{dz}\right|^2$, which is convergent.

The convergence of the series (2) is therefore proved, and the convergence is not just conditional, but absolute, since all the terms of the series are positive.

2° Suppose now that the point z is outside the fundamental circle.

If the point z is one of the points $-\frac{\delta_i}{\gamma_i}$, then one of the terms of the series is infinite and convergence is impossible. Suppose then that z does not coincide with any of the points $-\frac{\delta_i}{\gamma_i}$; I claim that the series (2) will again be convergent.

Consider another point z_1 inside the fundamental circle. The series

$$(2^a) \qquad \sum \left|\frac{df_i(z_1)}{dz_1}\right|^m = \sum \left|\gamma_i z_1 + \delta_i\right|^{-2m}$$

is convergent, as we have seen. I claim that it is the same for the series

$$(2) \qquad \sum \left|\frac{df_i}{dz}\right|^m = \sum \left|\gamma_i z + \delta_i\right|^{-2m}$$

In fact, we can find an upper bound R for $\left|z_1 + \frac{\delta_i}{\gamma_i}\right|$ and a lower bound r for $\left|z + \frac{\delta_i}{\gamma_i}\right|$; since the points $-\frac{\delta_i}{\gamma_i}$ have finite and bounded absolute value, they do not approach infinitely close to the point z.

We then have

$$\frac{|\gamma_i z + \delta_i|^{-2m}}{|\gamma_i z_1 + \delta_i|^{-2m}} < \frac{R^{2m}}{r^{2m}}$$

Each term of the series (2) is therefore smaller than $\left(\frac{R}{r}\right)^{2m}$ multiplied by the corresponding term of the series (2^a), which is convergent.

Then the series (2) is also convergent.

3° Now suppose that the point z is on the fundamental circle. The preceding demonstration is still applicable, provided that the point z is not approached infinitely closely by an infinity of points $-\frac{\delta_i}{\gamma_i}$. This is the case if the point z belongs to one of the sides of the second kind of a polygon R. The series (2) is then convergent. In the contrary case, the terms of the series (2), grow indefinitely, so that convergence does not take place.

The points of the fundamental circle are divided into two classes: those of the first belong to sides of the second kind on polygons R; the others, which do not satisfy this condition, are called **essential singular points of the group** G, so that the convergence condition for the series (2) can be enunciated as follows: **the point** z **must not coincide with any of the points** $-\frac{\delta_i}{\gamma_i}$, **nor with any of the essential singular points of the group** G.

The series (2) defines a function of z, but this function is not monogenic, as one easily sees from the form of the series. The sum of the series (2) depends equally on the group G, and if we suppose that the coefficients of the generators of this group are functions of a certain parameter t, then the sum of the series (2) will also be a function of t.

To enable me to express my thought more clearly, a small digression is necessary. Let us go back to Section XI of the memoir on Fuchsian groups, the Section entitled "Effective formation of Fuchsian groups", and to fix ideas, take Example I of that Section.

In that example, the Fuchsian groups under consideration are of the third family, generated by a normal polygon with 2n sides of the first kind and 2n sides of the second kind. We have seen that the coefficients of the n generators of this group are subject only to inequalities. It is therefore possible to express them by rational functions in 3n arbitrary parameters

$$u_1, u_2, \ldots, u_{3n}$$

which are required only to be real and to satisfy certain inequalities. The coefficients of all transformations in the group are then, like the generators, rational functions of the parameters u. Moreover, it is evident that all the different groups obtained by assigning different systems of values to the u are isomorphic.

The preceding can be extended to the more general case and, in order to state the coming results more easily, I shall introduce a new definition which will be useful in what follows. Consider two normal polygons R_0 and R'_0; I assume that they are expressed by the system of notation in Section VII of the memoir cited. R_0 and R'_0 have the same number of cycles of each category, and they are placed in correspondence with each other. I also assume that the sum of the angles of R_0 which belong to the same cycle of the first category is the same as the sum of the angles in the corresponding cycle of R'_0. I then say that the two polygons, and hence the two groups which they generate, are in the same **class**. It is clear that, in this case, the two groups are isomorphic.

Consider then an infinity of groups belonging to the same class C, and with n generators. We take any n transformations and see whether they can be used as generators for a discontinuous group belonging to class C. We find, in general, that their coefficients must satisfy certain equations and certain inequalities. These coefficients can then be expressed rationally as functions of p arbitrary parameters

$$u_1, u_2, \ldots, u_p$$

which are required only to be real and to satisfy certain inequalities. The only departure from Example I of Section XI is that in general we have

$$p < 3n$$

in place of $p = 3n$.

Two groups in the same class are, in general, in the same family. Nevertheless, there are exceptions. Take the Example I of Section XI, just cited. The groups considered are in general of the third family but, in certain limiting cases, they can reduce to groups of the second or fourth family. In general, a group of the second or fourth family can be considered to belong to a class formed from groups of the third family which do not reduce to the second or fourth family except for certain particular values of the parameter u. Likewise, a group of the sixth or the seventh family can be considered to belong to a class consisting of groups of the fifth family which do not reduce to the sixth or seventh family except for certain particular values of the parameters u. This remark will be useful to us in what follows.

Thus there are classes of Fuchsian groups which are all isomorphic to each other; the coefficients of their transformations are rational functions of certain real parameters u, subject to certain inequalities. If we form the series (2) for different groups belonging to the same class, then the sum of this series will evidently be a function of the u; I claim that it will be a **continuous** function of these parameters.

It is clear that each term of the series, being rational in the u, will be a continuous function of these parameters; but this does not suffice to make the same true for the sum of the series. If we consider, in fact, a series

$$S(x) = F_1(x) + F_2(x) + \dots + F_n(x) + \dots ,$$

whose terms are continuous functions of x and which is convergent, then the sum of this series can be a discontinuous function of x. However, we make another hypothesis. Suppose that, for

$$x_1 \geqslant x \geqslant x_0 \tag{3}$$

we have

$$|F_n(x)| < C_n$$

and that the series

$$C_1 + C_2 + \ldots + C_n + \ldots$$

is convergent; then we know that S(x) remains a continuous function of x as long as the latter variable satisfies the inequalities (3). This result is easily extended to the case of several variables. Thus to show that the series (2) is a continuous function of the u, it suffices to show that we can find an infinity of positive numbers Λ_i (independent of the u) such that the series $\Sigma\Lambda_i$ converges and such that the inequality

$$\left|\frac{df_i}{dz}\right|^m < \Lambda_i$$

is satisfied for all values of the parameters u between certain limits which, moreover, are as close as we please to the system of values for which we can prove the continuity of the sum of the series (2).

To demonstrate this continuity, I want to use certain considerations which at the same time will furnish a new proof of the convergence of this series — a useful outcome, in view of the importance of this result. I shall recall some of the definitions of Section II of the memoir on Fuchsian groups. In that Section, I have called two figures **congruent** if one can be transformed into the other by a linear transformation with real coefficients. Then, setting

$$z = x + y\sqrt{-1}$$

I have called the L of an arc the integral

$$\int \frac{|z|dz}{y}$$

taken along that arc, and the S of a plane area the integral

$$\iint \frac{dxdy}{y^2}$$

taken over the interior of that area.

The L of an arc and the S of an area are invariants of figures, i.e., two congruent arcs have the same L and two congruent areas have the same S.

Later, in Section XII of the memoir cited, I considered groups of transformations which were not real, but which left a certain **fundamental circle** invariant. By a completely natural extension, two figures are then said to be **congruent** when one is transformed into the other by a linear transformation preserving the fundamental circle. There are invariants for arcs and areas analogous to those we have found in the case of real transformations, and which we call, by extension, L and S. For example, in the case we are dealing with, the fundamental circle has center at the origin and radius one. We put

$$z = \rho e^{i\omega}$$

I call the L of an arc the integral

$$\int \frac{|dz|}{1-\rho^2}$$

taken along that arc, and the S of an area the integral

$$\iint \frac{\rho d\rho d\omega}{(1-\rho^2)^2}$$

taken over the interior of that area. One easily checks that two congruent arcs have the same L, while two congruent areas have the same S.

Consider a circle with center at the origin and radius ρ. Its S will be

$$\int_0^{2\pi} d\omega \int_0^{\rho} \frac{\rho d\rho}{(1-\rho^2)^2} = \frac{\pi\rho^2}{1-\rho^2}\,.$$

The L of its radius will be

$$R = \int_0^{\rho} \frac{d\rho}{1-\rho^2} = \frac{1}{2}\log\frac{1+\rho}{1-\rho}\,;$$

we call this quantity the R of the circle.

As a function of R:

$$\rho = \frac{e^{2R}-1}{e^{2R}+1}$$

$$S = \frac{\pi}{4}(e^{2R} + e^{-2R} - 2)$$

We now pass to a proof of convergence of the series (2), and suppose, to fix ideas, that the point z is inside the fundamental circle.

The proof extends without difficulty to the general case.

We enclose z by a contour C_0, which we can take to be so small that it lies entirely inside one of the polygons R, R_h for example. When the parameters u vary between the limits that we have fixed, the polygons R vary, but we can always assume that C_0 is so small that it remains constantly inside R_h. If we now consider the different transforms of the point z, i.e., the different points

$$\frac{\alpha_i z+\beta_i}{\gamma_i z+\delta_i}$$

then each of these points will be enclosed by a small contour C_i situated inside a certain polygon R_k, as we have seen above. All the contours C_i will be **congruent and disjoint from each other**.

I let σ be the S of C_0, and hence of all the C_i. If I now consider different circles cutting the fundamental circle orthogonally, and the arcs of these circles intercepted by C_0, then the L of these arcs lies below a certain bound which I call λ.

We now prove a few lemmas.

LEMMA IV. **The transforms of the point** z, **i.e., the points**

$$\frac{\alpha_i z+\beta_i}{\gamma_i z+\delta_i}$$

which are inside a circle C′ **with center at the origin and radius**

$$\rho' = \frac{e^{2R'}-1}{e^{2R'}+1}$$

are smaller in number than

$$\frac{\pi}{4\sigma}\left(e^{2(R'+\lambda)} + e^{-2(R'+\lambda)} - 2\right)$$

In fact, let N be that number.

If a point $\frac{\alpha_i z+\beta_i}{\gamma_i z+\delta_i}$ is inside a circle C′, then the corresponding contour will evidently be entirely inside the circle C″ with center at the origin and R which exceeds that of C′ by λ, i.e., whose R equals $R' + \lambda$.

Inside the circle C″, then, there are at least N contours C_i, whose total S is $N\sigma$.

But the S of the circle C″ is

$$\frac{\pi}{4}\left(e^{2(R'+\lambda)} + e^{-2(R'+\lambda)} - 2\right)$$

We therefore have

$$N < \frac{\pi}{4\sigma}\left(e^{2(R'+\lambda)} + e^{-2(R'+\lambda)} - 2\right)$$

Q.E.D.

LEMMA V. **We have, identically,**

$$\frac{1}{|\gamma_i z+\delta_i|^2} = \frac{1 - \left|\frac{\alpha_i z+\beta_i}{\gamma_i z+\delta_i}\right|^2}{1 - |z|^2}$$

In fact, consider an infinitely small contour C_0 enclosing the point z, and the transform C_i of this contour under the transformation

$$\left(z, \frac{\alpha_i z+\beta_i}{\gamma_i z+\delta_i}\right)$$

Letting ω_0 and ω_i be their areas, we have

$$\frac{\omega_i}{\omega_0} = \left| \frac{d \frac{\alpha_i z+\beta_i}{\gamma_i z+\delta_i}}{dz} \right|^2 = \frac{1}{|\gamma_i z+\delta_i|^4}$$

The S values of ω_0 and ω_i are

$$\iint \frac{dx\ dy}{(1-|z|^2)^2} = \frac{\omega_0}{(1-|z|^2)^2}$$

$$\iint \frac{dx\ dy}{\left(1-\left|\frac{\alpha_i z+\beta_i}{\gamma_i z+\delta_i}\right|^2\right)^2} = \frac{\omega_i}{\left(1-\left|\frac{\alpha_i z+\beta_i}{\gamma_i z+\delta_i}\right|^2\right)^2}$$

But these figures are congruent and have the same S, hence

$$\frac{\omega_i}{\omega_0} = \left[\frac{1-\left|\frac{\alpha_i z+\beta_i}{\gamma_i z+\delta_i}\right|^2}{1-|z|^2}\right]^2$$

whence

$$\left|\frac{1}{(\gamma_i z+\delta_i)^2}\right| = \frac{1-\left|\frac{\alpha_i z+\beta_i}{\gamma_i z+\delta_i}\right|^2}{1-|z|^2}$$

Q.E.D.

Consider two circles with center at the origin, one passing through the point z and the other through the point

$$\frac{\alpha_i z+\beta_i}{\gamma_i z+\delta_i}.$$

Let A be the R of the first circle and let R′ be the R of the second circle. We have

$$|z| = \frac{e^{2A}-1}{e^{2A}+1}, \qquad \left|\frac{\alpha_i z+\beta_i}{\gamma_i z+\delta_i}\right| = \frac{e^{2R'}-1}{e^{2R'}+1}$$

and finally

$$\frac{1}{|\gamma_i z+\delta_i|^2} = \frac{e^{2A}+e^{-2A}+2}{e^{2R'}+e^{-2R'}+2}.$$

THEOREM. **The series (2) is convergent.**

We construct an infinite sequence of circles with the origin as common center and R values increasing in arithmetic progression. Let $K_1, K_2, \ldots, K_n, \ldots$ be these circles, and let nr be the R of circle K_n.

We write the series (2) in the following form:

$$(2') \qquad \Sigma = U_1 + U_2 + U_3 + \ldots + U_n + \ldots$$

The term U_n of the series (2′) is obtained by grouping together all terms of the series (2) which correspond to points $\frac{\alpha_i z+\beta_i}{\gamma_i z+\delta_i}$ in the circular ring between the two circles K_{n-1} and K_n.

Since the terms of the series (2) are positive, such a grouping is legitimate, and the convergence of the series (2′) implies that of the series (2).

By virtue of Lemma IV, the number of terms of (2) grouped together in the term U is less than

$$\frac{\pi}{4\sigma}\left[e^{2(nr+\lambda)} + e^{-2(nr+\lambda)} - 2\right] < \frac{\pi}{4\sigma} e^{2(nr+\lambda)}$$

By virtue of Lemma V, each of them is smaller than

$$\left[\frac{e^{2A}+e^{-2A}+2}{e^{e(nr-r)}+e^{-2(nr-r)}+2}\right]^m < \left[\frac{e^{2A}+e^{-2A}+2}{e^{2(nr-r)}}\right]^m$$

We then have

$$U_n < \frac{\pi}{4\sigma}\left[e^{2A}+e^{-2A}+2\right]^m e^{2\lambda+2mr} e^{-2n(m-1)r}$$

Putting

$$\frac{\pi}{4\sigma}\left[e^{2A}+e^{-2A}+2\right]^m e^{2\lambda+2mr} = K$$

we have

$$U_n < \frac{K}{e^{2n(m-1)r}} \tag{4}$$

K is a constant independent of n and the second member of the inequality (4) is then, since $m > 1$, the nth term of a decreasing geometric progression. The series (2′), and hence the series (2), is therefore convergent.

Q.E.D.

Let us see the error involved in truncating the series (2) at the terms corresponding to points $\frac{\alpha_i z+\beta_i}{\gamma_i z+\delta_i}$ inside a circle with center at the origin and R value $(n-1)r$. The sum of the terms neglected is equal to

$$U_n + U_{n+1} + \dots$$

and consequently smaller than

$$\frac{Ke^{2n(1-m)r}}{1-e^{2r(1-m)}}$$

THEOREM. **The sum Σ of the series** (2) **is a continuous function of the parameters** u.

In fact, let Σ be the value of this sum for certain values

$$u_1, u_2, \dots, u_p$$

of the parameters u.

Let $\Sigma + \Delta\Sigma$ be the value of this same sum for the neighboring values

$$u_1 + \Delta u_1, u_2 + \Delta u_2, \dots, u_p + \Delta u_p$$

of the same parameters. I claim that we can take the Δu so small that

$$|\Delta\Sigma| < \epsilon$$

where ϵ is a given quantity.

Let Σ_0 be the sum of the first n—1 terms of the series (2′):

$$\Sigma_0 = U_1 + U_2 + \dots + U_{n-1}$$

Let

$$\Sigma_1 = U_n + U_{n+1} + \cdots$$

$$\Sigma = \Sigma_0 + \Sigma_1$$

Let $\Sigma_0 + \Delta\Sigma_0$, $\Sigma_1 + \Delta\Sigma_1$ be the sums of the corresponding terms in the series for $\Sigma + \Delta\Sigma$ so that

$$\Sigma + \Delta\Sigma = (\Sigma_0 + \Delta\Sigma_0) + (\Sigma_1 + \Delta\Sigma_1)$$

We have

$$\Sigma_1 < \frac{Ke^{2n(1-m)r}}{1-e^{2r(1-m)}} \qquad \Sigma_1 + \Delta\Sigma_1 < \frac{Ke^{2n(1-m)r}}{1-e^{2r(1-m)}}$$

We can then take n so large that

$$\Sigma_1 < \frac{\epsilon}{3} \qquad \Sigma_1 + \Delta\Sigma_1 < \frac{\epsilon}{3}$$

But, once n is chosen, Σ_0 will be a continuous function of the u; we can then take the Δu so small that

$$|\Delta\Sigma_0| < \frac{\epsilon}{3}$$

amd consequently

$$|\Delta\Sigma| < \epsilon$$

Q.E.D.

Now consider the following series

$$(5) \qquad \Theta(z) = \sum H\left(\frac{\alpha_i z+\beta_i}{\gamma_i z+\delta_i}\right)(\gamma_i z + \delta_i)^{-2m}$$

I suppose:

1° That H(z) represents a rational function of z, whose infinities are not on the fundamental circle, but which is otherwise arbitrary;

2° That the number m is an integer greater than 1. The function H(z) has a certain number of infinities

$$a_1, a_2, \ldots, a_p$$

If the point z coincides with one of the points

$$\frac{\alpha_i a_k+\beta_i}{\gamma_i a_k+\delta_i}$$

then one of the terms of the series is infinite and consequently the series cannot be convergent.

Suppose on the other hand that this is not the case. We can find a positive number M such that, for any i,

$$\left| H\left(\frac{\alpha_i z+\beta_i}{\gamma_i z+\delta_i}\right) \right| < M$$

and we can likewise choose M so large that this inequality continues to hold when we vary the parameters u within certain limits.

I now claim that the series $\Theta(z)$, which I call a **theta-Fuchsian series**, is convergent. In fact we have

$$\left| H\left(\frac{\alpha_i z+\beta_i}{\gamma_i z+\delta_i}\right)(\gamma_i z+\delta_i)^{-2m} \right| < M \left| \gamma_i z+\delta_i \right|^{-2m}$$

The absolute value of each term of the series (5) is therefore smaller than the corresponding term of a convergent series of positive terms. I.e., the series (5) is convergent, and the sum is independent of the order of the terms.

Moreover, we can show, as for the sum of the series (2), that the sum of the series (5) is a **continuous** function of the parameters.

II. Classification and general properties

Thus the series (5) is convergent, except at certain singular points; when convergent it defines a holomorphic function $\Theta(z)$. The function $\Theta(z)$ is essentially uniform, but it ceases to be holomorphic at the singular points where the series (5) ceases to converge.

These singular points are:

1^0 The points

$$\frac{\alpha_i a_k + \beta_i}{\gamma_i a_k + \delta_i}$$

i.e., the different transforms of the infinities of H(z) ; these points are the poles, in the neighborhood of which $\Theta(z)$ is meromorphic;

2^0 The points $\frac{-\delta_i}{\gamma_i}$, i.e., the different transforms of the point ∞. These points are again poles, in the neighborhood of which $\Theta(z)$ is meromorphic.

We demonstrate this double fact by remarking that, in the neighborhood of these points, one of the terms of the series (5) becomes infinite and, if we suppress this term, the remaining series converges.

3^0 Finally, we have the essential singular points of the group G, i.e., the points of the fundamental circle not belonging to a side of the second kind in one of the polygons R. These are also essential singular points of the function $\Theta(z)$.

Now here is the fundamental property of this function. We consider any transformation in the group G, for example

(6) $$\left\{z, \frac{\alpha_k z+\beta_k}{\gamma_k z+\delta_k}\right\}$$

and seek a relation between

$$\Theta\left[\frac{\alpha_k z+\beta_k}{\gamma_k z+\delta_k}\right] \qquad \text{and} \qquad \Theta(z)$$

The system of transformations

$$\left\{z, \frac{\alpha_i z+\beta_i}{\gamma_i z+\delta_i}\right\}$$

forms a group which, since it includes the transformation (6), will be identical with the system

$$\left[z, \frac{\alpha_i\left[\frac{\alpha_k z+\beta_k}{\gamma_k z+\delta_k}\right] + \beta_i}{\gamma_i\left[\frac{\alpha_k z+\beta_k}{\gamma_k z+\delta_k}\right] + \delta_i}\right] = \{z, f_i[f_k(z)]\}$$

if we make the abbreviations

$$\frac{\alpha_i z+\beta_i}{\gamma_i z+\delta_i} = f_i(z) \qquad \frac{\alpha_k z+\beta_k}{\gamma_k z+\delta_k} = f_k(z)$$

which permit us to write

$$\Theta(z) = \Sigma\, H\{f_i[f_k(z)]\} \left[\frac{d\{f_i[f_k(z)]\}}{dz}\right]^m$$

or

$$\Theta(z) = \Sigma\, H[f_i(f_k)] \left[\frac{df_i(f_k)}{df_k}\right]^m \left[\frac{df_k}{dz}\right]^m$$

But we also have

$$\Theta(f_k) = \Sigma\, H[f_i(f_k)] \left[\frac{df_i(f_k)}{df_k}\right]^m$$

and hence

$$\Theta(f_k) = \Theta(z) \left(\frac{df_k}{dz}\right)^{-m}$$

or

(7) $$\Theta\left(\frac{\alpha_k z+\beta_k}{\gamma_k z+\delta_k}\right) = \Theta(z)(\gamma_k z+\delta_k)^{2m}$$

We give the name **theta—Fuchsian function** to any uniform function of z enjoying property (7). We classify the theta—Fuchsian functions in the same way as the Fuchsian groups, by means of the properties of the corresponding normal polygon R_0.

We have seen that the polygons R_0 can be distributed among seven families, and that the first, second, fourth, sixth and seventh of these families are subdivided into two orders. But every group in the second order of the second, fourth, sixth and seventh family is identical to a group of the third or fifth family, or to a group in the first order of the sixth or seventh family (see Sections IX and XI of the memoir on Fuchsian groups). Hence we can always assume that the group G does not belong to the second order in the second, fourth, sixth or seventh family.

Having done this, I say that the function $\Theta(z)$ belongs to the first, third or fifth family if the group G does, and that $\Theta(z)$ belongs to the second, fourth, sixth or seventh family if G belongs to the first order of one of these families.

Similarly, I say that a theta—Fuchsian function is of genus p if the corresponding group is of that genus.

I can likewise extend, to theta–Fuchsian functions, the classification of Fuchsian groups into classes, which I spoke of in the previous paragraph.

Consider first the functions of the first, second, and sixth family. The corresponding polygons R have no sides of the second kind, so all points of the fundamental circle are essential singular points. The plane is divided into two parts, inside and outside this circle, by an essential singular line. The necessary conclusion, in conformity with the principles now admitted in the theory of functions, and illuminated by the works of Weierstrass, is that the expansion (5) represents two distinct functions, according as z is inside or outside the fundamental circle. The first of these functions exists only inside the circle and has all the outside as lacunary space; the second, on the contrary, exists only outside the fundamental circle. In what follows I shall always consider the first of these functions; in fact, the second of them, i.e., the one which exists only outside the fundamental circle, can easily be reduced to a theta–Fuchsian function existing only inside the fundamental circle, by changing z to $\frac{1}{z}$.

Consider then a theta–Fuchsian function existing only inside the fundamental circle and defined by a series such as (5); we can make two hypotheses:

We can assume that one or more of the infinities of H(z) are inside the fundamental circle; then the function $\Theta(z)$ has infinities (except in certain exceptional cases where several infinities of this function mutually annihilate each other) and we say that it is of the **first species**; we are then certain that the sum of the series (5) is not identically zero, because the sum can grow indefinitely.

On the other hand, we can assume that all infinities of H(z) are outside the fundamental circle; then the function $\Theta(z)$ has no infinities and we say that it is of the **second species**. The function $\Theta(z)$ can then be expanded in an ordinary series of increasing powers of z which remains convergent as long as z is inside the fundamental circle, i.e., while the function $\Theta(z)$ exists.

In this case there is nothing to prevent the sum of the series (5) being identically zero, and in fact we shall show later that all functions Θ of the **second species** can be expressed as linear combinations of a finite subset of this species.

We now pass to functions of the third, fourth, fifth and seventh family; the polygons R then have sides of the second kind, not all points of the fundamental circle are essential singular points, and instead of an essential singular line we have an infinity of isolated singular points. It follows that the series (5), instead of representing two distinct functions according as z is inside or outside the fundamental circle, represents one and the same transcendant which is everywhere holomorphic except at certain isolated points. The function $\Theta(z)$ is then a uniform transcendant existing throughout the plane, with an infinity of essential singular points.

We can ask what place it occupies in the classification of such functions given by Mittag–Leffler in his communication of April 3, 1882 in **Comptes rendus de l'Académie des Sciences de Paris**.

Since the essential singular points are infinite in number, they will approach infinitely close to certain points, singular points of the second species. The latter, in their turn, if they are infinite in number, will approach infinitely close to certain singular points of the third species, and so on.

I claim that this process never stops and that we therefore have an infinity of singular points of all species. In fact, if we have an infinity of singular points of the $(n-1)^{th}$ species, then we will have at least one singular point of the nth species; but if we have one we will have an infinity since all its transforms under the different transformations in the group G must also be singular points of the nth species.

Q.E.D.

Thus we are dealing with one of the functions which is of **second genus** in the terminology of Mittag–Leffler.

It seems at first that all the functions have infinities, since they exist in all the plane and they must have the poles of the rational function H(z), which must become infinite at some point of the plane. But it can happen that several infinities of the function $\Theta(z)$ mutually annihilate each other, so that we can construct, as in the preceding case, theta–Fuchsian functions of the second species. We shall see an example in Section VIII.

Among the singular points of our theta–Fuchsian functions, there are some which attract particular attention: they are the vertices of the polygon R which are of the second category and which belong to a cycle of the third subcategory (see Section V of the memoir on Fuchsian groups).

Let α be such a vertex; then the group G contains a parabolic transformation of the form

$$\left(\frac{1}{z-\alpha}, \frac{1}{z-\alpha} + \beta\right)$$

by definition of cycles of the third subcategory.

We put

$$\frac{2i\pi}{\beta} \frac{1}{z-\alpha} = t \qquad \frac{2i\pi}{\beta} \frac{1}{f_i(z)-\alpha} = \phi_i(t)$$

using the symbol $f_i(z)$ in the sense given above, i.e.,

$$f_i(z) = \frac{\alpha_i z+\beta_i}{\gamma_i z+\delta_i}$$

We now define a symbol H_1 as follows:

$$H_1(t) = H\left(\alpha + \frac{2i\pi}{\beta t}\right)\left(\frac{-2i\pi}{\beta t^2}\right)^m$$

It is clear that H_1 will be the algorithm of a rational function. We then find that, identically

$$\Theta(z) = \Sigma\, H[f_i(z)]\left[\frac{df_i(z)}{dz}\right]^m = \Sigma\, H_1[\phi_i(t)]\left[\frac{d\phi_i(t)}{dt}\right]^m\left(\frac{dt}{dz}\right)^m$$

But, since the series for $\Theta(z)$ is absolutely convergent, we can order the terms as we please. Here is how we shall order them:

Among the functions $\phi_i(t)$ we can choose an infinity

$$\theta_0(t) = t,\ \theta_1(t),\ \theta_2(t),\ \ldots$$

so that each function $\phi_i(t)$ can be expressed uniquely in the form

$$\phi_i(t) = \theta_k(t + 2hi\pi)$$

where h is a positive or negative integer; this permits us to write

$$\Theta(z) = \left[\frac{dt}{dz}\right]^m \sum_{k=0}^{\infty} \sum_{h=-\infty}^{\infty} H_1\,[\theta_k(t+2hi\pi)]\left[\frac{d\Theta_K(t+2hi\pi)}{dt}\right]^m$$

We consider a new algorithm $H'_k(t)$ defined as follows:

$$H'_k(t) = H_1[\theta_k(t)]\left[\frac{d\theta_k}{dt}\right]^m$$

whence

$$\Theta(z) = \left[\frac{dt}{dz}\right]^m \sum_{k=0}^{\infty} \sum_{h=-\infty}^{\infty} H''_k(t + 2hi\pi)\ .$$

It is necessary to carry out the summation with respect to h first; but $H'_k(t)$ is a rational function of t which tends to zero as t tends to infinity. We therefore have

$$\sum_{h=-\infty}^{\infty} H'_k(t + 2hi\pi) = H''_k(e^t)\ .$$

$H''_k(e^t)$ denotes a rational function of e^t which tends to zero as t tends to infinity. We then have

$$\Theta(z) = \left(\frac{dt}{dz}\right)^m \sum_k H''_k(e^t) .$$

The convergence of this series is evident, because we have obtained it by a certain grouping of terms of the series (5), which is absolutely and uniformly convergent. Moreover, the terms are rational in e^t and their infinities do not approach infinitely close to

$$e^t = 0 \quad \text{or} \quad e^t = \infty ,$$

so that we can find upper and lower bounds to the absolute values of e^t which make one of the functions H''_k infinite.

It follows from this that, in the neighborhood of the singular point $z = \alpha$, the function $\Theta(z)(z-\alpha)^{2m}$ is holomorphic in

$$e^{\frac{2i\pi}{\beta(z-\alpha)}} \quad \text{or} \quad e^{\frac{2i\pi}{\beta(\alpha-z)}}$$

according as the point α is approached from inside or outside the fundamental circle.

In other words, α is a **logarithmic singular point** of the function Θ.

Thus if we consider the different vertices of the polygons R we find:

1° That the vertices in a cycle of the first or third category are ordinary points of poles of the function Θ;

2° That the vertices in a cycle of the third subcategory are logarithmic singular points;

3° That the vertices in a cycle of the fourth subcategory are singular points of a higher nature.

III. Zeroes and infinities

We now have to study the distribution of zeroes and infinities of the function Θ. It is clear that if a point z is a zero or infinity of this function, then the same holds for all points corresponding to z, i.e., all the points $\frac{\alpha_i z+\beta_i}{\gamma_i z+\delta_i}$. Thus, each zero inside a polygon R_0 corresponds to a zero inside each of the polygons, which we do not regard as **really distinct** from the first. The number of **really distinct** zeroes and infinities of the function Θ will therefore be the number of zeroes and infinities contained in the interior of R_0, if the function exists only inside the fundamental circle, and in the interior of $R_0 + R'_0$ if it exists over the whole plane. Recall that R'_0 is the polygon symmetric to R_0 with respect to the fundamental circle.

Nevertheless, some special conventions are necessary to allow a simple statement of the result we shall reach concerning the number of zeroes and infinities. First, it is evident that a double zero or double infinity must be counted as two zeroes or two infinities; similarly for multiple zeroes and infinities. However, apart from zeroes in the interior of R_0, there may be some on the perimeter of this polygon. Suppose that we have one on a side ab of the first kind, and hence another, corresponding to the first, on the side equivalent to ab. These two zeroes are not really distinct and we should count them as only a single zero, or, if we wish, we could count each of them as a half–zero. Now suppose that a vertex of the first category is a zero of order p; there are, say, n vertices in the same cycle and each of them will be a zero of order p like the first. Suppose that the sum of angles corresponding to these vertices is $\frac{2\pi}{k}$; each vertex zero belongs to n − k different polygons, and if we divide it between the n − k polygons the share at a vertex of the polygon R_0

will be a zero of order $\frac{p}{n-k}$. It follows that the different vertices of the cycle represent only $\frac{p}{k}$ distinct zeroes.

It is easy to extend this convention to the case where one of the zeroes is a vertex of the second category which belongs to a cycle of the third subcategory. We have seen that if α is such a vertex then the function Θ can be put in the form

$$(z-\alpha)^{-2m}\Phi\left[e^{\frac{2i\pi}{\beta(\alpha-z)}}\right]$$

where Φ, being a holomorphic function of $e^{\frac{2i\pi}{\beta(\alpha-z)}}$, vanishes for

$$z = \alpha$$

i.e., for

$$e^{\frac{2i\pi}{\beta(\alpha-z)}} = 0$$

Then $u = 0$ will be a zero of the function $\Phi(u)$. Suppose that it is a zero of order p. We then say that the different vertices of the cycle containing α represent p distinct zeroes of the function Θ.

The preceding evidently applies to infinities.

We consider the infinities first. The series (5) becomes infinite when $H\left(\frac{\alpha_i z+\beta_i}{\gamma_i z+\delta_i}\right)$ or $\frac{1}{\gamma_i z+\delta_i}$ becomes infinite. Suppose first that the function Θ exists only inside the fundamental circle. Then $\frac{1}{\gamma_i z+\delta_i}$ cannot become infinite; moreover, to each infinity of $H(z)$ inside the fundamental circle there corresponds in general an infinity of one of the $H\left(\frac{\alpha_i z+\beta_i}{\gamma_i z+\delta_i}\right)$ which will be inside R_0.

Then:

The number of distinct infinities of Θ is equal to the number

of infinities of H **inside the fundamental circle.**

Now suppose that Θ exists over the whole plane.

To each infinity of H(z) inside the fundamental circle there corresponds an infinity of one of the $H\left[\frac{\alpha_i z+\beta_i}{\gamma_i z+\delta_i}\right]$ inside R_0.

To each infinity of H(z) outside the fundamental circle there corresponds an infinity of one of the $H\left[\frac{\alpha_i z+\beta_i}{\gamma_i z+\delta_i}\right]$ inside R'_0.

Inside each of the polygons R'_i, and consequently inside R'_0 we find one of the points $-\frac{\delta_i}{\gamma_i}$ which is an infinity of order 2m for Θ. However, there is one exception; the points $-\frac{\delta_i}{\gamma_i}$ are the different points corresponding to infinity; the polygon R′ which contains ∞ does not contain the point $-\frac{\delta_i}{\gamma_i}$; the point ∞ is not in general an infinity of Θ. Assuming, in order to avoid this difficulty, that the polygon R'_0 does not contain the point ∞ we can state the following result:

The number of infinities of Θ inside $R_0 + R'_0$, i.e., the **number of distinct infinities of Θ is equal to the number of infinities of Θ, plus** 2m.

We now pass to the investigation of the zeroes. Among them are some which attract attention immediately; I speak of the vertices of the first category which in certain cases are forced to be zeroes of a determinate order.

Let α be a vertex of the first category and let α' be its reflection in the fundamental circle. Suppose that α is part of a cycle and that the sum of the angles of this cycle is $\frac{2\pi}{k}$. One of the transformations in G will be

$$\left[\frac{z-\alpha}{z-\alpha'},\ e^{\frac{2i\pi}{k}}\frac{z-\alpha}{z-\alpha'}\right] = [z,\ f_1(z)]$$

We have

$$\Theta[f_1(z)] = \Theta(z)\left[\frac{df_i}{dz}\right]^{-m}$$

where

$$(8) \qquad \Theta(f_1)(f_1-\alpha')^{2m} = \Theta(z)(z-\alpha')^{2m}\left[\frac{d\left(\frac{f_1-\alpha}{f_1-\alpha'}\right)}{d\left(\frac{z-\alpha}{z-\alpha'}\right)}\right]^{-m}$$

But

$$\frac{f_1-\alpha}{f_1-\alpha'} = e^{\frac{2i\pi}{k}}\frac{z-\alpha}{z-\alpha'}$$

We also remark that we can expand $\Theta(z)(z-\alpha')^{2m}$ in powers of $\frac{z-\alpha}{z-\alpha'}$ so that

$$\Theta(z)(z-\alpha')^{2m} = \Theta_1\left[\frac{z-\alpha}{z-\alpha'}\right] = \Sigma\, A_p\left[\frac{z-\alpha}{z-\alpha'}\right]^p$$

Equation (8) then becomes

$$\Sigma\, A_p e^{\frac{2ip\pi}{k}}\left[\frac{z-\alpha}{z-\alpha'}\right]^p = \Sigma\, A_p e^{-\frac{2mi\pi}{k}}\left[\frac{z-\alpha}{z-\alpha'}\right]^p$$

This identity requires that we have

$$A_p = 0$$

or else

$$e^{\frac{2i\pi p}{k}} = e^{-\frac{2mi\pi}{k}}$$

i.e.,

$$p + m \equiv 0 \pmod{k} \tag{9}$$

Then the expansion of Θ_1 contains only terms whose order p satisfies the congruence (9). If k does not divide m, then $z = \alpha$ is a zero for Θ_1, and consequently for Θ.

This zero is of order at least equal to the remainder on division of $m(k-1)$ by k; and if the order of this zero differs from this remainder then it is a multiple of k. An evident exception is where α is an infinity of Θ.

Likewise, we have seen that the vertices which belong to a cycle of the third subcategory are, in general, zeroes of the function Θ.

We now have to find the number of really distinct zeroes of our function Θ and, to fix ideas, we suppose that Θ is a function of the first family. Let q be the number of distinct infinities, let p be the number of really distinct zeroes, i.e., the number sought. Now let p_0 be the number of zeroes inside R_0, leaving aside zeroes on the perimeter and at the vertices. We suppose, as can be arranged in general, that there are no zeroes on the perimeter except at vertices; we only have to consider zeroes on the sides as vertices separating collinear sides. Now suppose that the vertices are partitioned into a certain number of cycles of the first category: C_1, C_2, ..., C_i, ... Suppose that all the vertices of the cycle C_i are zeroes of order p_i, and that the sum of the angles of this cycle is $\frac{2\pi}{k_i}$, so that these zeroes must be counted, according to the above convention, as $\frac{p_i}{k_i}$ distinct zeroes. We must have

$$p_i + m \equiv 0 \pmod{k_i}$$

$$p = p_0 + \Sigma \frac{p_i}{k_i}$$

The problem is to evaluate p_0. To do this it is necessary to take the integral

$$\int \frac{\Theta'(z)dz}{\Theta(z)} \tag{10}$$

along the perimeter of R_0. The real part of this integral will be zero and the imaginary part will be written

$$2i\pi(p_0 - q)$$

The evaluation of the integral (10) presents a special difficulty here.

The function under the $\int$ sign becomes infinite at various points on the contour of integration, because we have seen that a certain number of vertices of R_0 are zeroes of Θ. We shall sidestep this difficulty by describing infinitely small arcs, of circles centered on each of these vertices, which connect the two sides which meet at the vertex considered; it is necessary to describe these arcs inside R_0, so as to leave the vertices outside the contour, because we want to evaluate p_0, i.e., the number of zeroes **inside** R_0.

Thus it is necessary to evaluate the integral (10):

1° Along the arcs of infinitely small circles described around the vertices;

2° Along the sides of the first kind;

3° Along the sides of the second kind.

Moreover, it suffices to calculate the imaginary part of the integral, since we know in advance that the real part is zero, and this imaginary part is nothing but the argument variation of Θ.

Suppose first that the function is of the first family, so that we have only vertices of the first category and sides of the first kind. Let 2n be the number of these sides.

First consider one of the small circular arcs around a vertex; suppose that this vertex belongs to a cycle C_i whose angle sum is

$\frac{2\pi}{k_i}$ and whose members are all zeroes of order p_i. Let λ be the angle at the vertex considered. The integral along the small arc of the corresponding circle will be $-\sqrt{-1}\ p_i\lambda$; along all the circular arcs around the different vertices in the cycle it will be

$$-\frac{2\pi\sqrt{-1}}{k_i}\ p_i$$

Finally, the integral (10), taken along all the infinitely small arcs around all the vertices of R_0 will be

$$-\ 2\pi\sqrt{-1}\ \Sigma\ \frac{p_i}{k_i}$$

It remains to evaluate our integral along the sides of the first kind. Suppose then that ab, a′b′ are two equivalent sides of R_0. It is necessary to calculate

$$J = \int_a^b \frac{\Theta' dz}{\Theta} + \int_{b'}^{a'} \frac{\Theta' dz}{\Theta} = \int_a^b \frac{\Theta' dz}{\Theta} - \int_{a'}^{b'} \frac{\Theta' dz}{\Theta}$$

Let $[z, f_i(z)]$ be the transformation in the group G which sends a′b′ to ab. From what we have seen above, we have

$$\Theta(f_i) = \Theta(z)\left(\frac{df_i}{dz}\right)^{-m}$$

where

$$\log\,\Theta(f_i) = \log\,\Theta(z) - m\,\log\,\frac{df_i}{dz}$$

Then

$$J = \int_{a'}^{b'} \left[\frac{\Theta'(f_i)}{\Theta(f_i)} - \frac{\Theta'(z)}{\Theta(z)}\right] dz = -m \int_{a'}^{b'} d\left[\log \frac{df_i}{dz}\right]^{*}$$

(* $\Theta'(f_i) = \dfrac{d\Theta[f_i(z)]}{dz}$ N.E.N.)

It therefore remains to find how the imaginary part of $\log \frac{df_i}{dz}$, or the argument of $\frac{df_i}{dz}$, varies as z varies from a′ to b′.

Recall that the sides of R_0, ab, a′b′, are circular arcs; I am going to draw the tangents to these circles at the points a,b,a′,b′; let ac, bc; a′c′,b′c′ be these tangents. Now let $\omega_1, \omega_2, \omega_3, \omega_4$ be the arguments of the complex quantities (c—a), (b—c), (c′—a′), (b′—c′).

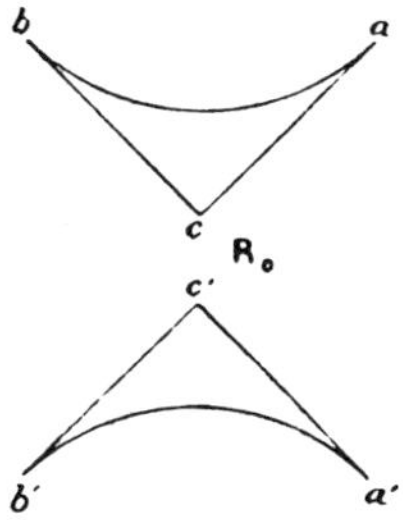

Figure 10

Suppose that we operate in the same way on all sides of the curvilinear polygon R_0; we obtain a rectilinear polygon P_0 of 4n sides, whose sides are tangent to sides of R_0 and whose vertices are those of R_0 together with the points c, c′. To designate the angles of the polygon P_0 which correspond to the vertices c and c′, I simply use the letters c and c′ again. Apart from the vertices c, the polygon P_0 has 2n angles which are shared with R_0, with sum $\Sigma \frac{2\pi}{k_i}$, whence the relation

$$\Sigma c + \Sigma \frac{2\pi}{k_i} = (4n-2)\pi$$

We can now write

$$\arg. \frac{df_i}{dz} = \omega_1 - \omega_3 \qquad \text{for } z = a$$

$$\arg.\ \frac{df_i}{dz} = \omega_2 - \omega_4 \qquad \text{for } z = b'$$

imaginary part of $J = -m(\omega_2 - \omega_1 - \omega_4 + \omega_3)\sqrt{-1}$

$$c = \omega_1 - \omega_2 + \pi$$

$$c' = \pi - \omega_3 + \omega_4$$

imaginary part of $J = m(c + c' - 2\pi)\sqrt{-1}$

The integral (10), taken along all the sides of the first kind, will then equal the imaginary part of ΣJ or,

$$\sqrt{-1}\ m\ \Sigma c - 2nm\pi\sqrt{-1}$$

where

$$(4n-2)m\pi\sqrt{-1} - 2\pi\sqrt{-1}\ \Sigma\ \frac{m}{k_i} - 2nm\pi\sqrt{-1} =$$

$$= (n-1)2m\pi\sqrt{-1} - 2\pi\sqrt{-1}\ \Sigma\ \frac{m}{k_i}$$

Thus the integral (10) reduces to

$$2\pi\sqrt{-1}\left[m(n-1) - \Sigma\ \frac{p_i+m}{k_i}\right]$$

whence we have the relation

$$p_0 = q + m(n-1) - \Sigma \frac{p_i+m}{k_i}$$

This number p_0 of zeroes inside R_0 is an integer because of the congruences

$$p_i + m \equiv 0 \pmod{k_i}$$

We now express the number p of really distinct zeroes, defined by the conventions above; we find

$$p = q + m\left[n-1 - \Sigma \frac{1}{k_i}\right]$$

Thus the excess of the number of distinct zeroes over the number of distinct infinities depends only on the number m, the number of sides of R_0, 2n, and the sum $\Sigma \frac{2\pi}{k_i}$ of angles of this polygon.

This sum satisfies the inequality

$$\Sigma \frac{2\pi}{k_i} < \pi(2n-2)$$

whence

$$\Sigma \frac{1}{k_i} < n - 1$$

and consequently $p > q$. The expression $(n - 1 - \Sigma \frac{1}{k_i})$ is proportional to the S of the polygon R_0.

Now suppose that the function Θ is of the second or sixth family. We always have that the sides are of the first kind; however, as well as the cycles C_1, C_2, ..., C_i, ... of the first category that we encountered in the first case, we have cycles C'_1, C'_2, ..., C'_t of the second category and the third subcategory. Consider the cycle C'_i; the different vertices of this cycle are zeroes; I suppose that they account for h_i distinct zeroes under the above convention. We then have

(11) $$p = p_0 + \Sigma \frac{p_i}{k_i} + \Sigma h_i$$

It is necessary to evaluate the integral (10):

1° Along the infinitely small circular arcs around the vertices of the first category;

2° Along the sides of the first kind;

3° Along the infinitely small circular arcs around the vertices of the second category.

The first part of the integral reduces, as above, to

$$-2\pi\sqrt{-1}\ \Sigma \frac{p_i}{k_i}$$

the second to

$$\Sigma J = 2\pi m\sqrt{-1}\left[n-1 - \Sigma \frac{1}{k_i}\right]$$

To evaluate the third, it will suffice to study how the argument of Θ varies in the neighborhood of vertices of the second category; to do this it is necessary to put the function Θ in the form

$$(z-\alpha)^{-2m}\Phi\left[e^{\frac{2i\pi}{\beta(\alpha-z)}}\right]$$

(where Φ is the algorithm of a holomorphic function), which is possible, as we have seen above. We then recognize that this third part of the integral reduces to $-2\pi\sqrt{-1}\ \Sigma h_i$.

The expression for the integral (10) then remains as

$$2\pi\sqrt{-1}\left[m(n-1) - \Sigma \frac{p_i+m}{k_i} - \Sigma h_i\right]$$

whence

$$p_0 = q + m(n-1) - \Sigma \frac{p_i+m}{k_i} - \Sigma h_i$$

and

$$p = q + m(n-1 - \Sigma \frac{1}{k_i}) .$$

We are led to the same expression as in the preceding case for the excess of the number of distinct zeroes over the number of distinct infinities. This excess is proportional to the number m and depends also on the number of sides of the polygon R_0 and the sum of its angles, indeed it is proportional to the product of m and the S of R_0.

Finally, suppose that the function Θ is of the third, fourth, fifth or seventh family, i.e., that it exists over the whole plane. We then have sides of the first and second kind, vertices of the first and third categories, and vertices of the second category belonging to cycles of the third subcategory. We have to find the numbers of distinct zeroes and infinities; these numbers will be defined as before. However, there is a remark to be made on the subject of the number of distinct infinities; in general this will be the number of infinities inside R_0 and R'_0, but a difficulty presents itself when the point ∞ lies in the region R'_0. In that case it can happen that ∞ is not an infinity of Θ, but that all the transforms of the point ∞ by transformations in the group G are infinities of Θ. The number of really distinct infinities will then be equal to the number of infinities inside $R_0 + R'_0$, **plus** 2m. To avoid this difficulty, I suppose that the polygon R_0 has been chosen so that ∞ does not lie in the region R'_0.

Having done this, we again take the integral (10); it will be necessary to evaluate it along the contour of R_0, then along the contour of R'_0, and add. Let 2n be the number of sides of the first kind; let l be the number of the second kind; also, we retain the above notation, so that p_0 and q represent the numbers of distinct zeroes and infinities inside $R_0 + R'_0$.

The integral has to be evaluated along:

1° The small circular arcs around vertices of the first category;

2° The small circular arcs around vertices of the third subcategory;

3° The sides of the first kind;

4° The sides of the second kind.

The first and second part of the integral make up

$$-2i\pi\left\{\Sigma \frac{p_i}{k_i} + \Sigma h_i\right\}.$$

The third part equals, as above, the imaginary part of ΣJ. We construct a polygon P_0 by taking tangents to the sides of the first kind at the different vertices of R_0. The resulting rectilinear polygon has $4n + l$ sides; its angles will be of two kinds. The first will be analogous to the angles we called c, c′ above; the others will be angles of R_0, but among them are those which sum to $\Sigma \frac{2\pi}{k_i}$, corresponding to vertices of the first category, others which are zero and correspond to vertices of the second category, and finally others, which are right angles and $2l$ in number, corresponding to vertices of the third category; whence the relation

$$\Sigma c + \Sigma \frac{2\pi}{k_i} + l\pi = (4n + l - 2)\pi .$$

Now we found above that

$$\text{imaginary part of } \Sigma J = m\sqrt{-1}(\Sigma c - 2n\pi)$$

and we then have

$$\text{imaginary part of } \Sigma J = 2\pi m\sqrt{-1}(n - 1 - \Sigma \frac{1}{k_i})$$

Finally, if we let $2\pi I\sqrt{-1}$ be the imaginary part of the integral taken along sides of the second kind, then we find that the integral (10) along R_0 is

$$2\pi\sqrt{-1}\left[m(n-1) - \Sigma \frac{p_i+m}{k_i} - \Sigma h_i + I\right]$$

Now it is necessary to take the integral (10) along R'_0. Suppose, following the above conventions, that the vertices of the first category in R'_0 account for $\Sigma \frac{p_i}{k_i}$ distinct zeroes, and that those of the second category account for $\Sigma h'_i$ distinct zeroes. Reasoning as above, we find the value of the integral along R'_0 to be

$$2\pi\sqrt{-1}\left[m(n-1) - \Sigma \frac{p'_i+m}{k_i} - \Sigma h'_i - I\right]$$

and so the sum of the two integrals is

$$2\pi\sqrt{-1}\left[2m(n-1) - \Sigma \frac{p_i+m}{k_i} - \Sigma \frac{p'_i+m}{k_i} - \Sigma h_i - \Sigma h'_i\right]$$

whence

$$p_0 = q + 2m(n-1) - \Sigma \frac{p_i+m}{k_i} - \Sigma \frac{p'_i+m}{k_i} - \Sigma(h_i + h'_i)$$

But if p denotes the number of distinct zeroes, as above, then we have

$$p = p_0 + \Sigma \frac{p_i}{k_i} + \Sigma \frac{p'_i}{k_i} + \Sigma h_i + \Sigma h'_i$$

whence

$$p = q + 2m\left(n-1 - \Sigma \frac{1}{k_i}\right)$$

Thus the excess of the number of distinct zeroes over the number of distinct infinities depends only on the number m, the number 2n of sides of the first kind, and the sum of the angles of the first category.

Above all, it is important to note that in all cases the numbers of distinct zeroes and infinities are finite.

IV. Fuchsian functions

If a function F(z) is uniform, if it is reproduced by all transformations in a Fuchsian group G, so that, for a transformation

$$\left(z, \frac{\alpha_i z+\beta_i}{\gamma_i z+\delta_i}\right)$$

in the group, we have identically

$$F\left(\frac{\alpha_i z+\beta_i}{\gamma_i z+\delta_i}\right) = F(z)$$

and finally, if the function F(z) has only a finite number of really distinct zeroes and infinities, then I say that it is a **Fuchsian function**. Do such functions exist? It is easy to construct one.

Consider two theta–Fuchsian functions corresponding to the same group G and with the same value for the integer m; their quotient will be a Fuchsian function. It is easy to generalize this procedure. We say that the number m is the **degree** of the series Θ. If we then consider a product of several such Θ series, the degree of this monomial will be the sum of the degrees of the factors. If we consider a polynomial in various Θ series, then we say that this polynomial is homogeneous and of degree k if all its terms are of the same degree k; the quotient of two homogeneous polynomials of degrees k and k′ will be a homogeneous rational function of degree $k - k'$. It is clear that every rational function of degree 0 with respect to different Θ series is a Fuchsian function. Later we shall see how every Fuchsian function can be expressed in this way.

What are the properties of Fuchsian functions?

1° **The singularities are the same as those of the theta–Fuchsian functions;** consequently, we have Fuchsian functions which exist only inside the fundamental circle and for which the whole circumference of the circle is an essential singular line, and others which exist throughout the plane and whose essential singular points are all on the fundamental circle, isolated, and infinite in number.

2° The number of distinct zeroes of a Fuchsian function F(z) is equal to the number of its distinct infinities; moreover, it is equal to the number of distinct zeroes of the Fuchsian function F(z) − a, i.e., the number of really distinct points for which F(z) takes the value a. Consequently, the number of points inside R_0 (or $R_0 + R'_0$) for which F(z) takes the same value is **constant** and **finite**.

3° Suppose two Fuchsian functions F(z) and $F_1(z)$ correspond to the same Fuchsian group G, and that the first takes the same values p times, and the second p_1 times, inside R_0; I then claim that there is an algebraic relation between F and F_1. In fact, to each value of F there correspond p_1 values of F_1; each symmetric function of these p_1 values is meromorphic in F for all values of F, finite of infinite. All these symmetric functions are then rational functions of F, hence F_1 itself is an algebraic function of F.

Q.E.D.

4° Now consider all Fuchsian functions which correspond to the same group G. We have just seen that there is an algebraic relation between any two of them. It follows that all of these functions can be expressed rationally in terms of two of them, which I shall call x and y. Moreover, x and y satisfy an algebraic relation

$$\psi(x,y) = 0 \tag{1}$$

What is the genus of the relation (1) or, equivalently, the genus of the corresponding Riemann surface? To resolve this question, it is necessary to go back to Section VIII of the memoir on Fuchsian groups, the Section entitled **Classification by genus**. In that section I supposed that we had cut out the region R_0 (or $R_0 + R'_0$), deformed it, and pasted the equivalent sides so as to obtain a closed surface. It is clear that from the viewpoint of the geometry of the situation the closed surface so obtained is no different from the Riemann surface under consideration. It follows that the genus of the relation (1) is precisely what we have called, in the memoir in question, **the genus of the group** G. Thus if the group G is of genus zero, it will be the same for the relation (1) and, consequently, all the Fuchsian functions will be expressible rationally in terms of **a single one** of them which I shall call x.

We can avoid these considerations borrowed from the geometry of the situation. This is what I have done in a work inserted in the **Memoires de l'Académie de Caen**, where I determined the genus of the relation (1) by the number of distinct cycles which can be described by an analytic point (x,y), but this leads to an excessively long discussion.

5^0 Consider the Fuchsian function we have come to call x, and form the following two functions

$$(2) \qquad \nu_1 = \sqrt{\frac{dx}{dz}} \qquad \nu_2 = z\sqrt{\frac{dx}{dz}}$$

It is clear that we have

$$\frac{1}{\nu_1}\frac{d^2\nu_1}{dx^2} = \frac{1}{\nu_2}\frac{d^2\nu_2}{dx^2} = \frac{2\frac{d^3x}{dz^3}\frac{dx}{dz} - 3\left(\frac{d^2x}{dz^2}\right)^2}{4\left(\frac{dx}{dz}\right)^4}$$

One easily checks that the third member of this double equation is a Fuchsian function of z; it is therefore, as we have come to see, a rational function of x and y, which I shall call $\phi(x,y)$.

The two integrals of the linear equation

$$\frac{d^2\nu}{dx^2} = \nu\phi(x,y) \tag{3}$$

are then

$$\nu = \nu_1 \qquad \nu = \nu_2$$

Thus the consideration of the Fuchsian function x permits the integration of the linear equation (3) whose coefficients are rational functions of the analytic point (x,y). We see that the independent variable z is expressed as a Fuchsian function of z, i.e., of the ratio of the integrals. When we know this Fuchsian function we can deduce the integrals themselves with the aid of the formula (2).

In the particular case where the group G is of genus zero, equation (3) can be written

$$\frac{d^2\nu}{dx^2} = \nu\phi(x) \tag{3'}$$

where $\phi(x)$ is a rational function of x alone; in this case, in fact, we have seen that every Fuchsian function is expressed rationally in x.

We now investigate what are the singular points of equation (3) and how the integrals behave in the neighborhood of each of them. In general, the integrals will be holomorphic functions of x; there are two exceptions which give us two kinds of singular point:

1° At singular points of the relation (1), y ceases to be a holomorphic function of x, and it is the same for the integrals ν_1 and ν_2; but ν_1 and ν_2 remain **holomorphic functions of the analytic point** (x,y). Thus we do not have a true singular point in this case.

2° At the points (x,y) which correspond to the different vertices of the polygon R_0. Within this second species of singular points we distinguish:

1° Those which correspond to vertices of the first category.

In the neighborhood of such a singular point the integrals of the equation (3) are regular, to use the expression of Fuchs. If we now form the determinant equation corresponding to this singular point, the difference between the roots of this equation is the reciprocal of an integer. In fact, the singular point corresponds to a vertex of R_0 which is part of a cycle of the first category, and the sum of the angles in this cycle equals $\frac{2\pi}{p}$, where p is an integer; we easily see that the difference between the roots of the determinant equation is precisely $\frac{1}{p}$.

In the particular case where the vertex of R_0 in question belongs to a cycle of the first subcategory, we have $p = 1$, by definition of this subcategory. Consequently, the difference between the roots of the determinant equation is unity. Thus such a point is not singular.

2° Now consider a singular point which corresponds to a vertex of R_0 belonging to a cycle of the second category. This cycle will be of the third subcategory because, as we have seen above, we can always assume that we have no cycle of the fourth subcategory. We form the determinant equation corresponding to such a singular point; the roots of this equation are equal and the integrals of equation (3) will be **logarithmic** but **regular**.

3° Finally, it is necessary to consider the analytic points (x,y) which correspond to vertices of the third category, but we easily recognize that these are not singular points.

In summary, the integrals of equation (3) are everywhere regular and (if we leave aside the singular points of the relation (1)) the number of singular points of the differential equation (3) is equal to the number of cycles of the second or third subcategory present in the polygon R_0 (or in the two polygons R_0 and R'_0).

V. First family; genus zero

After having studied the properties of Fuchsian functions in general, we have to concern ourselves, separately, with the different families of functions and the genera into which they are divided. We commence with a special study of the most simple functions: those of the first family and genus zero.

I first recall the form of the polygon R_0 for such functions. The sides are all of the first kind and 2n in number; the sides of rank p and $2n + 1 - p$ are equivalent and, consequently, congruent. The number of cycles is n+1; two of them contain only a single vertex. The first the vertex of rank 1, and the second the vertex of rank n+1. The n−1 other cycles consist of two vertices, namely the vertices of rank p and $2n + 2 - p$. (This is Example II of Section VII of the memoir on Fuchsian groups.)

A remarkable particular case is that where the polygon R_0 is symmetric with respect to the arc of the circle which cuts the fundamental circle orthogonally and which joins the vertices of rank 1 and n+1. I then say simply that the polygon is symmetric.

In the case we are concerned with the genus is 0 and consequently all the Fuchsian functions are rationally expressible in terms of one of them, which I call x or f(z). This condition does not uniquely determine the function $x = f(z)$. If, in fact, all the Fuchsian functions are expressible rationally in terms of x, then they are also expressible rationally in terms of $\frac{ax+b}{cx+d}$, where a, b, c, d, are any constants. Thus it is necessary, in order to define the function $x = f(z)$ completely, to impose three further conditions. Here are the ones we choose. Let α_1, α_2, α_{n+1} be the vertices of rank 1, 2 and n+1; we suppose that we have

$$f(\alpha_1) = 0 \quad f(\alpha_2) = 1 \quad f(\alpha_{n+1}) = \infty$$

The function f(z) will then be completely determined. With regard to this function, I make the following remark: if the polygon R_0 is symmetric, then the function f(z) is real and positive on the whole perimeter of this polygon.

I let α_i denote the vertex of rank i; together with α_{2n+2-i}, this vertex forms a cycle with angle sum equal to $\frac{2\pi}{\beta_i}$, where β_i is an integer. Finally, I put

$$a_i = f(\alpha_i) = f(\alpha_{2n+2-i})$$

whence

$$a_1 = 0 \quad a_2 = 1 \quad a_{n+1} = \infty$$

For $z = \alpha_i$, the first $\beta_i - 1$ derivatives of f(z) vanish, so that the expansion of f(z) is of the form

$$f(z) = a_i + b_1(z-\alpha_i)^{\beta_i} + b_2(z-\alpha_i)^{\beta_i+1} + \dots$$

For $z = \alpha_{n+1}$, f(z) is an infinity of order β_{n+1}, so that the expansion of f(z) is of the form

$$f(z) = \frac{b_1}{(z-\alpha_{n+1})^{\beta_{n+1}}} + \frac{b_2}{(z-\alpha_{n+1})^{\beta_{n+1}+1}} + \dots$$

We have seen that if we put

$$\nu = \sqrt{\frac{dx}{dz}}$$

then the function ν so defined satisfies an equation

$$\text{(3}'\text{)} \qquad \frac{d^2\nu}{dx^2} = \nu\phi(x)$$

where $\phi(x)$ is rational in x. What is the form of the function $\phi(x)$ here? We easily find that

$$\phi(x) = \frac{P(x)}{(x-\alpha_1)^2(x-\alpha_2)^2\ldots(x-\alpha_n)^2} = \frac{P(x)}{Q^2(x)}$$

where $Q(x)$ is the product $(x-\alpha_1)(x-\alpha_2)\ldots(x-\alpha_n)$ and $P(x)$ is a polynomial in x of degree $2n - 2$.

The polynomial $P(x)$ must satisfy the following conditions:

$$P(\alpha_i) = Q'^2(\alpha_i)\left[\frac{1}{4} - \frac{1}{4\beta_i^2}\right]$$

We use the notation $Q'(x)$ for the derivative of $Q(x)$. Moreover, the coefficient of x^{2n-2} in $P(x)$ must be equal to

$$-\left[\frac{1}{4} - \frac{1}{4\beta_{n+1}^2}\right]$$

There we have n+1 conditions which the polynomial $P(x)$ must satisfy if the equation (3′) is to be integrable in the way we have claimed. These conditions are not sufficient. In fact, since the polynomial $P(x)$ is of degree $2n - 2$, $n - 2$ parameters in this polynomial remain arbitrary; but in these $n - 2$ parameters, which are **complex** quantities, I must distinguish real and imaginary parts, so they are equivalent to $2n - 4$ **real** parameters. But how many arbitrary parameters do we have in the group G? The group G is defined by a polygon R_0 of 2n sides. To define such a polygon, $4n - 3$ conditions are necessary in general. However, our polygon is not arbitrary; equivalent sides must be congruent, which takes n conditions; the angle sums of the different cycles must equal

$$\frac{2\pi}{\beta_1}, \frac{2\pi}{\beta_2}, \ldots, \frac{2\pi}{\beta_{n+1}}$$

respectively, which takes n+1 conditions. There remain, then, $2n - 4$ arbitrary parameters in our group G. In our function $\phi(x)$, on the contrary, we have $2n - 4$ parameters which remain arbitrary in P(x), in addition we have n–2 arbitrary complex parameters in Q(x) namely $\alpha_3, \alpha_4, \ldots, \alpha_n$; these are equivalent to $2n - 4$ real parameters. Thus in the function $\phi(x)$ we have $2n - 4$ arbitrary parameters more than in the group G. Hence, for the equation (3′) to be integrated in the way I have claimed, it is necessary, in addition to the conditions stated above, for $2n - 4$ other conditions to be satisfied. These conditions are transcendental and very complicated; their study will be found in another memoir.* (*Sur les groupes des équations linéaires, **Acta Mathematica IV**, pp 201–312 N.E.N.)

Now suppose that we are given $\alpha_3, \alpha_4, \ldots, \alpha_n$, i.e., the singular points of the equation (3′); Q(x) will be determined and the $2n - 4$ parameters of P(x) will remain. The number of parameters remaining arbitrary is therefore precisely the number of conditions to be satisfied. In a question where only algebraic functions or simple transcendants were involved we could conclude that the parameters could always be disposed to satisfy these conditions, and consequently that **there is always an equation of the form** (3′), **integrable by Fuchsian functions alone, with singular points**

$$\alpha_1, \alpha_2, \ldots, \alpha_n, \alpha_{n+1}$$

given in advance. But here we cannot be content with such a rough guess, and this highly important theorem requires a special proof, which I shall give in a later memoir.* (*Translator's note: the memoir cited above.)

In the particular case where the polygon R_0 is symmetric, the singular points $\alpha_1, \alpha_2, \ldots, \alpha_{n+1}$, and hence the coefficients of P(x), are real.

In the more particular case again, where $n = 2$, equation (3′) easily reduces to the hypergeometric equation of Gauss. Moreover, we have

$$2n - 4 = 0$$

so that the number of transcendent conditions mentioned above is zero. Then, for x to be a Fuchsian function of the ratio of the integrals, it is necessary and sufficient that

$$P(0) = -\frac{1}{4}\left[1 - \frac{1}{\beta_1^2}\right]$$

$$P(1) = -\frac{1}{4}\left[1 - \frac{1}{\beta_2^2}\right]$$

and that the coefficient of x^2 in P(x)

$$= -\frac{1}{4}\left[1 - \frac{1}{\beta_2^2}\right]$$

where β_1, β_2, β_3 are integers.

We now study the theta–Fuchsian functions.

Let $\Theta(z)$ be a theta–Fuchsian function enjoying the characteristic property

$$\Theta\left[\frac{\alpha_k z+\beta_k}{\gamma_k z+\delta_k}\right] = \Theta(z)(\gamma_k z+\delta_k)^{2m}$$

If we take the Fuchsian function $f(z) = x$ and its derivative $f'(z)$, then it is clear that the function

$$\frac{\Theta(z)}{[f'(z)]^m}$$

will be a Fuchsian function, and consequently a rational function of x. We then have the following general expression for the functions $\Theta(z)$:

$$\text{(1)} \qquad \left(\frac{dx}{dz}\right)^m F(x)$$

where F is the algorithm of a rational function.

We have seen that there are two species of theta–Fuchsian functions: those which have infinities inside the fundamental circle, and those which have not.

What will be the general form of those in the second species? The expression (1) may not become infinite for any value of x. In that case F(x) can only become infinite when $\frac{dx}{dz}$ is zero, i.e., when x is one of the singular points $x = \alpha_1$, $x = \alpha_2$, ..., $x = \alpha_n$.

It follows that we must have

$$\text{(2)} \qquad \Theta(z) = \left(\frac{dx}{dy}\right)^m \frac{\theta_p(x)}{(x-\alpha_1)^{\lambda_1}(x-\alpha_2)^{\lambda_2}\ldots(x-\alpha_n)^{\lambda_n}}$$

where $\theta_p(x)$ denotes a polynomial of order p. If we now say that $\Theta(z)$ must not become infinite for $x = \alpha_1$, $x = \alpha_2$, ..., $x = \alpha_n$, nor for $x = \infty$, we get

$$\lambda_i < m\left(1 - \frac{1}{\beta_i}\right)$$

$$\text{(3)} \qquad p < \Sigma\, \lambda_i - m\left(1 + \frac{1}{\beta_{n+1}}\right)$$

These n + 1 inequalities are compatible.

These inequalities do not permit p to be given values as large as we please, hence there is always a number q such that $p < q$ but the inequalities (3) permit p to attain the bound $q - 1$.

If the number m is given, then the number q will be determined, and it is clear that there will always be a linear identity between q theta–Fuchsian functions of the second species.

If we bring back the expression we have given for those transcendants in Section I, we see that we have a rational function H(z) comprising an infinite number of arbitrary parameters. It follows that we have an infinity of rational functions H for which the corresponding theta–Fuchsian series are identically zero.

Then there is an infinity of linear identities between the theta–Fuchsian series. We shall study them later.

We return to the expression (2). Until now we have assumed that the numbers m, λ_1, λ_2, ... which appear in it are integers. If they are not, then the function defined by this expression will not enjoy the fundamental property of theta–Fuchsian functions, but it can remain uniform. For this it is sufficient for the numbers

$$m(\beta_i - 1) - \lambda_i \beta_i \qquad (i = 1, 2, \ldots, n)$$

and

$$m(\beta_{n+1} + 1) - \Sigma\, \lambda_i \beta_{n+1}$$

to be integers.

It is clear that the n + 1 equations

$$(4) \qquad \begin{aligned} m(\beta_i - 1) - \lambda_i \beta_i &= \epsilon_i \\ m(\beta_{n+1} + 1) - \Sigma\, \lambda_i \beta_{n+1} &= \epsilon_{n+1} \end{aligned}$$

where the numbers ϵ are given integers, always give values for the λ and for m, since these are linear equations with non–zero determinant.

Suppose then that we give all the ϵ in the equations (4) the value zero, except for ϵ_k which we give the value 1. After this we carry the values of m and the λ drawn from the equations (4) to the expression (2), and replace the θ_p in this same expression by a constant. We have defined a function X_k which is holomorphic everywhere inside the fundamental circle, and which has no zeroes except the point a_k and the points corresponding to it. Moreover, all these zeroes are simple.

These functions $X_1, X_2, \ldots, X_n, X_{n+1}$, whose role is very important, have been discovered by Halphen in the particular case n=2. We can find certain interesting relations between them and x. Consider, in fact, the series X_1 and X_{n+1}. Let m_1 and m_{n+1} be the corresponding values of the number m. These values will be fractional.

Let Y be any uniform function capable to being put in the form

$$(2') \qquad Y = \left[\frac{dx}{dz}\right]^M \frac{K}{(x-\alpha_1)^{\lambda_1}(x-\alpha_2)^{\lambda_2}\ldots(x-\alpha_n)^{\lambda_n}}$$

where K denotes a numerical factor. This function has no zeroes or infinities except at $\alpha_1, \alpha_2, \ldots, \alpha_{n+1}$ and the points which correspond to them. Suppose that these zeroes are of order $\gamma_1, \gamma_2, \ldots, \gamma_{n+1}$ respectively; it goes without saying that if α_i, for example, is an infinity, then γ_i will be negative. We have

$$Y = KX_1^{\gamma_1}X_2^{\gamma_2} \ldots X_{n+1}^{\gamma_{n+1}}$$

where K is a numerical factor. In particular, we have

$$x = K_1X_1^{\beta_1}X_{n+1}^{-\beta_{n+1}}$$

and

$$x - \alpha_i = K_iX_i^{\beta_i}X_{n+1}^{-\beta_{n+1}}$$

where the K are always numerical factors: whence the identities

$$K_1X_1^{\beta_1} = K_iX_i^{\beta_i} + a_iX_{n+1}^{\beta_{n+1}}$$

We have now expressed the theta–Fuchsian functions in two

entirely different ways: by the series (5) (Section I) and by the formula (1) of this section.

We therefore know **a priori** that there is an infinity of identities of the form

$$\Sigma\, H\left(\frac{\alpha_i z+\beta_i}{\gamma_i z+\delta_i}\right)(\gamma_i z+\delta_i)^{-2m} = \left(\frac{dx}{dz}\right)^m F(x)$$

where H and F are the algorithms of two rational functions.

The problem now is to effectively write down these identities and in particular to recognize in which cases the function F is identically zero.

However, to arrive at the completest possible solution, it is necessary to have recourse to considerations of a somewhat different order.

Until now we have always regarded the number m as positive; suppose now that it is an integer, but negative. Consider in particular the following function:

$$\Lambda(z) = \left(\frac{dx}{dz}\right)^{-h} \frac{[f(z)-\alpha_1]^{\mu_1}\,[f(z)-\alpha_2]^{\mu_2}\cdots[f(z)-\alpha_n]^{\mu_n}}{[f(z)-f(z_1)][f(z)-f(z_2)]\cdots[f(z)-f(z_p)]}$$

I assume that h is a positive integer and that the exponents $\mu_1, \mu_2, \ldots, \mu_n$ and the number p of factors in the denominator, which are also positive integers, satisfy the following inequalities:

$$\mu_i > h\left(1-\frac{1}{\beta_i}\right)$$

(5)

$$\Sigma\,\mu_i < h\left(1+\frac{1}{\beta_{n+1}}\right) + p$$

Because of these inequalities, the function $\Lambda(z)$ cannot become infinite except when one of the factors in the denominator vanishes. It follows that the infinities of this function are all simple and that they are covered by the formulae

$$\text{(6)} \qquad \frac{\alpha_i z_1+\beta_i}{\gamma_i z_1+\delta_i}, \frac{\alpha_i z_2+\beta_i}{\gamma_i z_2+\delta_i}, \ldots, \frac{\alpha_i z_p+\beta_i}{\gamma_i z_p+\delta_i}$$

where

$$\left[z, \frac{\alpha_i z+\beta_i}{\gamma_i z+\delta_i}\right]$$

represents one of the transformations in the group G.

We put

$$F(z) = [f(z)-\alpha_1]^{\mu_1}[f(z)-\alpha_2]^{\mu_2}\ldots [f(z)-\alpha_n]^{\mu_n}$$

and

$$\frac{1}{[f(z)-f(z_1)][f(z)-f(z_2)]\ldots[f(z)-f(z_p)]} =$$

$$= \frac{A_1}{f(z)-f(z_1)} + \frac{A_2}{f(z)-f(z_2)} + \ldots + \frac{A_p}{f(z)-f(z_p)}$$

so that

$$\text{(7)} \qquad \begin{aligned} &A_1 + A_2 + \ldots + A_p = 0 \\ &A_1 f(z_1) + A_2 f(z_2) + \ldots + A_p f^z(z_p) = 0 \\ &A_1 f^z(z_1 + A_2 f^z(z_2) + \ldots + A_p f^z(z_p) = 0 \\ &\ldots\ldots\ldots\ldots\ldots\ldots\ldots\ldots \\ &A_1 f^{p-2}(z_1 + A_2 f^{p-2}(z_2) + \ldots + A_p f^{p-2}(z_p) = 0 \\ &A_1 f^{p-1}(z_1 + A_2 f^{p-1}(z_2) + \ldots + A_p f^{p-1}(z_p) = 1 \end{aligned}$$

The residue corresponding to the infinity

$$\frac{\alpha_i z_k + \beta_i}{\gamma_i z_k + \delta_i}$$

will be written

$$\frac{A_k F(z_k)}{[f'(z_k)]^{h+1}} (\gamma_i z_k + \delta_i)^{-2(h+1)}$$

It follows that we have identically

$$\Lambda(z) = \Sigma_i \Sigma_k \frac{A_k F(z_k)}{[f'(z_k)]^{h+1}} \frac{(\gamma_i z_k + \delta_i)^{-2(h+1)}}{z - \dfrac{\alpha_i z_k + \beta_i}{\gamma_i z_k + \delta_i}} + G(z) \tag{8}$$

where G(z) denotes a function holomorphic inside the fundamental circle. More generally, if $\phi(z)$ is a rational function of z with no infinities inside the fundamental circle, then we have

$$\Lambda(z)\phi(z) = \sum_{i=0}^{\infty} \sum_{k=1}^{p} \frac{A_k F(z_k)}{[f'(z_k)]^{h+1}} \times$$

(9)

$$\times \frac{\phi\left[\dfrac{\alpha_i z_k + \beta_i}{\gamma_i z_k + \delta_i}\right]}{(\gamma_i z_k + \delta_i)^{2(h+1)}} \frac{1}{z - \dfrac{\alpha_i z_k + \beta_i}{\gamma_i z_k + \delta_i}} + G(z)$$

I claim that the holomorphic function G(z) in the formulae (8) and (9) is identically zero.

First proof

Consider a circle C with radius r and center at the origin, hence concentric with the fundamental circle.

We envision the different polygons R_i which are, entirely or in part, inside this circle.

The set of these polygons forms a polygonal figure S depending only on the radius r of the circle C. We consider the integral

$$\int \frac{\Lambda(z)\phi(z)dz}{z - x}$$

taken along the contour of this figure S. The theorem I have claimed can be taken as proved once I establish that this integral tends to 0 as r tends to 1, i.e., to the radius of fundamental circle.

To do this I shall show:

1° That the perimeter of S remains finite, whatever the value of r;

2° That the absolute value of the function under the $\int$ sign remains smaller than a certain quantity M when the variable remains on the contour of integration;

3° That M tends to 0 as r tends to 1.

By virtue of Lemma IV of Section I, the number of polygons R_i partially or totally inside the circle C is smaller than

$$N = \frac{\pi}{4\sigma}(e^{2(R+\lambda)} + e^{-2(R+\lambda)} - 2)$$

where σ and λ are constants and R is defined by the equation

$$r = \frac{e^{2R}-1}{e^{2R}+1}$$

Then, **a fortiori**, the number of sides of S will be smaller than 2nN. Consider one of these sides belonging to a polygon R_i, let l_i be the length of that side, and let l_0 be the length of the corresponding side of R_0. Finally, letting $\left(z, \frac{\alpha_i z+\beta_i}{\gamma_i z+\delta_i}\right)$ be the transformation in G which corresponds to R_i, we have

$$l_i < l_0 \, . \, H$$

where H is the greatest value taken by

$$\left|\frac{1}{(\gamma_i z+\delta_i)^2}\right|$$

as the variable z describes the side l_0.

Suppose that the polygon R_0 is entirely contained in the circle with center at the origin and radius

$$\frac{e^{2A}-1}{e^{2A}+1}$$

By virtue of Lemma V of Section I we have

$$H < \frac{e^{2A} + e^{-2A} + 2}{e^{2R} + e^{-2R} + 2}$$

Let L be the longest side of R_0. The perimeter of S will be smaller than

$$2nN \, . \, H \, . \, L$$

and, **a fortiori**, smaller than

$$\frac{2n\pi L}{\sigma}\left[e^{2A} + e^{-2A} + 2)\right]\frac{2e^{2(R+\lambda)}}{e^{2R}}$$

It will therefore be finite.

Q.E.D.

We must evidently assume that x is not on any of the contours of integration. The function $\frac{\phi(z)}{z-x}$ is then a well-defined rational function which does not become infinite for any value taken by the variable on the different contours of integration. We can then easily find an upper bound M_1 to its absolute value.

We can also assume that the function $\Lambda(z)$ does not become infinite along the perimeter of R_0, and we can find an upper bound M_2 to its absolute value as z describes the contour of R_0. But we have identically

$$\Lambda\left[\frac{\alpha_i z+\beta_i}{\gamma_i z+\delta_i}\right] = \Lambda(z)\,\frac{1}{(\gamma_i z+\delta_i)^{2h}}$$

We conclude that, when z describes the contour of S, the absolute value of $\Lambda(z)$ is smaller than

$$M_2H^h$$

and that the function under the $\int$ sign is smaller than

$$M = M_1M_2H^h$$

Moreover, as r tends to 1, M tends to 0, because M_1 and M_2 are constants and H tends to 0.

The stated theorem is therefore proved.

Second Proof

We put

$$\Phi(z,\alpha) = \sum \frac{\phi\left(\frac{\alpha_i a+\beta_i}{\gamma_i a+\delta_i}\right)}{z-\frac{\alpha_i a+\beta_i}{\gamma_i a+\delta_i}} \frac{1}{(\gamma_i a+\delta_i)^{2(h+1)}}$$

If we regard z as a constant in this expression, then $\Phi(z,a)$ will be a theta–Fuchsian function of a. Suppose first that z and a are both inside the fundamental circle; this theta–Fuchsian function does not become infinite except at a = z and points corresponding to it, i.e., when f(a) = f(z). The residue corresponding to the infinity z = a will equal $-\phi(z)$. We then have

$$(10) \qquad \Phi(z,a) = \frac{[f'(a)]^{n+1}}{f(z)-f(a)} \frac{\theta_q[f(a)]}{F_1(a)} \frac{F_1(z)\phi(z)}{\theta_q[f(z)][f'(z)]^h}$$

In this expression (10), $F_1(a)$ denotes the following product:

$$[f(a)-a_1]^{\lambda_1}[f(a)-a_2]^{\lambda_2}\ldots\ [f(a)-a_n]^{\lambda_n}$$

$\theta[f(a)]$ denotes a polynomial of degree q in f(a) whose coefficients can be functions of z.

Finally, the numbers $\lambda_1, \lambda_2, \ldots, \lambda_n$ and q must satisfy the inequalities (3)

$$\lambda_i < (h+1)\left(1 - \frac{1}{\beta_i}\right)$$

(11)

$$q < \Sigma\, \lambda_i - (h+1)\left(1 + \frac{1}{\beta_{n+1}}\right) + 1$$

In fact, we easily see that this must be the form of any theta—Fuchsian function of a with no infinity distinct from the infinity $a = z$.

Comparing the inequalities (3) and (11), we find

$$\lambda_i \leqslant \mu_i \qquad q < p$$

I now propose to establish equation (9) while showing that G(z) is identically zero, i.e., to show the identity

$$\phi(z)\Lambda(z) = \sum_{k=1}^{p} \frac{A_k F(z_k)}{[f'(z_k)]^{h+1}}\, \Phi(z,z_k)$$

or

$$\phi(z)\Lambda(z) = \sum_k \frac{A_k}{f(z)-f(z_k)} \frac{F(z_k)}{F_1(z_k)} \times$$

$$\times\ \theta_q[f(z_k)] \frac{F_1(z)\phi(z)}{\theta_q[f(z)][f'(z)]^h}$$

We can simplify the right hand side by assuming that

$$\lambda_i = \mu_i$$

If this is not already the case, we multiply $F_1(a)$ and $\theta_q[f(a)]$ by

$$[f(a)-a_1]^{\mu_1-\lambda_1}[f(a)-a_2]^{\mu_2-\lambda_2} \dots [f(a)-a_n]^{\mu_n-\lambda_n}$$

The degree of the polynomial $\theta_q[f(a)]$ will then be increased, but it will remain less than p.

Thus we can always assume

$$\lambda_i = \mu_i \qquad \text{and, consequently} \qquad F(z) = F_1(z)$$

while continuing to assume

$$p > q$$

The identity to be proved then simplifies and can be written

$$\sum_k \frac{A_k}{f(z)-f(z_k)} \frac{F(z)\phi(z)}{[f'(z)]^h} =$$
$$= \sum_k \frac{A_k}{f(z)-f(z_k)} \frac{\theta_q[f(z_k)]}{\theta_q[f(z)]} \frac{F(z)\phi(z)}{[f'(z)]^h}$$

If we now put

$$\frac{\theta_q[f(z_k)]-\theta_q[f(z)]}{f(z) - f(z_k)} = P[f(z_k)]$$

then P represents a polynomial in $f(z_k)$ whose degree equals $q - 1$, and consequently cannot exceed $p - 2$. The identity to be proved then reduces to

$$\Sigma\, A_k P[f(z_k)] = 0$$

and in this form it is evident, by virtue of the relations (7). The stated theorem is thus proved again.

This second proof leads us very naturally to an important theorem concerning theta–Fuchsian functions of the second species. Let

$$(12) \qquad \Sigma\, H\left[\frac{\alpha_i z+\beta_i}{\gamma_i z+\delta_i}\right](\gamma_i z+\delta_i)^{-2(h+1)}$$

be a series in which H is the algorithm of a rational function with no infinities inside the fundamental circle. We have seen that this series represents a theta–Fuchsian function without infinities, and that it can be equated with an expression of the form

$$(13) \qquad [f'(z)]^{h+1}\,\frac{\theta_{q-1}[f(z)]}{F(z)}$$

where θ_{q-1} is a polynomial of degree $q - 1$ in f(z) and q is the greatest integer staisfying the inequalities (11).

Each function of the form (13) can be expressed linearly in terms of q particular such functions, hence each function of the form (13) can be expressed linearly in terms of q others. But we can make two hypotheses.

Either, every function of the form (13) can be put in the form (12), in which case not every function of the form (12) can be expressed linearly in terms of $q - 1$ particular such functions.

Or else it is not possible to put each function (13) in the form (12), in which case every function (12) can be expressed linearly in terms of $q - 1$ others.

This second hypothesis has to be rejected.

Suppose, on the contrary, that it holds, and let

$$\Theta_1,\ \Theta_2,\ \ldots,\ \Theta_{q-1}$$

be the q − 1 functions of the form (12) which can be used to express all others as linear combinations.

Now let

$$z_1, z_2, \ldots, z_q$$

be any q values of z. We can always find q numbers

$$A_1, A_2, \ldots, A_q$$

satisfying the conditions

$$A_1\Theta_1(z_1) + A_2\Theta_1(z_2) + \ldots + A_q\Theta_1(z_q) = 0$$

$$A_1\Theta_2(z_1) + A_2\Theta_2(z_1) + \ldots + A_q\Theta_2(z_q) = 0$$

$$\cdots\cdots\cdots\cdots\cdots\cdots$$

$$A_1\Theta_{q-1}(z_1) + A_2\Theta_{q-1}(z_2) + \ldots + A_q\Theta_{q-1}(z_q) = 0$$

We then have, whatever the function H in the expression (12) may be,

$$(14) \qquad \sum_{k=1}^{q} \sum_{i} A_k H\left[\frac{\alpha_i z_k + \beta_i}{\gamma_i z_k + \delta_i}\right](\gamma_i z_k + \delta_i)^{-2(h+1)} = 0$$

Let $\left[z, \frac{\alpha z+\beta}{\gamma z+\delta}\right]$ be any transformation in the group G; we easily prove the following identity by taking the function $\Phi(z,z_k)$ defined above and setting $\phi(z) = 1$:

$$(15)\qquad \theta\left[\frac{\alpha z+\beta}{\gamma z+\delta},\ z_k\right] - (\gamma z+\delta)^{-2h}\theta(z,z_k) =$$

$$= \Sigma_i(\gamma_i z_k+\delta_i)^{-2h-2}\frac{\left\{\dfrac{(\gamma z+\delta)^{2h+1} - \left[\gamma\left(\dfrac{\alpha_i z_k+\beta_i}{\gamma_i z_k+\delta_i}\right)+\delta\right]^{2h+1}}{z-\dfrac{\alpha_i z_k+\beta_i}{\gamma_i z_k+\delta_i}}\right\}}{(\gamma z+\delta)^{2h}\left[\gamma\left(\dfrac{\alpha_i z_k+\beta_i}{\gamma_i z_k+\delta_i}\right)+\delta\right]^{2h+1}} =$$

$$= \Sigma_i(\gamma_i z_k+\delta_i)^{-2h-2}H\left[\frac{\alpha_i z_k+\beta_i}{\gamma_i z_k+\delta_i}\right]$$

where $H(z)$ is the algorithm of a rational function with no infinities inside the fundamental circle.

We now put

$$\Lambda(z) = \sum_k A_k\Phi(z,z_k)$$

and from the formulae (14) and (15) we deduce

$$\Lambda\left[\frac{\alpha z+\beta}{\gamma z+\delta}\right] - (\gamma z + \delta)^{-2h}\Lambda(z) = 0$$

Since this holds whatever transformation $(z, \frac{\alpha z+\beta}{\gamma z+\delta})$ we have chosen in the group G, we necessarily have

$$\Lambda(z) = \left(\frac{dx}{dz}\right)^{-h}F(x)$$

where $F(x)$ is rational in X.

But $\Lambda(z)$ has no more than q distinct infinities z_1, z_2, ..., z_q, whereas it is easy to see that a function of the form $\left[\frac{dx}{dz}\right]^{-h}F(x)$ must always have more than q distinct infinities.

Hence the hypothesis made at the beginning is absurd.

Hence each expression of the form (13) can always be put in the form (12).

Hence each function of the form (1) can always be expressed by a series of the form (5) (I, p.148), provided $m > 1$.

Now, supposing that we are given a series of the form (5) (I), how can we effectively put it in the form (1)?

Given a series $\Theta(z)$ of the form (5) of Section I, it is a question of putting it in the form

$$\left[\frac{dx}{dz}\right]^{m} F(x)$$

Initially, we know the infinities of $\Theta(z)$ and the corresponding residues, but these do not suffice to completely determine $F(x)$; a certain number of undetermined coefficients remain in this rational function. We could first calculate a sufficient number of values of the series $\Theta(z)$ to get equations which permit these coefficients to be determined. It is simpler to operate as follows. Consider the function

$$\Theta(z)\left[\frac{dx}{dz}\right]^{-m} = F(x)$$

It becomes infinite, not only when $\Theta(z)$ is infinite, but also when $\frac{dx}{dz}$ is zero, i.e., when

$$z = \alpha_1,\ z = \alpha_2,\ \ldots,\ z = \alpha_n$$

We then expand the series $\Theta(z)\left[\frac{dx}{dz}\right]^{-m}$ in increasing powers of $z - \alpha_i$ and calculate the coefficients of all the negative powers. When these coefficients are known, and hence also the first coefficients in the expansion of x in powers of $z - \alpha_i$, the function $F(x)$ will also be known.

But to find these coefficients themselves, it is necessary to calculate a certain number of values of the series $\Theta(z)$ and analogous series. It is the case, however, that this difficulty can be avoided.

We take the function

$$\Lambda(z) = \left(\frac{dx}{dz}\right)^{-h} \frac{F(z)}{[f(z)-f(z_1)][f(z)-f(z_2)]\ldots[f(z)-f(z_p)]}$$

that we have studied above. We have proved the following identity:

$$(8) \qquad \Lambda(z) = \sum_i \sum_k \frac{A_k F(z_k)}{[f'(z_k)]^{h+1}} \frac{(\gamma_i z_k+\delta_i)^{-2(h+1)}}{z - \dfrac{\alpha_i z_k+\beta_i}{\gamma_i z_k+\delta_i}}$$

Differentiating this identity 2h + 1 times with respect to z, we obtain

$$(16) \qquad \frac{1}{(2h+1)!} \frac{d^{2h+1}\Lambda}{dz^{2h+1}} =$$

$$= - \Sigma_i \Sigma_k \frac{A_k F(z_k)}{[f'(z_k)]^{h+1}} \frac{(\gamma_i z_k+\delta_i)^{-2(h+1)}}{\left(z - \dfrac{-\alpha_i z_k+\beta_i}{\gamma_i z_k-\delta_i}\right)^{2(h+1)}}$$

the right hand side can be written

$$- \sum_i \sum_k \frac{A_k F(z_k)}{[f'(z_k)]^{h+1}} \frac{(\gamma_i z-\alpha_i)^{-2(h+1)}}{\left(z_k - \dfrac{-\delta_i z+\beta_i}{\gamma_i z-\alpha_i}\right)^{2(h+1)}}$$

We remark that if we have the transformation $\left(z, \dfrac{\alpha_i z+\beta_i}{\gamma_i z+\delta_i}\right)$ in the group G, then we likewise have the inverse transformation $\left(z, \dfrac{-\delta_i z+\beta_i}{\gamma_i z-\alpha_i}\right)$. I.e., the preceding expression can be written

$$-\sum_{i}\sum_{k} \frac{A_k F(z_k)}{[f'(z_k)]^{h+1}} \, \frac{(\gamma_i z+\delta_i)^{2(h+1)}}{\left(z_k - \dfrac{\alpha_i z+\beta_i}{\gamma_i z+\delta_i}\right)^{2(h+1)}}$$

Thus the right hand side of the identity (16) is a series of the form (5) (Section I); it remains to put the left hand side in the form (1). This is a calculation which presents no difficulties. I indicate the result in the cases h = 1 and h = 2.

Let

$$\Lambda(z) = \left(\frac{dx}{dz}\right)^{-h} \Lambda_0(x)$$

and let $\Lambda_1, \Lambda_2, \ldots$ be the successive derivatives of Λ_0 with respect to x. We now take equation

$$(3') \qquad \frac{d^2 \nu}{dx^2} = \nu \phi(x)$$

and let $\phi_1, \phi_2, \ldots$ be the successive derivatives of ϕ with respect to x. In the case h = 1 we have

$$(17) \qquad \frac{d^3 \Lambda}{dz^3} = (\Lambda_3 - 2\Lambda_0\phi_1 - 4\Lambda_1\phi)\left(\frac{dx}{dz}\right)^2$$

and in the case h = 2,

$$(18) \qquad \frac{d^5 \Lambda}{dz^5} = (\Lambda_5 - 20\Lambda_3\phi - 30\Lambda_2\phi_1 - 18\Lambda_1\phi_2 - 4\Lambda_0\phi_3 + 64\Lambda_1\phi^2 + 64\Lambda_0\phi\phi_1)\left(\frac{dx}{dz}\right)$$

We can find an infinity of relations analogous to (16) by various procedures; for example by differentiating this relation with respect to the various parameters $z_1, z_2, \ldots, z_p$.

I now want to pose another problem which I can resolve only partially.

Consider the series (5) (Section I)

$$\sum H\left(\frac{\alpha_i z+\beta_i}{\gamma_i z+\delta_i}\right)(\gamma_i z+\delta_i)^{-2m} ;$$

it depends on the rational function H, and I therefore represent it by the notation

$$\Theta[z,H]$$

How must the rational function H **be chosen in order that** Θ **be identically zero?**

No doubt the complete solution of this problem will be very complicated, but it can be assisted, in each particular case, by the following considerations:

1° If we have, identically,

$$\Theta(z,H_1) = \Theta(z,H_2) = 0$$

then we likewise have

$$\Theta(z,H_1+H_2) = 0$$

2° If the rational function H can be expanded in a convergent series of the following form

$$\text{(S)} \qquad k_1H_1 + k_2H_2 + \ldots + k_nH_n + \ldots$$

where $k_1, k_2, \ldots, k_n$ are constant coefficients and $H_1, H_2, \ldots, H_n$ are rational functions such that

$$\Theta(z,H_n) = 0$$

and if the series (S) is convergent whenever z is inside the fundamental circle, then we have

$$\Theta(z,H) = 0$$

identically.

3° If we have, identically,

$$\Theta(z,H) = 0$$

and if we put

$$H_1(z) = H\left(\frac{\alpha z+\beta}{\gamma z+\delta}\right)(\gamma z+\delta)^{-2m}$$

where $\left(z, \frac{\alpha z+\beta}{\gamma z+\delta}\right)$ is one of the transformations in G, then we have, identically,

$$\Theta(z,H_1) = 0$$

In fact, if we make the abbreviations

$$\frac{\alpha_i z+\beta_i}{\gamma_i z+\delta_i} = f_i(z) \qquad \frac{\alpha z+\beta}{\gamma z+\delta} = f_1(z) \qquad \frac{-\delta z+\beta}{\gamma z-\alpha} = f_1^{-1}(z)$$

then the series (5) can be put in the form

$$\Theta(z,H) = \Sigma\ H[f_1[f_i[f_1^{-1}(z)]]]\left[\frac{df_1[f_i[f_1^{-1}(z)]]}{dz}\right]^m$$

If we now replace z by $f_1(z)$ in the identity

$$\Theta(z,H) = 0$$

then the result is

$$0 = \Sigma\, H[f_1[f_i(z)]] \left[\frac{df_1[f_i(z)]}{df_i(z)}\right]^m \left[\frac{df_i}{dz}\right]^m \left[\frac{dz}{df_1(z)}\right]^m =$$

$$= \Theta(z,H_1)\left[\frac{dz}{df_1(z)}\right]^m$$

which proves the identity stated.

4° Returning to the notation used at the beginning of this Section, let α_r be one of the vertices of the polygon R_0, and let α'_r be its reflection in the fundamental circle. One of the transformations in G may be written as

$$\left(\frac{z-\alpha_r}{z-\alpha'_r}\ ,\ e^{\frac{2i\pi}{\beta_r}}\ \frac{z-\alpha_r}{z-\alpha'_r}\right)$$

where β_r is a known integer. If we put

$$H(z) = \left(\frac{z-\alpha_r}{z-\alpha'_r}\right)^p \frac{1}{(z-\alpha'_r)^{2m}}$$

then we have, identically,

$$\Theta(z,H) = 0$$

$$\text{unless } p + m \equiv 0 \qquad (\mathrm{mod}\ \beta_r)$$

In fact, if we consider the different transformations (z,f_i) in the group G, we can choose an infinity

$$(z,\phi_0),\ (z,\phi_1),\ (z,\phi_2),\ \ldots,\ (z,\phi_n),\ \ldots$$

such that each of the functions $f_i(z)$ is defined in one and only one way by a relation of the following form:

$$\frac{f_i-\alpha_r}{f_i-\alpha'_r} = e^{\frac{2hi\pi}{\beta_r}} \frac{\phi_k-\alpha_r}{\phi_k-\alpha'_r}$$

where h is an integer less than or equal to β_r. We then have

$$\Theta(z,H) = \Sigma_i \frac{(f_i-\alpha_r)^p}{(f_i-\alpha'_r)^{p+2m}} \left[\frac{df_i}{dz}\right]^m$$

where

$$\Theta(z,H) = \sum_{k=0}^{\infty} \sum_{h=1}^{\beta_r} \frac{(\phi_k-\alpha_r)^p}{(\phi_k-\alpha'_r)^{p+2m}} \left[\frac{d\phi_k}{dz}\right]^m e^{\frac{2k(p+m)i\pi}{\beta_r}}$$

But if we do not have

$$p + m \equiv 0 \qquad (\text{mod } \beta_r)$$

then we have

$$\sum_{h=1}^{\beta_r} e^{\frac{2h(p+m)i\pi}{\beta_r}} = 0$$

Hence we have

$$\Theta(z,H) = 0$$

Q.E.D.

5^0 Again we take up the identity

$$\phi(z)\Lambda(z) = \sum_k \frac{A_k F(z_k)}{[f'(z_k)]^{h+1}} \Phi(z,z_k) \tag{9}$$

where

$$\Phi(z,z_k) = \sum_i \frac{\phi\left(\dfrac{\alpha_i z_k+\beta_i}{\gamma_i z_k+\delta_i}\right)}{z - \dfrac{\alpha_i z_k+\beta_i}{\gamma_i z_k+\delta_i}} \frac{1}{(\gamma_i z_k+\delta_i)^{2(h+1)}}$$

If we set, in turn,

$$\phi(z) = 1 \quad \text{and} \quad \phi(z) = \frac{1}{z-b}$$

where b is a constant outside the fundamental circle, then we obtain two identities of the form (9).

Multiplying the first by $\frac{1}{z-b}$ and subtracting the second, we obtain

$$\sum_i \frac{A_k F(z_k)}{[f'(z_k)]^{h+1}} \Psi(b,z_k) = 0 \tag{19}$$

by putting

$$\Psi(b,a) = \sum_i \frac{1}{b - \dfrac{\alpha_i a+\beta_i}{\gamma_i z+\delta_i}} \frac{1}{(\gamma_i a+\delta_i)^{2(h+1)}}$$

Again, putting

$$\Theta(b,a) = -\frac{d^{2h+1}\Psi}{db^{2h+1}}\frac{1}{(2h+1)!}$$

we obtain, by differentiating (19) 2h + 1 times with respect to b,

$$\sum_k \frac{A_k F(z_k)}{[f'(z_k)]^{h+1}}\Theta(b,z_k) = 0 \tag{20}$$

But $\Theta(b,z_k)$ is a theta–Fuchsian function of b. The left hand side of the identity (20) is therefore a series of the form (5), (Section I). Hence we obtain in this way a series of this form which is identically zero when b, which is the independent variable here, remains **outside** the fundamental circle.

What we seek, however, is a series of the form (5) (Section I) which remains identically zero when the independent variable remains **inside** the fundamental circle. Fortunately it is easy to pass from one case to the other. In fact we put

$$b = \frac{1}{z}$$

We have

$$\frac{\alpha_i b+\beta_i}{\gamma_i b+\delta_i} = \frac{1}{\dfrac{\alpha'_i z+\beta'_i}{\gamma'_i z+\delta'_i}}$$

where α'_i, β'_i, γ'_i, δ'_i denote the complex conjugates of α_i, β_i, γ_i, δ_i, so that the transformations $\left[z, \dfrac{\alpha'_i z+\beta'_i}{\gamma'_i z+\delta'_i}\right]$ form a group G′ conjugate to G.

Likewise, putting

$$b_i = \frac{\alpha_i b+\beta_i}{\gamma_i b+\delta_i} \qquad z_i = \frac{\alpha'_i z+\beta'_i}{\gamma'_i z+\delta'_i}$$

we have

$$\frac{db_i}{db} = \frac{z^2}{z_i^2}\frac{dz_i}{dz}$$

We then have

$$\Theta(b,a) = \sum_i \frac{1}{(b_i-a)^{2h+2}} \left(\frac{db_i}{db}\right)^{h+1} =$$

$$= \sum_i \frac{z^{2h+2}}{(1-az_i)^{2h+2}} \left(\frac{dz_i}{dz}\right)^{h+1}$$

or finally,

$$\Theta(b,a) = z^{2h+2}\Theta'(z,a)$$

putting

$$\Theta'(z,a) = \sum_i \frac{1}{(1-az_i)^{2h+2}} \left(\frac{dz_i}{dz}\right)^{h+1}$$

We arrive at the following identity:

$$\sum_k \frac{A_k F(z_k)}{[f'(z_k)]^{h+1}} \Theta'(z,z_k) = 0$$

The functions Θ' are theta–Fuchsian functions of z, so we indeed have an identity of the form sought. But the group for the functions Θ' is not the group G, it is the conjugate group G′. Passing from the group G′ to the group G by changing $\sqrt{-1}$ to $-\sqrt{-1}$, we obtain

$$\sum_k \left[\frac{A_k F(z_k)}{[f'(z_k)]^{h+1}}\right]' \Theta'_1(z,z'_k) = 0$$

In this formula,

$$\left[\frac{A_k F(z_k)}{[f'(z_k)]^{h+1}}\right]'$$

and z'_k denote the complex conjugates of the quantities

$$\frac{A_k F(z_k)}{[f'(z_k)]^{h+1}}$$

and z_k, and

$$\Theta'_1(z,a) = \sum \frac{1}{\left(1-a\,\dfrac{\alpha_i z+\beta_i}{\gamma_i z+\delta_i}\right)^{2h+2}} \frac{1}{(\gamma_i z+\delta_i)^{2h+2}}$$

so that, if we put

$$H(z) = \sum_k \left[\frac{A_k F(z_k)}{[f'(z_k)]^{h+1}}\right]' \frac{1}{(1-z'_k z)^{2h+2}}$$

we have

$$\Theta(z,H) = 0$$

identically as long as z remains inside the fundamental circle.

6^0 Suppose that we have, identically

$$\Theta(z,H) = 0$$

and that H depends on an arbitrary parameter λ; then we likewise have

$$\Theta\left(z, \frac{dH}{d\lambda}\right) = 0$$

By combining the preceding six principles in various ways, we obtain an infinity of relations of the form

$$\Theta(z,H) = 0$$

VI. First family; arbitrary genus

We have dwelt on the functions of the first family and genus zero because they are the simplest of all, and because the general principles, once shown in this particular case, extend without difficulty to functions of all the families and all genera. We lay less stress on the other Fuchsian functions and confine ourselves to making known their most remarkable aspects.

When the polygon R_0, while remaining in the first family, is not of the simple form studied in the previous paragraph, and is not susceptible to reduction, then the corresponding Fuchsian functions are of the first family and genus greater than zero.

We begin with a simple example. Take a polygon R_0 with 4n sides and such that opposite sides are equivalent. Then all the vertices form a single cycle, so that the sum of all the angles of the polygon R_0 is an aliquot part of 2π. Suppose first that it is precisely equal to 2π. We can then express all the Fuchsian functions derived from the polygon R_0 in terms of two of them which we call x and y, and between which there is an algebraic relation

$$(1) \qquad \psi(x,y) = 0$$

which is of genus n.

Recall equation (3) of Section IV. It is

$$(3) \qquad \frac{d^2\nu}{dx^2} = \nu\phi(x,y)$$

and we have, identically

$$\phi(x,y) = \frac{1}{2}\frac{x'''x'}{(x')^4} - \frac{3}{4}\frac{(x'')^2}{(x')^4}$$

where x', x'', x''' denote the successive derivatives of x with respect to z.

What is the form of the function $\phi(x,y)$? To simplify, I shall assume that the function y of x defined by equation (1) has no singular points of any but the simplest kind; i.e., whenever we have

$$\frac{d\psi}{dy} = 0$$

we also have

$$\frac{d\psi}{dx} \gtrless 0 \qquad \frac{d^2\psi}{dy^2} \gtrless 0$$

(which corresponds, on the algebraic curve $\psi(x,y) = 0$ to a tangent parallel to the y axis not having contact with the curve or order higher than first), or else

$$\frac{d\psi}{dx} = 0 \quad \left[\frac{d^2\psi}{dxdy}\right]^2 - \frac{d^2\psi}{dx^2}\frac{d^2\psi}{dy^2} \lessgtr 0 \quad \frac{d^2\psi}{dy^2} \gtrless 0$$

(which corresponds to an ordinary double point on the algebraic curve $\psi(x,y) = 0$).

We suppose in addition that if equation (1) is of order r then for x infinite we have r finite and distinct values of the ratio $\frac{y}{x}$ (which corresponds to r distinct asymptotic directions of the curve $\psi(x,y) = 0$).

These hypotheses do not restrict generality; among the Fuchsian functions corresponding to the group G, we can in fact choose the functions x and y in infinitely many ways so that the others can be expressed rationally in terms of x and y and the choice can always be made so as to satisfy the above hypotheses.

We now return to our function $\phi(x,y)$. It is clear that it remains finite for infinite values of x; it is likewise finite whenever x' is not zero. But x' does not vanish unless, at the same time,

$$\frac{d\psi}{dy} = 0 \qquad \frac{d\psi}{dx} \gtrless 0$$

Then the only infinities of the function $\phi(x,y)$ are the singular points of the equation (1) which are only ordinary singular points and not double points of the curve (1).

We consider one of these infinities, and let α, β, γ be the corresponding values of x, y, z. We easily see that the expression of $x - \alpha$ begins with a term in $(z-\gamma)^2$, while that of $y - \beta$ begins with a term in $z - \gamma$, and we conclude that

$$\lim (y-\beta)^2 \left[\frac{dx}{dy}\right]^2 \phi(x,y) = -\frac{3}{4} \qquad (\text{for } x = \alpha,\ y = \beta)$$

These conditions are not sufficient to completely determine the function $\phi(x,y)$. The latter must also satisfy a number of transcendental conditions whose deeper study will be the subject of a later memoir. We also show in another memoir that if these conditions are not satisfied, but if the coefficients of $\phi(x,y)$ satisfy certain inequalities, then x is no longer a Fuchsian function of z, it is true, but it is still a uniform (Kleinian) function.

On how many distinct parameters does our group G depend?

To define a polygon of 4n sides, 8n − 3 conditions are necessary; however, our polygon R_0 is not arbitrary. It is subject to 2n + 1 conditions, namely that opposite sides be congruent and that the sum of the angles equal 2π. Thus 6n − 4 arbitrary parameters remain. But we have seen, in Section IX of the memoir on Fuchsian groups, that the same group G can be regarded as derived from an infinity of different polygons, and in fact, in the case we are concerned with, we can replace the polygon R_0 by another R'_0, with the same number of sides, disposed in the same manner, but with any point of the plane as vertex. Thus there is a double infinity of polygons R_0 corresponding to the same group G, and the number of parameters of G is two less than the number of parameters of R_0. **The group G therefore depends on 6n − 6 parameters.**

On how many of these parameters does the relation (1) depend? Of course, I do not consider two relations

$$\psi(x,y) = 0 \qquad \psi_1(\xi,\eta) = 0$$

to be distinct if we can pass from one to the other by a uniform transformation, i.e., by replacing ξ and η by rational functions of x and y. The theory of Abelian functions informs us that the number of these parameters is 3n − 3 for a relation of genus n. But these are 3n − 3 **complex** parameters, which correspond to 6n − 6 **real** parameters, i.e., the number of parameters on which relation (1) depends is precisely the number of parameters of the group G. Does it follow that we can always dispose the parameters of the group G so as to obtain whatever algebraic relation we want between x and y? This is what I shall prove in a later memoir.

In the particular case n = 1 we are no longer dealing with

Fuchsian functions in the strict sense. In fact, in the curvilinear polygons R_0 whose sides are arcs of circles orthogonal to the fundamental circle, the angle sum is always smaller than that for a rectilinear polygon with the same number of sides. For a quadrilateral, the angle sum must be less than 2π. But here we have assumed that the angle sum of our curvilinear polygon with 4n sides is precisely 2π. If we want to take $n = 1$, then the polygon R_0 must become rectilinear, the radius of the fundamental circle must become infinite and the quadrilateral R_0 reduces to a simple rectilinear parallelogram. The group G then reduces to transformations of the form $(z, z+\alpha)$, and the Fuchsian functions derived from it reduce to doubly periodic functions. It is in this sense that we can say that elliptic functions are special cases of Fuchsian functions.

Suppose then that $n > 1$. The polygon R_0 can present different kinds of symmetry. First, it can be symmetric with respect to a circle cutting the fundamental circle orthogonally; in this case the coefficients of the functions $\psi(x,y)$ and $\phi(x,y)$ are real. Now suppose that it is symmetric with respect to a point, possibly after suitable change of the variable z, or by application of the rule of Section IX in the memoir on Fuchsian groups; in this case relation (1) reduces to the hyperelliptic relation

$$y^2 = (x-a_1)(x-a_2)\ldots(x-a_{2n})$$

However, the converse is not true.

Now consider an Abelian integral of the first species

$$\int g(x,y)dx \tag{4}$$

and replace x and y by their values in z; we obtain

$$\int g(x,y)dx = \int g(z)\frac{dx}{dz}\,dz = G(z)$$

where g(z) denotes a Fuchsian function of z and G(z) denotes a uniform function of z, holomorphic inside the fundamental circle but ceasing to exist outside it.

If

$$\left(z, \frac{\alpha_i z+\beta_i}{\gamma_i z+\delta_i}\right)$$

is one of the transformations in the Fuchsian group then we have, identically,

$$G\left(\frac{\alpha_i z+\beta_i}{\gamma_i z+\delta_i}\right) = G(z) + K_i$$

where K_i is one of the periods of the Abelian integral. Thus, corresponding to each transformation in our group, we have one of the periods of the integral (4).

However, we can look at this from another point of view. Let ab be one of the sides of the polygon R_0; then

$$\int_a^b g(z) \frac{dx}{dz}\, dz = G(b) - G(a)$$

Thus $G(b) - G(a)$ will be one of the periods of the integral (4), and hence each side of R_0 corresponds to one of these periods.

If cd is the side equivalent to ab we evidently have

$$G(b) - G(a) = G(c) - G(d)$$

It follows that two equivalent sides will correspond to the same period, taken in opposite senses, and the 2n pairs of equivalent sides of our polygon will correspond to a fundamental system of 2n periods of the integral (4). But these periods will not be the normal periods.

However, among the polygons equivalent to R_0, by virtue of the rule in Section IX of the memoir on Fuchsian groups, there is one with remarkable features in this regard. Consider a polygon R_0 of 4n sides, whose vertices are

$$a_1, b_1, c_1, d_1, a_2, b_2, c_2, d_2, \ldots, a_n, b_n, c_n, d_n$$

following the perimeter in the positive sense.

For greater symmetry of notation, I denote the last vertex indifferently by d_n or d_0. I assume that the side $d_{i-1} a_i$ is equivalent to $c_i b_i$ and that the side $a_i b_i$ is equivalent to $d_i c_i$. In other words, the side of rank 4p + 1 is equivalent to the side of rank 4p + 3, and the side of rank 4p + 2 is equivalent to the side of rank 4p + 4. One sees without difficulty that all the vertices of such a polygon belong to the same cycle and that the polygon R'_0 is of genus n. I assume, in addition, that the angle sum is 2π; then we see easily that it can be reduced to a polygon such as R_0 by application of the rule of Section IX.

We now consider two Abelian integrals of the first species,

$$\int g(x,y)dx = \int g(z)\frac{dx}{dz}\, dz = G(z) \tag{4}$$

$$\int g_1(x,y)dx = \int g_1(z)\frac{dx}{dz}\, dz = G_1(z) \tag{5}$$

We put

$$\int_{d_{i-1}}^{a_i} g(z)\frac{dx}{dz}\,dz = -\int_{b_i}^{c_i} g(z)\frac{dx}{dz}\,dz = G(a_i) - G(d_{i-1}) = A_i$$

$$\int_{a_i}^{b_i} g(z)\frac{dx}{dz}\,dz = -\int_{c_i}^{d_i} g(z)\frac{dx}{dz}\,dz = G(b_i) - G(a_i) = B_i$$

$$\int_{d_{i-1}}^{a_i} g_1(z)\frac{dx}{dz}\,dz = -\int_{b_i}^{c_i} g_1(z)\frac{dx}{dz}\,dz = G_1(a_i) - G_1(d_{i-1}) = A'_i$$

$$\int_{a_i}^{b_i} g_1(z)\frac{dx}{dz}\,dz = -\int_{c_i}^{d_i} g_1(z)\frac{dx}{dz}\,dz = G_1(b_i) - G_1(a_i) = B'_i$$

Thus the periods A_i, B_i, A'_i, B'_i of the integrals (4) and (5) correspond to the different sides of the polygon R'_0.

If we now take the integral

$$\int G_1(z)g(z)\frac{dx}{dz}\,dz$$

along the perimeter of R_0 then it must be zero and, by expanding the integral, we find without difficulty the well-known identity

$$\Sigma_i(A_iB'_i - B_iA'_i) = 0$$

which proves that the periods A_i, B_i, ... are the normal periods.

I shall not further emphasize these considerations, which show the extent to which one can benefit from Fuchsian functions in the study of Abelian integrals.

I now suppose that the angle sum of R_0 is no longer 2π, but $\frac{2\pi}{\beta}$, where β is an integer.

Let us see how the change in this hypothesis affects the preceding. The relation (1) will always be of genus n. The function $\phi(x,y)$ will enjoy the same properties as before, with the only difference that it will become infinite for $x = a$, $y = b$, i.e., for the values of x and y which correspond to vertices of the polygon R_0.

We then have

$$\lim(x-a)^2\phi(x,y) = -\frac{1}{4}\left(1 - \frac{1}{\beta^2}\right) \qquad (\text{for } x = a,\ y = b)$$

The singular points of equation (3) are therefore those of the relation (1), and the point a,b.

Now consider a polygon R_0 of the first family, but otherwise arbitrary; let 2n be the number of sides and p the number of cycles. Suppose that if we calculate the angle sums of the different cycles we find

$$\frac{2\pi}{\beta_1}, \frac{2\pi}{\beta_2}, \ldots, \frac{2\pi}{\beta_p}$$

respectively for the first, second, ..., p^{th} of these cycles. The Fuchsian functions derived from this polygon R_0 are expressed rationally in terms of two of them, which I call x and y, and between which there is an algebraic relation

$$(1) \qquad \psi(x,y) = 0$$

of genus

$$p = \frac{n-p+1}{2}$$

Let (a_1,b_1), (a_2,b_2), ..., (a_p,b_p) be the values taken by x and y when the variable z comes to one of the vertices of the first, second,

..., p^{th} cycle.

I let (c_i, d_i) be the analytic points for which

$$\frac{d\psi}{dy} = 0 \qquad \frac{d\psi}{dx} \gtrless 0$$

I let (e_i, f_i) be the analytic points for which

$$\frac{d\psi}{dx} = \frac{d\psi}{dy} = 0$$

Moreover, I retain the previous hypotheses on the function $\psi(x,y)$, which I suppose to be of order r. The number of points c_i, d_i is then

$$r(r-1) + 2P - (r-1)(r-2) = 2r + 2P - 2$$

and that of the points e_i, f_i is

$$\frac{(r-1)(r-2)}{2} - P$$

Equation (3) is

$$\frac{d^2\nu}{dx^2} = \nu\phi(x,y)$$

The function $\phi(x,y)$, which is rational in x and y becomes infinite:

1° For $x = c_i$, $y = d_i$; we then have

$$\lim (y-d_i)^2 \left[\frac{dx}{dy}\right]^2 \phi(x,y) = -\frac{3}{4} \qquad (\text{for } x = c_i,\ y = d_i)$$

2° For $x = a_i$, $y = b_1$; we then have

$$\lim(x-a_i)^2 \phi(x,y) = \frac{1-\beta_i^2}{4\beta_i^2} \qquad (\text{for } x = a_i,\ y = b_i)$$

The number of conditions we must impose on the group G in order to have a given relation (1) between x and y and given values for the quantities (a_i,b_i) and the integers β_i is $6P - 6 + 2p$. The number of parameters on which the group G depends is also $6P - 6 + 2p$. It does not follow immediately that we can dispose these $6P - 6 + 2p$ parameters in such a way as to satisfy these $6P - 6 + 2p$ conditions. But I shall show, in a later memoir, that this is effectively the case, and that there **always** exist two Fuchsian functions x and y which correspond to the same group G, between which there is a **given** algebraic relation

$$\psi(x,y) = 0$$

and which, for p **given** systems of values

$$(x = a_1,\ y = b_1),\ (x = a_2,\ y = b_2),\ \ldots,\ (x = a_p,\ y = b_p)$$

see their derivatives of the first

$$\beta_1 - 1,\ \beta_2 - 1,\ \ldots,\ \beta_p - 1$$

orders vanish (where β_1, β_2, ..., β_p are **given** integers).

We now study the theta–Fuchsian functions. Their general form will be

(6) $$\left(\frac{dx}{dz}\right)^m F(x,y)$$

where F denotes a rational function of x and y.

Under what conditions is this theta–Fuchsian function of the second species, i.e., holomorphic inside the unit circle? The function F(x,y) can always be written:

$$\frac{P(x,y)}{\left(\frac{d\psi}{dy}\right)^m (x-a_1)^{\lambda_1}(x-a_2)^{\lambda_2}\dots(x-a_p)^{\lambda_p}}$$

where $\lambda_1, \lambda_2, \dots, \lambda_p$ are the greatest integers satisfying the inequalities

$$\lambda_i \leqslant m\left[1 - \frac{1}{\beta_i}\right]$$

We then easily recognize the necessary and sufficient conditions for the function (6) to be of the second species:

1° The rational function P(x,y) must reduce to a polynomial

2° The p(r—1) points of intersection of the curve $\psi(x,y) = 0$ with the lines $x = a_1$, $x = a_2$, ..., $x = a_p$ (excepting the singular points $(x = a_1, y = b_1)$, $(x = a_2, y = b_2)$, ..., $(x = a_p, y = b_p)$ must belong to the curve $P(x,y) = 0$; the latter must have a contact of order $\lambda_i - 1$ with the curve $\psi(x,y) = 0$ at the points where this curve meets the line $x = a_i$ (always excepting the singular point $x = a_i$, $y = b_i$).

This is equivalent to

$$(r-1)(\lambda_1+\lambda_2+\dots+\lambda_p) = (r-1)\Sigma\, \lambda_i$$

conditions on the polynomial P(x,y).

3° The degree of the polynomial P(x,y) must be greater than or equal to

$$\Sigma\, \lambda_i + m(r-3)$$

4° Now consider the points $x = e_i$, $y = f_i$, i.e., the double points of the curve $\psi(x,y) = 0$; they must belong to the curve $P(x,y) = 0$ and they must also (since $m > 1$) be double points of the latter curve. Finally, the two branches of the curve $P = 0$ must have respective contacts of order $m - 2$ with the two branches of the curve $\psi = 0$. This is equivalent to $2m - 1$ conditions on the polynomial $P(x,y)$ for each of the double points $x = e_i$, $y = f_i$, i.e., to

$$(2m-1)\left[\frac{(r-1)(r-2)}{2} - P\right]$$

conditions in all.

A polynomial of degree $\Sigma \lambda + m(r-3)$ depends on

$$\tfrac{1}{2}[\Sigma \lambda + m(r-3) + 1][\Sigma \lambda + m(r-3) + 2]$$

parameters. But if we take account of the relation (1) and if we have

$$\Sigma \lambda + m(r-3) \geqslant r \tag{7}$$

then this number is reduced to

$$r \Sigma \lambda + (m - \tfrac{1}{2})r(r-3)$$

But our polynomial is subject to $(r-1)\Sigma \lambda$ conditions on the one hand, and to $(2m-1)\left[\frac{(r-1)(r-2)}{2} - P\right]$ on the other. Thus there remain

$$\Sigma \lambda + (2m-1)(P-1) \tag{8}$$

arbitrary parameters in this polynomial However, this is only a lower bound on the number of parameters really remaining in our polynomial. It may happen that

1^0 Condition (7) is not satisfied. Then the expression for the number of arbitrary parameters remaining in our polynomial is no longer

$$\Sigma \lambda + (2m-1)(P-1)$$

but

$$\tfrac{1}{2}[\Sigma \lambda + m(r-3) + 1][\Sigma \lambda + m(r-3) + 2] -$$

$$- (r-1)\Sigma \lambda - (2m-1)\left[\frac{(r-1)(r-2)}{2} - P\right]$$

The second expression differs from the first by

$$\tfrac{1}{2}[\Sigma \lambda + m(r-3) - r + 1][\Sigma \lambda + m(r-3) - r + 2]$$

It follows that the two expressions are no different if we have

$$\Sigma \lambda + m(r-3) - r = -1 \text{ or } -2$$

Consequently, the expression (8) does not cease to represent the number of parameters remaining arbitrary unless we have

$$\Sigma \lambda + (m-1)r - 3m < -2$$

But the latter inequality, combined with $m \geqslant 2$, $r \geqslant 3$, implies the following equations:

$$r = 3 \qquad \Sigma \lambda = 0$$

This case, where we have $r = 3$, $P = 1$, $\Sigma \lambda = 0$ and where, consequently, all the λ are zero, is precisely that of the elliptic functions, which we do not have to deal with.

2° We can also ask whether the

$$(r-1)\Sigma \lambda + (2m-1)\left[\frac{(r-1)(r-2)}{2} - P\right]$$

conditions we have imposed on our polynomial are all distinct. Here is how we can avoid this difficulty. Let N be the number of distinct zeroes of the expression (6); the number $N + 1$ will evidently be an upper bound to the number of parameters remaining arbitrary in this expression. But this bound can again be reduced. In fact, the expression (6) vanishes when the variable z comes to one of the vertices of R_0; these are the zeroes which are not arbitrary. Then if N' is the number of really distinct zeroes which coincide with the vertices of R_0 we must take our upper bound to be $N - N' + 1$. This is not all. If we consider a rational expression in x and y, admitting q zeroes and q infinities, then the theory of Abelian integrals informs us that we cannot choose the q zeroes and q infinities arbitrarily, since knowledge of the q infinities and $q - P$ of the zeroes suffices to determine the last P zeroes. But the $N - N'$ zeroes which remain in our expression (6) are those of the rational function

$$\frac{P(x,y)}{[F'(y)]^m (x-a_1)^{\lambda_1} (x-a_2)^{\lambda_2} \cdots (x-a_p)^{\lambda_p}}$$

whose infinities we know. It follows that $N - N' - P$ of them suffice to determine all the others, and we finally have to take $N - N' - P + 1$ as our upper bound.

But the principles of Section 3 give us

$$N = m(n - 1 - \Sigma \frac{1}{\beta_i})$$

$$N' = \Sigma \frac{m(\beta_i - 1) - \lambda_i \beta_i}{\beta_i} = mp - m \Sigma \frac{1}{\beta_i} - \Sigma \lambda$$

$$N - N' - P + 1 = \Sigma \lambda + (2m-1)(P-1)$$

i.e., our upper bound coincides with the lower bound found above. There is an exception in the case $\Sigma\lambda = 0$, $P = 1$, i.e., in the elliptic function case which we do not have to discuss.

It follows that all the theta–Fuchsian functions of the second species (not only those which can be expressed by a series of the form (5) (Section I), but also those which cannot be expressed by such a series) can be expressed linearly in terms of $\Sigma \lambda + (2m-1)(P-1)$ particular ones.

We now consider a function of the form

$$\Lambda(z) = \left(\frac{dx}{dz}\right)^{1-m} F(x,y)$$

where F denotes a rational function of x and y, and seek the minimum number of distinct infinities of such an expression.

We assume that the vertices of R_0 are not infinities of our function $\Lambda(z)$. If the number of infinities is to be a minimum, it is necessary that the number of zeroes be also; consequently it is necessary that the vertices of R_0 be zeroes of the lowest possible order, and that there be as few zeroes as possible outside these vertices. To achieve this we put, as is always possible

$$F(x,y) = \frac{(x-a_1)^{\lambda_1}(x-a_2)^{\lambda_2}\ldots(x-a_p)^{\lambda_p}\left(\frac{d\psi}{dy}\right)^{m-1}}{P(x,y)}$$

where the λ are precisely the integers defined above. We recognize that, for the number of distinct zeroes to be a minimum, it is necessary and sufficient:

1° that P(x,y) be a polynomial of degree

$$\Sigma \lambda + (m-1)(r-3)$$

2° that the curve $P = 0$ pass through all points of intersection of the curve $\psi = 0$ with the lines

$$x = a_1, x = a_2, \ldots, x = a_p$$

other than the singular points ($x = a_i$, $y = b_i$) and that the two curves have a contact of order $\lambda_i - 1$ at these points;

3° finally, that the curve $P = 0$ admit the same double points as $\psi = 0$, and in such a way that the branches which cross at these double points have contacts of order $m - 2$ with those of $\psi = 0$.

In fact is is possible to satisfy these conditions, since the polynomial P(x,y) contains

$$r \Sigma \lambda + \left(m - \frac{3}{2}\right) r(r-3)$$

parameters, as we have seen, and we have to satisfy

$$(r-1)\Sigma \lambda + (m - \tfrac{3}{2})[(r-1)(r-2) - 2P]$$

conditions.

Then there remain

$$\Sigma \lambda + (2m-3)(P-1)$$

arbitrary parameters in our polynomial P(x,y). Moreover, we easily see that, if our conditions are satisfied, then the number of distinct zeroes, and consequently the number of distinct infinities, is a minimum, since the number of distinct zeroes which coincide with vertices of R_0 is also as small as possible, and there are no zeroes outside these vertices. Applying the rule of Section III, we find that the minimum number of infinities is

$$J = \Sigma \lambda + (2m-2)(P-1)$$

Since the polynomial P(x,y) depends on only $\Sigma \lambda + (2m-3)(P-1)$ parameters, it follows that our J distinct infinities cannot be chosen arbitrarily, but that knowledge of $J - P$ of them suffices to determine the P others. We can then, in general, determine the function Λ in such a way that it has J given infinities and no others.

What is the smallest number J' such that we can always determine the function Λ with J' given infinities and no others? The function Λ then admits $J' - J$ zeroes and $J' - J$ distinct infinities more than the functions of the same form with only J infinities and, consequently, it contains $2J' - 2J$ more parameters than they do, or in all

$$2J' - 2J + \Sigma \lambda + (2m-3)(P-1)$$

where, replacing J by its value

$$2J' - \Sigma \lambda - (2m-1)(P-1)$$

we must then have

$$2J' - \Sigma \lambda - (2m-1)(P-1) \geqslant J' + 1$$

or

(9) $$J' \geqslant \Sigma \lambda + (2m-1)(P-1) + 1$$

This then is the desired lower bound for the number J'.

Now we shall provide the resolution of an important question which arises very naturally. We have seen that all expressions of the form (6) which have no infinities can be expressed linearly in terms of $\Sigma\,\lambda + (2m-1)(P-1)$ particular ones. The theta–Fuchsian series of the form (5) (Section I) and the second species can all be put in the form (6) and hence can be expressed linearly in terms of $\Sigma\,\lambda + (2m-1)(P-1)$ among them. But we do not yet know whether all expressions of the form (6) which have no infinities can be expressed by a series (5) (Section I). If this were false, then all series of the form (5) (Section I) and the second species would be expressible linearly in terms of

$$\Sigma\,\lambda + (2m-1)(P-1) - 1$$

particular ones. I am going to show that this hypothesis is impossible.

In fact, we have seen in the preceding section (p.193 onwards) that if all series of the form (5) (Section I) and the second species are expressible linearly in terms of $q - 1$ particular ones, then we can construct a function

$$\Lambda(z) = \left[\frac{dx}{dz}\right]^{1-m} F(z)$$

where $F(z)$ represents a Fuchsian function of z admitting q **given** infinities and no others. Then, if the hypothesis made at the beginning is tenable, we can construct a function $\Lambda(z)$ admitting

$$J' = \Sigma\,\lambda + (2m-1)(P-1)$$

given infinities and no others. The number J' then would not satisfy the inequality (9), which is contrary to what we have seen above.

Consequently, each expression of the form (6) which has no infinities can be put in the form of a series (5) (Section I) of the second species; hence **every expression of the form** (6), with or without infinities, **can be put in the form of a series** (5) (Section I) of the first or second species.

We move on to an important property of functions of the form $\Lambda(z)$. These functions always have infinitely many infinities; let $z_1, z_2, \ldots, z_i, \ldots$ be these infinities, which we suppose to be simple in order to fix ideas, and let $A_1, A_2, \ldots, A_i, \ldots$ be the corresponding residues. Let $\phi(z)$ be a rational function with no infinities inside the fundamental circle. We have, identically

$$\Lambda(z)\phi(z) = \sum_i \frac{A_i \phi(z_i)}{z-z_i}$$

This is proved in the manner of the preceding section; the first proof can be repeated without changing a word; the second requires slight modifications in order to be applied.

The different terms on the right hand side of the identity can be grouped in a more advantageous manner. Suppose that q is the number of distinct infinities of $\Lambda(z)$, let $z_1, z_2, \ldots, z_q$ be these infinities, and let $A_1, A_2, \ldots, A_q$ be the corresponding residues. Then we can write

$$(10) \qquad \Lambda(z)\phi(z) = \sum_k A_k \sum_i \frac{\phi\left(\dfrac{\alpha_i z_k+\beta_i}{\gamma_i z_k+\delta_i}\right)}{\left(z - \dfrac{\alpha_i z_k+\beta_i}{\gamma_i z_k+\delta_i}\right)(\gamma_i z_k+\delta_i)^{2m}}$$

In this formula,

$$\left(z, \frac{\alpha_i z+\beta_i}{\gamma_i z+\delta_i}\right)$$

are the different transformations in the group G. Differentiating the relation (10) $2m-1$ times with respect to z, we arrive at a formula quite analogous to (16) in the previous section. The left hand side is an expression of the form (6), and the right hand side is a series of the form (5) (Section I).

VII. Second family; genus zero

We consider a polygon R_0 whose 2n vertices are all situated on the fundamental circle, and whose sides are circular arcs cutting the fundamental circle orthogonally. All the angles of our polygon are then formed by two tangential circular arcs and consequently are zero. I suppose that the sides of rank p and $2n + 1 - p$ are equivalent; the number of cycles will then be $n + 1$, two of which consist of single vertices, the first the vertex of rank 1 and the second the vertex of rank $n + 1$. The other $n - 1$ cycles each consist of two vertices, namely the vertices of rank p and $2n + 2 - p$.

I let S_p denote the transformation which changes the side of rank p into that of rank $2n + 1 - p$. We therefore have n transformations $S_1, S_2, \ldots, S_n$ as generators of our group G.

I next define transformations $S'_2, \ldots, S'_n$ by the following equations

$$S'_2 = S_2 S_1^{-1},\ S'_3 = S_3 S_2^{-1},\ \ldots,\ S'_n = S_n S_{n-1}^{-1}$$

and, for greater symmetry of notation, I put

$$S_1 = S'_1,\ S_{n+1} = S'_{n+1}$$

I now let α_i denote the vertex of rank i; it is clear that α_i will be one of the fixed points of S'_i. I suppose that **the** n + 1 **transformations** $S'_1, S'_2, \ldots, S'_{n+1}$ **are parabolic**. This will not be possible if the 2n points α_i are chosen arbitrarily on the fundamental circle. They must satisfy the relation

$$(1) \qquad (\alpha_2-\alpha_1)(\alpha_4-\alpha_3) \ldots (\alpha_{2p}-\alpha_{2p-2}) \ldots (\alpha_{2n}-\alpha_{2n-1}) =$$

$$= (\alpha_3-\alpha_2)(\alpha_5-\alpha_4) \ldots (\alpha_{2p+1}-\alpha_{2p}) \ldots (\alpha_{2n}-\alpha_1)$$

Under these conditions, our polygon R_0 is of genus zero and of the first order in the second family; our group G and the Fuchsian functions derived from it are of genus zero and of the second family. As in Section V we can choose from these Fuchsian functions one which I call $x = f(z)$, in terms of which all others are rationally expressible. To complete its definition I suppose that

$$f(\alpha_1) = 0 \qquad f(\alpha_2) = 1 \qquad f(\alpha_{n+1}) = \infty$$

As in Section V, I also put

$$a_i = f(\alpha_i) = f(\alpha_{2n+2-i})$$

whence

$$a_1 = 0 \qquad a_2 = 1 \qquad a_{n+1} = \infty$$

If we put

$$\nu = \sqrt{\frac{dx}{dz}}$$

then ν satisfies a linear equation

$$(3') \qquad \frac{d^2\nu}{dx^2} = \nu\phi(x)$$

where $\phi(x)$ is rational in x. What is the form of the function $\phi(x)$ here? We easily find that

$$\phi(x) = \frac{P(x)}{[Q(x)]^2}$$

where Q(x) is the product $(x-a_1)(x-a_2)\ldots(x-a_n)$ and P(x) is a polynomial of degree 2n − 2. The polynomial P(x) must satisfy the following two conditions:

$$P(a_i) = \frac{1}{4} Q'^2(\alpha_i)$$

and

$$\text{coefficient of } x^{2n-2} \text{ in } P(x) = -\frac{1}{4}$$

These conditions are not sufficient for x to be a Fuchsian function of the ratio of the integrals of the equation (3′). It is necessary, in addition, to subject P(x) to 2n − 4 transcendental conditions.

We shall show, in another memoir, that we can always choose the group G so that the numbers a_3, a_4, ..., a_n take any given values.

In the particular case where the polygon R_0 is symmetric in the circle joining α_1 and α_{n+1} which cuts the fundamental circle orthogonally, the coefficients of the rational function $\phi(x)$ are all real.

If, in addition, n = 2, then f(z) reduces to the modular function k^2.

We see how the integrals of the equation (3′) behave in the neighborhood of one of the singular points a_1, a_2, ..., a_n. The integrals will be **regular and logarithmic**, to use the terminology of Fuchs and his disciples. The two roots of the determinant equation will be $\frac{1}{2}$; so that the integrals can be put in the form

$$\nu_1 = (x-a_i)^{\frac{1}{2}}[P_i+Q_i\log(x-a_i)]$$

and

$$\nu_2 = (x-a_i)^{\frac{1}{2}} Q_i$$

where P_i and Q_i are holomorphic functions in $x - a_i$, of which the first vanishes for $x = a_i$ and the second does not.

For $x = \infty$ we likewise have

$$\nu_1 = x^{\frac{1}{2}}(P_{n+1} + Q_{n+1} \log x), \qquad \nu_2 = x^{\frac{1}{2}} Q_{n+1}$$

where P_{n+1} and Q_{n+1} are holomorphic functions of $\frac{1}{x}$, of which the first vanishes for $x = \infty$ and the second does not.

We now express z as a function of x in the neighborhood of each of these singular points. In the neighborhood of $x = a_i$ we must have

$$z = \frac{\lambda Q_i + \mu[Q_i \log(x-a_i) + P_i]}{\lambda' Q_i + \mu'[Q_i \log(x-a_i) + P_i]}$$

where λ, μ, λ', μ' are suitably chosen constants. Suppose that the transformation S'_i is written

$$\left(\frac{1}{z-\alpha_2}, \frac{1}{z-\alpha_i} + \beta_i\right)$$

and put

$$e^{\frac{2\pi\sqrt{-1}}{(z-\alpha_i)\beta_i}} = t_i$$

then express the fact that z reduces to α_i for $x = a_i$ and undergoes the transformation S'_i when $\log(x-a_i)$ is changed to $\log(x-a_i) + 2\pi\sqrt{-1}$. We see that the preceding relation can be written

$$\frac{2\pi\sqrt{-1}}{(z-\alpha_i)\beta_i} = P'_i + \log(x-a_i)$$

where P'_i is a holomorphic function of $x - a_i$. We deduce from this that

$$t_i = e^{\frac{2\pi\sqrt{-1}}{(z-\alpha_i)\beta_i}} = Q'_i$$

where Q_i is a holomorphic function of $(x-\alpha_i)$; the expansion of Q'_i in powers of $(x-a_i)$ begins with a term in $(x-a_i)$ whose coefficients can never be zero. We deduce from this that

$$x = a_i + \Phi_i(t_i)$$

where Φ_i denotes a holomorphic function of t_i whose expansion in powers of that variable begins with a first degree term whose coefficient is never zero. Such is the form of the expression for x as a function of z in the neighborhood of a singular point $x = a_i$. Likewise, in the neighborhood of $z = \alpha_{n+1}$, $x = \infty$ we have

$$\frac{1}{x} = \Phi_{n+1}(t_{n+1})$$

where Φ_{n+1} denotes a holomorphic function of t_{n+1} whose expansion in powers of that variable begins with a first degree term which is never zero. We can also write the preceding relation in the form

$$x = \frac{1}{t_{n+1}} \Psi(t_{n+1})$$

where Ψ denotes a holomorphic function of t_{n+1} which does not vanish with that variable. As for the derivative $\frac{dx}{dz}$, it can be put in an analogous form. In the neighborhood of $z = \alpha_i$ we have

$$\frac{dx}{dz} = \frac{1}{(z-\alpha_i)^2} t_i \Phi'_i(t_i)$$

where Φ'_i is a holomorphic function of t_i which does not vanish with that variable. Likewise, in the neighborhood of $z = \alpha_{n+1}$, we have

$$\frac{dx}{dz} = \frac{1}{(z-\alpha_{n+1})^2} \frac{\Psi'(t_{n+1})}{t_{n+1}}$$

where Ψ' denotes a holomorphic function of t_{n+1} which does not vanish with that variable.

We now move on to the study of the theta–Fuchsian functions.

A function represented by a series of the form (5) (Section I) can always be put in the following form:

$$\left(\frac{dx}{dz}\right)^m F(x) \tag{2}$$

where F denotes a rational function of x. But the converse is not true. In fact, we have seen in Section 2 that each series of the form (5) (Section I) can, in the neighborhood of $z = \alpha_i$, be put in the form

$$\frac{1}{(z-\alpha_i)^{2m}} R(t_i)$$

where R denotes a holomorphic function of t_i which vanishes with that variable. It follows that, for $x = a_i$, the function $(x-a_i)^m F(x)$ must vanish and, when x increases indefinitely, $x^m F(x)$ must tend to zero. Such are the **necessary** conditions which the expression (2) must satisfy in order to be represented by a series of the form (5) (Section I).

We call these the conditions A.

Now suppose that the function (2) admits no infinities inside the fundamental circle. F(x) must then be of the form

$$\frac{\Pi(x)}{(x-a_1)^{\lambda_1}(x-a_2)^{\lambda_2}\cdots(x-a_n)^{\lambda_n}}$$

where $\Pi(x)$ denotes a polynomial in x. But, as a result of the conditions A, the numbers $\lambda_1, \lambda_2, \ldots, \lambda_n$ are at most equal to $m-1$ and the degree of $\Pi(x)$ is at most equal to $n(m-1)-m-1$. Consequently, all theta–Fuchsian series of the form (5) (Section I) and the second species can be expressed linearly in terms of $n(m-1)-m$ particular ones.

However we still do not know whether every expression of the form (2) with no infinities and satisfying the conditions A can be put in the form of a series (5) (Section I) not, consequently, whether every series of that form and the second species can be expressed linearly in terms of $n(m-1)-m-1$ particular ones.

To recognize whether this is so, we consider a function of the form

$$\Lambda(z) = \left(\frac{dx}{dz}\right)^{1-m} F(x)$$

where F denotes a rational function, and seek the number J' of its distinct infinities, supposing that it tends to zero as z tends to α_i. We can write

$$F(x) = \frac{(x-a_1)^{m-1}(x-a_2)^{m-1}\cdots(x-a_n)^{m-1}\Pi(x)}{\Phi(x)}$$

where $\Pi(x)$ and $\Phi(x)$ denote two polynomials in x. For $\Lambda(z)$ to tend to zero as z tends to α_i, it is necessary and sufficient that

$$\Phi(a_1),\ \Phi(a_2),\ \ldots,\ \Phi(a_n)$$

be non—zero and that the degree of the numerator exceed that of the denominator by at most $m - 1$. But the degree of $\Phi(x)$ is the number J' of distinct infinities of $\Lambda(z)$. We therefore have

$$J' \geqslant n(m-1) - m + 1$$

I then claim that it is impossible to express each series of the form (5) (Section I) and the second species linearly in terms of $n(m-1) - m - 1$ particular ones, because if it were possible we could construct a function of the form $\Lambda(z)$ with $n(m-1) - m$ distinct infinities, which is impossible from what we have seen. We therefore conclude (as in the two preceding sections) that every expression of the form (2) satisfying the conditions A can be expressed by a series of the form (5) (Section I).

The formula (10) of the preceding section also applies to the functions we are now dealing with. We prove it as in the two preceding sections; the first proof demands a few slight modifications, the second can be applied without changing anything. All the consequences we deduced from that formula in the two preceding sections, among others the formula (16) of Section V, are again true in the present case.

One easily sees how to extend the above principles to functions of the second and sixth families and arbitrary genus.

We can then move on to the study of Fuchsian functions which exist throughout the whole plane.

VIII. Third family

Among these, the simplest are those of the third family which we shall discuss first.

Suppose that we are given n pairs of circles

$$(c_1 \text{ and } c'_1),\ (c_2 \text{ and } c'_2),\ \ldots,\ (c_n \text{ and } c'_n)$$

I assume that these 2n circles are all disjoint from each other, that they cut the fundamental circle orthogonally and that, consequently, they are symmetric with respect to this circle. I suppose that we have n transformations S_1, S_2, ..., S_n where S_i changes c_i into c'_i while preserving the fundamental circle. The group generated by these transformations S_i will be a Fuchsian group G whose generating polygon S_i consists of the part of the plane inside the fundamental circle and outside the various circles c and c′. All the vertices of R_0 are of the third category. As for the polygon R'_0 symmetric to R_0 with respect to the fundamental circle, this will be the part of the plane outside the fundamental circle and outside the circles c and c′. The group G, and consequently the Fuchsian functions derived from it, will be of the third family and of genus n.

All these Fuchsian functions can be expressed rationally in terms of two among them which I call x and y, between which there is an algebraic relation of genus n:

(1) $$\psi(x,y) = 0$$

If in addition we put

$$\nu = \sqrt{\frac{dx}{dz}}$$

then the function ν will satisfy equation (3) of Section IV, namely:

(3) $$\frac{d^2\nu}{dx^2} = \nu\phi(x,y)$$

where ϕ is rational in x and y.

The group G depends on only 3n − 3 distinct real parameters; one cannot dispose them so as to make the relation $\psi(x,y) = 0$

arbitrary. In fact, how many arbitrary parameters remain in an algebraic relation of genus n

$$\psi(x,y) = 0$$

when we agree not to regard this relation as distinct from one derived from it by replacing x and y by rational functions of x′ and y′? We know that there remain 3n − 3 parameters which we call the **moduli** of the relation (1) and which are, so to speak, the invariants of this relation with respect to the operations of replacing x and y by rational functions of x′ and y′. In proposing to obtain a **given** algebraic relation between two Fuchsian functions x and y we must therefore impose 3n −3 **complex** conditions, which are equivalent to 6n − 6 **real** conditions. This number exceeds by 3n − 3 the number of parameters at our disposal. Thus the left hand side of the relation (1) is subject to 3n − 3 real conditions; it is easy to find what they are.

In fact, we can assume that we have made an operation which consists in replacing z by its reflection in the unit circle. This amounts to changing $\sqrt{-1}$ into $-\sqrt{-1}$ and therefore is a linear change of variable. It changes R_0 into R'_0 and conversely, so that the group G and the system of Fuchsian functions derived from it do not change. Then the 3n − 3 moduli of the relation (1) also do not change. But when we change $\sqrt{-1}$ to $-\sqrt{-1}$ the moduli are obliged to change into their complex conjugates; afterwards, when we make a linear change of variable, they do not change. Consequently, the 3n − 3 **moduli of the relation** (1) do not differ from their complex conjugates, **they are therefore real**.

It follows that we can always choose x and y among the Fuchsian functions derived from the group G so that **all the coefficients of the relation** (1) **are real**. Then the coefficients of $\phi(x,y)$ will be real also.

I assume in addition that we have chosen x and y, as is always possible, in such a way that the relation (1) satisfies the conditions we imposed on it at the beginning of Section VI (p.208).

When this is done, the function $\phi(x,y)$ will satisfy the following conditions:

1^{0} When x becomes infinitely large of first order, it becomes infinitely small of fourth order;

2^{0} The only infinities of the function $\phi(x,y)$ will be the singular points of the function y of x defined by relation (1), and will not include the double points of the algebraic curve $\psi(x,y) = 0$. If one of these points is $x = a$, $y = b$ we will have

$$\lim(y-b)^2 \left(\frac{dx}{dy}\right)^2 \phi(x,y) = -\frac{3}{4}$$

for $x = a$, $y = b$.

These conditions, which we observe are the same as those we found in the first example, examined in Section VI, do not suffice to determine the rational function $\phi(x,y)$. It is also subject to n transcendental conditions. Here is the form in which these transcendental conditions can be presented:

The function $\phi(x,y)$ must be chosen in such a way that when the analytic point (x,y) is made to describe n suitably chosen cycles (corresponding to n suitably chosen periods of the Abelian integral

$$\int g(x,y)dx$$

of the first species), we see the integrals of equation (3) return to their initial values.

We now move on to the study of the theta–Fuchsian series

$$\Theta(z) = \sum_i H\left(\frac{\alpha_i z+\beta_i}{\gamma_i z+\delta_i}\right)(\gamma_i z+\delta_i)^{-2m}$$

i.e., to the study of series of the form (5) (Section I), and we first find how it can happen that the function defined by this series lacks any infinity apart from the essential singular points.

In general, the series $\Theta(z)$ admits, as we have seen in Section III, the following infinities:

1° The points corresponding to the point ∞, i.e., the points

$$z = -\frac{\delta_i}{\gamma_i}$$

2° The infinities of the function H(z) and the points corresponding to them.

Suppose, to fix ideas, that the function H(z) does not become infinite at any of the points $-\frac{\delta_i}{\gamma_i}$. First, what is the condition that the points $-\frac{\delta_i}{\gamma_i}$ not be infinities of $\Theta(z)$? For this, it is necessary and sufficient that the degree of the denominator of the rational function H(z) surpass that of the numerator by 2m.

Now consider an infinity a of the function H(z) itself. In the series $\Theta(z)$, the term

$$H(z)$$

(corresponding to $\alpha_i = 1$, $\delta_i = 1$, $\beta_i = 0$, $\gamma_i = 0$, i.e., to the identity transformation) becomes infinite for $z = a$. Thus if $\Theta(z)$ is to remain finite for $z = a$, it is necessary that another term of this series become infinite, in such a way that the sum of these two terms (which both increase indefinitely as z tends to a) tends on the contrary to a finite limit. For this it is necessary that the infinities of H(z) include $z = b$ such that

$$b = \frac{\alpha a+\beta}{\gamma a+\delta}$$

where $\left(z, \frac{\alpha z+\beta}{\gamma z+\delta}\right)$ is one of the transformations in the group G. It is then clear that the term

$$H\left(\frac{\alpha z+\beta}{\gamma z+\delta}\right)(\gamma z+\delta)^{-2m}$$

becomes infinite for $z = a$. But it is necessary that the sum of this term with $H(z)$ remain finite for $z = a$. Suppose that A and B are the residues of $H(z)$ which correspond to $z = a$ and $z = b$. We have

$$H(z) = \frac{A}{z-a} + H'(z)$$

$$H\left[\frac{\alpha z+\beta}{\gamma z+\delta}\right](\gamma z+\delta)^{-2m} = \frac{B}{z-a}(\gamma a+\delta)^{2-2m} + H'_1(z)$$

where $H'(z)$ and $H'_1(z)$ are rational functions of z which remain finite for $z = a$. Then we must have

$$A + B(\gamma a+\delta)^{2-2m} = 0$$

Such are the two necessary and sufficient conditions for $\Theta(z)$ to remain finite at $z = a$, and consequently at all points corresponding to it.

Having established this, here is what must be done to construct a function $\Theta(z)$ which admits no infinity. We take a rational function $H(z)$ with 2p infinities $a_1, b_1, a_2, b_2, a_3, b_3, \ldots, a_p, b_p$ and with $A_1, B_1, A_2, B_2, A_3, B_3, \ldots, A_p, B_p$ as the corresponding residues.

Moreover, I assume that we have

$$b_1 = \frac{\alpha_1 a_1+\beta_1}{\gamma_1 a_1+\delta_1},\ b_2 = \frac{\alpha_2 a_2+\beta_2}{\gamma_2 a_2+\delta_2},\ \ldots\ldots\ b_p = \frac{\alpha_p a_p+\beta_p}{\gamma_p a_p+\delta_p}$$

where

$$\left[z, \frac{\alpha_1 z+\beta_1}{\gamma_1 z+\delta_1}\right] \quad \left[z, \frac{\alpha_2 z+\beta_2}{\gamma_2 z+\delta_2}\right], \ \ldots\ldots \ \left[z, \frac{\alpha_p z+\beta_p}{\gamma_p z+\delta_p}\right]$$

are p transformations in the group G. I also subject the 2p residues to two systems of conditions. First,

$$A_1 + B_1 + A_2 + B_2 + \dots + A_p + B_p = 0$$

$$A_1 a_1 + B_1 b_1 + A_2 a_2 + B_2 b_2 + \dots + A_p a_p + B_p b_p = 0$$

$$A_1 a_1^2 + B_1 b_1^2 + A_2 a_2^2 + B_2 b_2^2 + \dots + A_p a_p^2 + B_p b_p^2 = 0$$

....

$$A_1 a_1^{2m-2} + B_1 b_1^{2m-2} + \dots + A_p a_p^{2m-2} + B_p b_p^{2m-2} = 0$$

in order that the degree of the denominator of H(z) surpass that of the denominator by 2m. Then

$$A_k + B_k(\gamma_k a_k + \delta_k)^{2-2m} = 0 \qquad (k = 1,2, \dots, p)$$

in order that a_k and b_k not be infinities of $\Theta(z)$.

These $p + 2m - 1$ conditions will be compatible provided that

$$p > 2m - 1$$

and if they are satisfied, the series $\Theta(z)$ formed with the aid of the rational function

$$H(z) = \sum_{k=1}^{p} \left[\frac{A_k}{z-a_k} + \frac{B_k}{z-b_k}\right]$$

will not have any infinity. I then claim that it is of the second species, so that we have here, as in the three preceeding paragraphs, the distinction between the two species of the theta–Fuchsian series.

Every series of the form $\Theta(z)$, whatever species it belongs to, can be put in the form

(2) $$\left[\frac{dx}{dz}\right]^m F(x,y)$$

where F(x,y) denotes a rational function of x and y. Those functions of the form (2) which do not admit infinities can be expressed linearly, as we have seen in Section VI, in terms of (2m—1)(n—1) particular ones. It follows that all series $\Theta(z)$ of the second species can be expressed linearly in terms of (2m—1)(n—1) particular ones. It remains to show, as in the preceding three paragraphs, that it is impossible to express them linearly in terms of (2m—1)(n—1) — 1 particular ones.

To abbreviate, we put

$$q = (2m-1)(n-1)$$

Suppose that all the series $\Theta(z)$ of the second species can be expressed linearly in terms of q — 1 particular ones which I call $\Theta_1, \Theta_2, \ldots, \Theta_{q-1}$; I am going to show that this hypothesis is absurd.

In fact, if we choose q points $z_1, z_2, \ldots, z_q$ **at random** we can always find q numbers $A_1, A_2, \ldots, A_q$ such that

$$A_1\Theta_1(z_1) + A_2\Theta_1(z_2) + \ldots + A_q\Theta_1(z_q) = 0$$

$$A_1\Theta_2(z_1) + A_2\Theta_2(z_2) + \ldots + A_q\Theta_2(z_q) = 0$$

....

$$A_1\Theta_{q-1}(z_1) + A_2\Theta_{q-1}(z_2) + \ldots + A_q\Theta_{q-1}(z_q) = 0$$

and consequently such that

(5) $$A_1\Theta(z_1) + A_2\Theta(z_2) + \ldots + A_q\Theta(z_q) = 0$$

where $\Theta(z)$ denotes an arbitrary series of the form (5) (Section I) and the second species.

Let p be an integer smaller than 2m. I can always form a function $\theta_p(z)$ which satisfies the following conditions: it does not admit any infinity other than the points $z = -\frac{\delta_i}{\gamma_i}$, and these infinities are of order $2m - p$. Under these conditions, the point $z = \infty$ is not an infinity of our function $\theta_p(z)$, it is a zero or order p.

I suppose that

$$\lim z^p \theta_p(z) = 1 \qquad \text{for } z = \infty$$

$$\lim z^{2m-1-p}[z^p \theta_p(z) - 1] = 0 \qquad \text{for } z = \infty$$

These conditions do not suffice to completely define the function $\theta_p(z)$, since if $\Theta(z)$ is a series of the form (5) (Section I) and the second species, then $\theta_p(z) + \Theta(z)$ satisfies as well as $\theta_p(z)$. However, we choose our function $\theta_p(z)$ to be any series of the form (5) (Section I) which satisfies the stated conditions.

We consider an arbitrary series of the form (5) (Section I) and the first species

$$\Psi(z) = \sum_i H\left[\frac{\alpha_i z + \beta_i}{\gamma_i z + \delta_i}\right](\gamma_i z + \delta_i)^{-2m}$$

Suppose that the polynomial part of the rational function

$$z^{2m} H(z)$$

is

$$B_0 + B_1 z + B_2 z^2 + \dots + B_{2m} z^{2m}$$

We easily see that the point $z = \infty$ is a zero of order at least $2m$ for the theta-Fuchsian function

$$\Psi(z) - B_0\theta_0(z) - B_1\theta_1(z) - B_2\theta_2(z) - \dots - B_{2m}\theta_{2m}(z)$$

and consequently that this theta–Fuchsian function does not become infinite for $z = -\frac{\delta_i}{\gamma_i}$.

Having established this, we consider the following function

$$\Phi(z,a) = \sum_i \frac{1}{z - \frac{\alpha_i a+\beta_i}{\gamma_i z+\delta_i}} (\gamma_i a+\delta_i)^{-2m}$$

As a function of z, it has no infinities other than the points

$$z = \frac{\alpha_i a+\beta_i}{\gamma_i a+\delta_i}$$

i.e., the points corresponding to a.

As a function of a, it is a theta–Fuchsian series and its infinities are:

1^0 The points $a = \frac{\alpha_i z+\beta_i}{\gamma_i a+\delta_i}$, i.e., the points corresponding to z;

2^0 The points $a = -\frac{\delta_i}{\gamma_i}$, i.e., the points corresponding to $a = \infty$. These points are infinities of order $2m - 1$. As for the point $a = \infty$ itself, it is a first order zero of our series.

If we look for the polynomial part of the rational function

$$\frac{a^{2m}}{z-a}$$

we find

$$-(a^{2m-1} + za^{2m-2} + z^2a^{2m-3} + \ldots + z^{2m-2}a + z^{2m-1})$$

We conclude that the theta–Fuchsian function

$$\Omega(z,a) = \Phi(z,a) + z^{2m-1}\theta_0(a) + z^{2m-2}\theta_1(a) + \ldots + z\theta_{2m-2}(a) + \theta_{2m-1}(a)$$

does not have infinities at the points $a = -\frac{\delta_i}{\gamma_i}$.

Consequently, it has no infinities other than the points

$$a = \frac{\alpha_i z + \beta_i}{\gamma_i z + \delta_i}$$

the residues corresponding to which will be

$$-(\gamma_i z + \delta_i)^{2m-2}$$

Let $\left[z, \frac{\alpha z + \beta}{\gamma z + \delta}\right]$ be an arbitrary transformation in our group G. The function

$$(\gamma z + \delta)^{2m-2} \Omega\left[\frac{\alpha z + \beta}{\gamma z + \delta}, a\right]$$

will, like $\Omega(z,a)$, be a theta–Fuchsian function of a; it has no infinities other than the points

$$a = \frac{\alpha_i z + \beta_i}{\gamma_i z + \delta_i}$$

with the same residues. Then we can write:

$$(6) \qquad (\gamma z + \delta)^{2m-2} \Omega\left[\frac{\alpha z + \beta}{\gamma z + \delta}, a\right] - \Omega(z,a) = \Theta(a)$$

where $\Theta(a)$ denotes a theta–Fuchsian function of a, capable of being put in the form (5) (Section I), and of the second species. Now we put

$$(7) \qquad \Lambda(z) = A_1 \Omega(z,z_1) + A_2 \Omega(z,z_2) + \ldots + A_q \Omega(z,z_q)$$

Comparing equations (5), (6) and (7), we easily find the following identity:

$$(\gamma z + \delta)^{2m-2} \Lambda\left[\frac{\alpha z + \beta}{\gamma z + \delta}\right] = \Lambda(z)$$

which holds for all transformations $\left(z, \frac{\alpha z+\beta}{\gamma z+\delta}\right)$ in the group G. It follows that $\Lambda(z)$ can be put in the form

$$(8) \qquad \left(\frac{dx}{dz}\right)^{1-m} F(x,y)$$

where F(x,y) denotes a rational function of x and y. But the function $\Lambda(z)$ admits q distinct infinities $z_1, z_2, \ldots, z_q$ which have been chosen at random, and no others at a finite distance. We can therefore construct a function of the form (8) admitting q **given** infinities and no others. But we have seen in Section VI that this is impossible. Thus the hypothesis made at the beginning is absurd.

We conclude, as in the preceding three sections:

1° that not all series of the form (5) (Section I) and the second species can be expressed linearly in terms of $q - 1$ particular ones;

2° that every expression of the form (2) which has no infinities can be put in the form (5) (Section I);

3° that every expression of the form (2) can be expressed by a series of the form (5) (Section I) (provided $m > 1$);

4° that every theta-Fuchsian function can be expressed in infinitely many ways as the quotient of two series of the form (5) (Section I).

Consider the function

$$\Lambda(z) = \left(\frac{dx}{dz}\right)^{1-m} F(x,y)$$

Suppose that it has infinitely many infinities $z_1, z_2, \ldots, z_k, \ldots$ with residues $A_1, A_2, \ldots, A_k, \ldots$ Suppose in addition that we have

$$\Lambda(z) = B_h z^h + B_{h+1} z^{h-1} + \ldots + B_1 z + \Lambda_1(z)$$

where $\Lambda_1(z)$ tends to a finite limit when z increases indefinitely. We have, identically,

$$\Lambda(z) = B_h z^h + B_{h+1} z^{h-1} + \ldots + B_1 z + \sum_{k=1}^{\infty} \frac{A_k}{z-z_k} \tag{9}$$

This identity can be proved in two ways, like the corresponding identity in Section V.

A remarkable special case is that of $n = 1$.

In this case the transformations in the group G reduce to

$$\left(\frac{az+b}{cz+d}, K^p \frac{az+b}{cz+d}\right)$$

where a, b, c, d are constants, K is a positive real number greater than 1, and the exponent p can take all integer values, positive or negative, in the different transformations.

Here is how we can form the Fuchsian functions in the present case. Let $f(\xi)$ be a doubly periodic function with periods $2i\pi$ and log K. The transcendant $f\left[\log\left(\frac{az+b}{cz+d}\right)\right]$ will then be a Fuchsian function.

For example, put

$$x = \mathrm{sn}\left(\log \frac{az+b}{cz+d}\right), \quad y = \mathrm{cn}\left(\log \frac{az+b}{cz+d}\right)\mathrm{dn}\left(\log \frac{az+b}{cz+d}\right)$$

The expressions of the form

$$\left(\frac{dx}{dz}\right)^m F(x,y) \tag{2}$$

can be written

$$\frac{f\left(\log \frac{az+b}{cz+d}\right)}{(az+b)^m(cz+d)^m} \tag{10}$$

where $f(\xi)$ always denotes a doubly periodic function.

On the other hand, the series $\Theta(z)$ of the form (5) (Section I) will become

$$(11) \qquad \sum_i \frac{K^{mi}\phi\left(K^i \frac{az+b}{cz+d}\right)}{(cz+d)^{2m}}$$

where ϕ is the algorithm of a rational function.

If we equate the expressions (10) and (11) and set

$$\frac{az+b}{cz+d} = e^{\xi}$$

then we get

$$f(\xi) = \sum_i K^{mi} e^{m\xi} \phi(K^i e^{\xi})$$

Thus we obtain the expansion of a doubly periodic function of ξ as a series whose terms are rational functions of e^{ξ}. If in place of ξ we take $\eta = \xi\sqrt{-1}$ as independent variable, then the left hand side becomes a doubly periodic function of η and, in the series on the right hand side, all terms will be rational functions of sin η and cos η. Thus we rediscover by a circuitous route a result at which Jacobi arrived directly, and from which he derived so many beautiful consequences.

Now we see what becomes of the formula (9) in the present case.

We have

$$\Lambda(z) = (az+b)^{m-1}(cz+d)^{m-1}f(\xi)$$

where $f(\xi)$ is always a doubly periodic function.

On the other hand, on the right hand side we have:

1° A polynomial of degree h in z; but it is easy to see, in general case, i.e., at least when

$$f(\log \frac{a}{c}) = \infty$$

that the degree h cannot exceed $2m - 2$. We can then regard it as a homogeneous polynomial of degree $2m - 2$ in $(az+b)$ and $(cz+d)$. If we divide it by $(az+b)^{m-1}(cz+d)^{m-1}$ it will take the following form:

$$P_{m-1}(e^{\xi}) + P'_{m-1}(e^{-\xi})$$

where P_{m-1} and P'_{m-1} denote polynomials of degree $m - 1$ in e^{ξ} and in $e^{-\xi}$.

2^0 A set of terms $\sum \frac{A_k}{z-z_k}$ which we can group as follows: let $z_1, z_2, \ldots, z_q$ be the q distinct infinities of $\Lambda(z)$ and suppose that there are no others; let $A_1, A_2, \ldots, A_q$ be the corresponding residues; let $\xi_1, \xi_2, \ldots, \xi_q$ be the corresponding values of ξ. We group the set of all terms of the form $\frac{A_k}{z-z_k}$ which correspond to the infinities which are not distinct from $z = z_p$. Their sum will be

$$A_p \sum_i \frac{ce^{\xi}-a}{(cK^i e^{\xi_p}-a)^{2m-1}} \frac{K^{im}}{e^{\xi}-K^i e^{\xi_p}} (ce^{\xi_p}-a)^{2m}$$

Since we have, on the other hand,

$$(az+b)^{m-1}(cz+d)^{m-1} = e^{(m-1)\xi}(ce^{\xi}-a)^{2-2m}$$

we find the following expression for the doubly periodic function $f(\xi)$:

$$f(\xi) = P_{m-1}(e^{\xi}) + P'_{m-1}(e^{-\xi}) + \sum_{p=1}^{q} \left[A_p \left[ce^{\xi_p}-a\right]^{2m} \times \right.$$

$$\left. \times \sum_{i=-\infty}^{\infty} \left[\frac{ce^{\xi}-a}{cK^i e^{\xi_p}-a}\right]^{2m-1} \frac{K^{mi} e^{(1-m)\xi}}{e^{\xi}-K^i e^{\xi_p}} \right]$$

IX. Fifth family; genus zero

I believe it is unnecessary to multiply these examples any further. Those that I have studied up to now suffice in fact to show how we can apply the general principles to each particular case.

I now want, without treating completely the questions which concern us, to say a few words about certain remarkable functions in the fifth family.

Consider a polygon R_0 bounded in the following fashion; it has 2n sides of the first kind and one side of the second kind; I suppose that its vertices are α_0, α_1, α_2, ..., α_n, α_{n+1}, α_{n+2}, ..., α_{2n}, following the perimeter in the positive sense. I suppose that the side $\alpha_n\alpha_{n+1}$ is of the second kind and that the others are of the first kind, with the pairs $\alpha_0\alpha_1$ and $\alpha_0\alpha_{2n}$, $\alpha_1\alpha_2$ and $\alpha_{2n}\alpha_{2n-1}$, $\alpha_2\alpha_3$ and $\alpha_{2n-1}\alpha_{2n-2}$, ..., $\alpha_{p+1}\alpha_{p+2}$ and $\alpha_{2n-p}\alpha_{2n-p-1}$, ..., and finally $\alpha_{n-1}\alpha_n$ and $\alpha_{n+2}\alpha_{n+1}$ being equivalent and therefore congruent. If we omit the side from α_n to α_{n+1}, all vertices will be of the first category and partitioned into n closed cycles. The vertex α_0 will be one of these cycles, the others consist respectively of α_1 and α_{2n}, α_2 and α_{2n-1}, α_3 and α_{2n-2}, ..., and finally α_{n-1} and α_{n+2}; thus the curvilinear angles α_0, $\alpha_1 + \alpha_{2n}$, $\alpha_2 + \alpha_{2n-1}$, ..., $\alpha_{n-1} + \alpha_{n+2}$ must be aliquot parts of 2π. As for the angles α_n and α_{n+1}, they are forced to be right angles because the sides $\alpha_{n-1}\alpha_n$ and $\alpha_{n+1}\alpha_{n+2}$ arrive orthogonally at the fundamental circle, of which the side $\alpha_n\alpha_{n+1}$ is an arc. We then construct a polygon R'_0 symmetric to R_0 in the fundamental circle. I let α'_i denote the vertex of R'_0 which is the reflection of α_i in the fundamental circle.

We easily see that the polygon R_0 is of the fifth family and genus zero, and that it serves to define a Fuchsian group and an infinity of Fuchsian functions. The latter can all be expressed rationally in terms of one among them which I call

$$x = f(z)$$

Since we can choose one such among our Fuchsian functions in infinitely many ways, the function $x = f(z)$ will not be completely determined unless we adjoin certain additional conditions.

Let β and γ be two points on the side $\alpha_n\alpha_{n+1}$ of the second kind; I suppose that we have

$$f(\alpha_n) = f(\alpha_{n+1}) = 0 \qquad f(\beta) = 1 \qquad f(\gamma) = \infty$$

I then put

$$f(\alpha_i) = a_i \qquad f(\alpha'_i) = a'_i$$

Since the polygon $R_0 + R'_0$ is symmetric with respect to the fundamental circle, we can see, by reasoning like that of the previous section, that the numbers a_i and a'_i are complex conjugates.

We now put

$$\nu = \sqrt{\frac{dx}{dz}}$$

and the function ν will satisfy a differential equation of the form (3') (Section IV). because the corresponding Fuchsian group is of genus zero. As we easily see, this equation can be written:

$$(3') \qquad \frac{d^2\nu}{dx^2} = \nu \frac{P(x)}{[Q(X)]^2}$$

where $Q(x)$ is the product

$$(x-a_0)(x-a'_0)(x-a_1)(x-a'_1) \;\ldots.\; (x-a_{n-1})(x-a'_{n-1})$$

and P(x) is a polynomial of degree 4n — 4 in x, with all its coefficients real.

X. Resumé

In studying the preceding examples, I have encountered different results which are common to all Fuchsian functions. I do not believe it is useful to give a proof in the most general case, since it would not differ from those given in the various particular cases and it would be a repetition without interest. I therefore confine myself to giving it here in the form of a resumé.

Using an arbitrary rational function H(z), we form the theta—Fuchsian series

$$(1) \qquad \sum_i H\left(\frac{\alpha_i z+\beta_i}{\gamma_i z+\delta_i}\right)(\gamma_i z+\delta_i)^{-2m} = \Theta[z,H(z)]$$

This series can always be put in the form

$$(2) \qquad \left(\frac{dx}{dz}\right)^m F(x,y)$$

where F(x,y) denotes a rational function of two Fuchsian functions x and y, in terms of which all others can be expressed rationally and between which there is an algebraic relation

$$(3) \qquad \psi(x,y) = 0$$

For an expression of the form (2) to be expressible in the form (1), it is necessary that it vanish when z comes to the vertices of the second category in the polygon R_0.

Under these conditions, every expression of the form (2) **can be put in the form** (1), **provided that** m **is an integer greater than** 1.

It follows that every Fuchsian function can be expressed as a quotient of two series of the form (1) in infinitely many ways.

The series of the form (1) are of two species: those which have infinities and those which do not. The latter can be expressed linearly in terms of finitely many particular ones. Thus there are infinitely many linear relations between theta–Fuchsian series of the second species.

Here is one of these relations which can serve as a point of departure for finding, if not all the others, at least a great number of them.

Let $\left[z, \frac{\alpha z+\beta}{\gamma z+\delta}\right]$ be any transformation in the group G. We have, identically,

$$\Theta[z,H(z)] = \Theta\left[z,H\left(\frac{\alpha z+\beta}{\gamma z+\delta}\right)(\gamma z+\delta)^{-2m}\right](\gamma z+\delta)^{2m}$$

Now consider a function of the following form:

$$\Lambda(z) = \left(\frac{dx}{dz}\right)^{1-m} F(x,y)$$

where F(x,y) as always denotes a rational function and m is an integer greater than 1. I suppose in addition that this function vanishes when z comes to one of the vertices of the second category in R_0.

Let q be the number of distinct infinities of $\Lambda(z)$, let $z_1, z_2, \ldots, z_q$ be the distinct infinities, let $A_1, A_2, \ldots, A_q$ be the corresponding residues, and let $\phi(z)$ be an arbitrary rational function of z with no infinity inside the fundamental circle.

We have, identically,

$$(4) \qquad \Lambda(z)\phi(z) = \sum_{k=1}^{q} \sum_{i=0}^{\infty} \frac{A_k \phi\left(\dfrac{\alpha_i z_k+\beta_i}{\gamma_i z_k+\delta_i}\right)}{z - \dfrac{\alpha_i z_k+\beta_i}{\gamma_i z_k+\delta_i}} (\gamma_i z_k+\delta_i)^{-2m}$$

if the Fuchsian functions exist only inside the fundamental circle and

$$(4') \qquad \Lambda(z)\phi(z) = \sum_{k=1}^{q} \sum_{i=0}^{\infty} \frac{A_k}{z - \dfrac{\alpha_i z_k - \beta_i}{\gamma_i z_k + \delta_i}} (\gamma_i z_k + \delta_i)^{-2m} + P(z)$$

(where P(z) is a polynomial in z) if the Fuchsian functions exist throughout the whole plane.

It follows that every Fuchsian function can be put in the form of a quotient of two series such as (4) or (4′) in infinitely many ways.

XI. Historical

If we leave aside the doubly periodic functions, the first Fuchsian functions to have been noticed are the modular functions. They reveal themselves, so to speak, in the study of elliptic transcendants, and their main properties, in particular uniformity and the possession of an essential singular line, have been noted for a long time. Moreover, the works on them and the remarkable results in recent times by Hermite, Dedekind, Fuchs and Klein are too well known to need emphasis.

But another category of Fuchsian functions came to light in 1872. These arise from the hypergeometric equation of Gauss; they are only a particular case of those we have studied in Section V, and are obtained when the polygon R_0 considered in that Section is symmetric and reduces to a quadrilateral.

In a memoir in **Volume 75 of Borchardt's journal**, Schwarz announced without proof that these functions are uniform and have a circle as essential singular line. With the same stroke, this announces the discontinuity of the corresponding group, as I have previously remarked in the historical section of the memoir of Fuchsian groups. Unfortunately, being diverted from this subject by other studies, Schwarz confines himself to a few lines on these transcendants, and does not push his researches further.

In another order of ideas, Schwarz has obtained other results which relate directly to our subject. In various memoirs inserted in **Volumes 70** and **74** of **Borchardt's Journal** and in the **Monatsberichte** of the Academy of Berlin, Schwarz has rigorously proved the so—called Dirichlet principle and the possibility of **mapping** onto the (disc bounded by a) circle any plane figure, in particular a polygon bounded by circular arcs. If he had known the conditions for groups to be discontinuous, he would have been able to proceed in this way to prove the existence of Fuchsian functions in the particular case where the polygon R_0 is symmetric.

I could have made use of these results, but I have preferred to follow another path:

1° Because I could otherwise have proved the existence of Fuchsian functions only when the polygon R_0 was symmetric;

2° Because the series expansions I have used not only give a proof of existence of the function, but also its analytic expression.

MEMOIR ON KLEINIAN GROUPS*

(*Completed 19 May 1883, published 8 September 1883. N.E.N.)

Acta Mathematica 3 (1883), 49—92

I. Complex transformations

In a previous memoir* (*Theory of Fuchsian groups N.E.N.) I have studied the discontinuous groups formed by linear transformations with real coefficients. In the present work I intend to expose certain results concerning groups of linear transformations with complex coefficients. These transformations are divided naturally into four categories, as we shall see.

Let

$$\left(z, \frac{\alpha z+\beta}{\gamma z+\delta}\right)$$

be any transformation, where I always suppose that

$$\alpha\delta - \beta\gamma = 1$$

If we have

$$(\alpha+\delta)^2 = 4$$

then the transformation can be put in the form

$$\left(\frac{1}{z-a}, \frac{1}{z-a} + K\right)$$

where a and K are constants. We say that it is **parabolic**.

If we have

$$(\alpha+\delta)^2 \lessgtr 4$$

then the transformation can be put in the form

$$\left(\frac{z-a}{z-b}, K\frac{z-a}{z-b}\right)$$

where a, b, K are constants.

If we have

$$(\alpha+\delta)^2 \text{ real positive and } > 4$$

then K is real positive and the transformation is **hyperbolic**.

If we have

$$(\alpha+\delta)^2 \text{ real positive and } < 4$$

then K is imaginary or negative and of an absolute value 1; the transformation is **elliptic**.

Finally, if $(\alpha+\delta)^2$ is imaginary or negative, then K is likewise imaginary or negative and the transformation is **loxodromic**.

A common property of all linear transformations is that they transform circles into circles. In fact, if we follow the example of Hermite and represent the complex conjugates of quantities u, v, ... by u_0, v_0, ..., then the transformation

$$\left(z, \frac{\alpha z+\beta}{\gamma z+\delta}\right) \tag{1}$$

can be written

$$\left(z_0, \frac{\alpha_0 z_0+\beta_0}{\gamma_0 z_0+\delta_0}\right)$$

Moreover, the general equation of a circle can be written

$$Azz_0 + Bz + B_0z_0 + C = 0 \tag{2}$$

where A and C are essentially real* (*For the radius of the circle to be real, it is also necessary that $BB_0 - AC > 0$. One easily checks that the analogous expression formed from the coefficients of equation (3) is $(BB_0-AC)(\alpha\delta-\beta\gamma)(\alpha_0\delta_0-\beta_0\gamma_0) = BB_0 - AC$ and consequently positive. N.E.N.) and it is clear that (1) (Translator's Note: which Poincaré views as the coordinate change which renames z as $\frac{\alpha z+\beta}{\gamma z+\delta}$) returns us to the circle (2) from the circle whose equation is

$$A(\alpha z+\beta)(\alpha_0z_0+\beta_0) + B(\alpha z+\beta)(\gamma_0z_0+\delta_0) +$$

$$+ B_0(\gamma z+\delta)(\alpha_0z_0+\beta_0) + C(\gamma z+\delta)(\gamma_0z_0+\delta_0) = 0$$

or

$$\begin{aligned} & zz_0(A\alpha\alpha_0+B\alpha\gamma_0+B_0\gamma\alpha_0+C\gamma\gamma_0) + \\ & + z(A\alpha\beta_0+B\alpha\delta_0+B_0\gamma\beta_0+C\gamma\delta_0) + \\ & + z_0(A\beta\alpha_0+B\beta\gamma_0+B_0\delta\alpha_0+C\delta\gamma_0) + \\ & + (A\beta\beta_0+B\beta\delta_0+B_0\delta\beta_0+C\delta\delta_0) = 0 \end{aligned} \tag{3}$$

We now write down the equation for inversion in the circle (2), i.e., the operation which changes any point z into its transform by reciprocal radius vectors, taking the pole of the transformation to be the center of the circle (2) and the parameter of the transformation to be the square of the radius of this circle. Here is that equation.

Letting t be the transform of z under the inversion, we have

$$t + \frac{B_0}{A} = \frac{BB_0 - AC}{A(Az_0 + B)}$$

Let C_1 and C_2 be any two circles, and let I_1 and I_2 denote the respective inversions in these two circles. If we subject an arbitrary point to the inversion I_1, then the inversion I_2, then we can represent the resulting operation by the notation $I_1 I_2$. It will be a linear transformation, as one easily convinces oneself. This transformation will be parabolic if the two circles C_1 and C_2 are tangential, elliptic if they cross, and hyperbolic if they do not meet.

Suppose that we observe the resultant of not just two, but of several successive inversions. If the number of inversions is even, then the result will be a linear transformation; if the number is odd, the result will be a more complex operation which can be regarded as a linear transformation followed by an inversion. Moreover, **each linear transformation can be regarded as the resultant of an even number of inversions in infinitely many ways.**

Thus the group obtained by combining the different imaginable inversions in various ways contains all linear transformations.

We put

$$z = \xi + \eta\sqrt{-1}$$

where ξ and η are the coordinates of a point in the plane respresenting z.

Now consider an arbitrary point, not in the $\xi\eta$ plane, but in space, and let ξ, η, ζ be its coordinates; I assume ζ is positive, so the point considered is above the $\xi\eta$ plane. We have seen that the transformation (1) changes an arbitrary point ξ, η of the $\xi\eta$ plane to another point of the same plane. We are going to extend the definition of the transformation (1) in such a way that it applies, not only to a point of the $\xi\eta$ plane, but to any point of space. As we have seen, the transformation (1) can be viewed as the result of a certain number of successive inversions with respect to certain circles in the $\xi\eta$ plane which I call C_1, C_2, ..., C_n. Let Σ_1, Σ_2, ..., Σ_n be the spheres which have the same centers and the same radii as these circles. Consider the operation which consists in effecting n successive inversions with respect to the spheres Σ_1, Σ_2, ..., Σ_n. This operation, if applied to a point of the $\xi\eta$ plane, does not differ from the transformation (1). Thus we have redefined the transformation (1) so as to make it applicable to a point in space outside the $\xi\eta$ plane.

An inversion with respect to one of the spheres Σ_1, Σ_2, ..., Σ_n transforms spheres into spheres, transforms the $\xi\eta$ plane into itself, preserves angles, and transforms each infinitesimally small figure into a similar infinitesimally small figure; thus all these properties hold for the resultant, i.e., for the generalized transformation (1).

Suppose that an inversion with respect to the circle C_1, for example, changes a certain circle K of the $\xi\eta$ plane into another circle K_1 of the same plane. Let S and S_1 be the spheres with the same centers and radii as K and K_1; it is clear that the inversion with respect to the sphere Σ_1 will change S into S_1. Then if the transformation (1) changes the circle K into a certain circle K_n, and if S and S_n are the spheres with the same centers and radii as K and K_n, then the generalized transformation (1) will change S into S_n. In fact, the transformation (1) is equivalent to n inversions with respect to the circles C_1, C_2, ..., C_n respectively; these inversions change the circle K into K_1, then K_2, ... then finally into K_n. Let S_i be the sphere with the same center and radius as K_i. The generalized transformation (1) will be equivalent to n inversions with respect to spheres Σ_1, Σ_2, ..., Σ_n respectively, and these inversions will change the sphere S successively into S_1, S_2, ..., and finally S_n.

Q.E.D.

To justify the preceding definition, it is necessary to establish the following:

The transformation (1) can be regarded **in infinitely many ways** as the result of an even number of inversions with respect to circles in the $\xi\eta$ plane. Thus the circles C_1, C_2, ..., C_n are not perfectly determinate. It is necessary to see that the generalized transformation (1) is nevertheless a perfectly determinate operation. Suppose in fact that the transformation (1) can be regarded:

1° On the one hand, as the resultant of n inversions with respect to the circles $C_1, C_2, \ldots, C_n$;

2° On the other hand, as the resultant of p inversions with respect to p other circles $C'_1, C'_2, \ldots, C'_p$ in the $\xi\eta$ plane.

Let Σ'_i be the sphere with the same center and radius as C'_i.

Now let P be any point of space. If we apply to it the inversions in the spheres $\Sigma_1, \Sigma_2, \ldots, \Sigma_n$ successively, then we obtain a certain point Q. If we apply to P the inversions in the spheres $\Sigma'_1, \Sigma'_2, \ldots, \Sigma'_p$ successively, then we obtain a certain point Q′. The two operations, with we call (P,Q), (P,Q′), both satisfy the definition of the generalized transformation (1), hence it is necessary to show that the two points Q and Q′ coincide. Well, we can pass three spheres S, S′, S″ through the point P, with their centers on the $\xi\eta$ plane and cutting it in three great circles K, K′, K″.

The transformation (1) will change these three great circles into three other circles K_1, K'_1, K''_1 in the $\xi\eta$ plane. Let S_1, S'_1, S''_1 be the spheres with the same centers and radii as these circles; the operation (P,Q), and likewise the operation (P,Q′), will change S, S′, S″ into S_1, S'_1, S''_1. The point Q, and likewise the point Q′, is then found as the intersection of the three spheres S_1, S'_1, S''_1. Thus these two points coincide. Hence the generalized transformation (1) is a perfectly determinate operation.

Q.E.D.

It remains to find the equations of this operation. To define a point P of space, we employ the following three coordinates

$$z = \xi + i\eta \qquad z_0 = \xi - i\eta \qquad \rho^2 = \xi^2 + \eta^2 + \zeta^2 = zz_0 + \zeta^2$$

Since ζ is assumed positive, these three coordinates suffice to completely determine the point P.

Let ρ'^2, z' and z'_0 be the three coordinates of the point Q to which P is sent by the generalized transformation (1). The occurrence of P on the sphere with the same center and radius as the circle (3) is expressed by

$$\begin{aligned}
(4)\qquad & \rho^2(A\alpha\alpha_0 + B\alpha\gamma_0 + B_0\gamma\alpha_0 + C\gamma\gamma_0) + \\
& + z(A\alpha\beta_0 + B\alpha\delta_0 + B_0\gamma\beta_0 + C\gamma\delta_0) + \\
& + z_0(A\beta\alpha_0 + B\beta\gamma_0 + B_0\delta\alpha_0 + C\delta\gamma_0) + \\
& + (A\beta\beta_0 + B\beta\delta_0 + B_0\delta\beta_0 + C\delta\delta_0) = 0
\end{aligned}$$

If the point P is on the sphere with the same center and radius as the circle (3), then the point Q must be found on the sphere with the same center and radius as the circle (2) obtained from (3) by the transformation (1); whence the equation

$$(5)\qquad A\rho'^2 + Bz' + B_0z'_0 + C = 0$$

If we regard A, B, B_0 and C as unknowns, these two equations must be equivalent. We then have

$$\rho'^2 = \frac{\rho^2\alpha\alpha_0+z\alpha\beta_0+z_0\beta\alpha_0+\beta\beta_0}{\rho^2\gamma\gamma_0+z\gamma\delta_0+z_0\delta\gamma_0+\delta\delta_0}$$

$$z' = \frac{\rho^2\alpha\delta_0+z\alpha\delta_0+z_0\beta\gamma_0+\beta\delta_0}{\rho^2\gamma\gamma_0+z\gamma\delta_0+z_0\delta\gamma_0+\delta\delta_0}$$

$$z'_0 = \frac{\rho^2\gamma\alpha_0+z\gamma\beta_0+z_0\delta\alpha_0+\delta\beta_0}{\rho^2\gamma\gamma_0+z\gamma\delta_0+z_0\delta\gamma_0+\delta\delta_0}$$

These are the equations of the generalized transformation (1).

We now move on to the general properties of this transformation.

If the transformation (1) is elliptic then it leaves unchanged all points of the circle C which passes through the two fixed points, has its center in the middle of the line joining the two fixed points, and its plane perpendicular to the $\xi\eta$ plane. It likewise leaves invariant infinitely many circles whose characteristic property is that the spheres passing through them are orthogonal to the circle C just defined. We call C the **fixed circle** of the elliptic transformation.

If the transformation (1) is hyperbolic, then there are only two points of the space which are not changed by the transformation: they are the two fixed points in the $\xi\eta$ plane. The transformation leaves invariant all the spheres which pass through these two points.

If the transformation (1) is parabolic, then there is only a single point left unchanged by the operation, moreover all the spheres which are tangential, at the fixed point, to a certain line in the $\xi\eta$ plane are left invariant.

Finally, suppose that the transformation (1) is loxodromic; it leaves invariant the circle C whose diameter is the line adjoining the fixed points and whose plane is normal to the $\xi\eta$ plane. However, with the exception of the fixed points, it changes all the points of this circle.

In summary, the only transformations which have fixed points outside the $\xi\eta$ plane are elliptic.

We have previously seen that the generalized transformation (1) transforms each infinitely small figure into a similar infinitely small figure. Let us find the ratio of similitude. If we consider a single inversion, then it is evident that the homologous dimensions of two infinitely small figures, one of which is a transform of the other, will be in the same ratio as the ζ coordinates of the centers of gravity of these figures. It will then be the same when we consider the resultant of several inversions instead of one, i.e., when we consider the generalized transformation (1).

Thus if we let ds, dw, dv respectively denote an infinitely small arc, area, or volume, if ζ is the distance of that arc, area, or volume from the $\xi\eta$ plane, and if we let ds′, dw′, dv′ denote their respective transforms under the generalized transformation (1), and if ζ' is the distance of ds', dw' or dv' from the $\xi\eta$ plane we have

(6) $$\frac{ds}{\zeta} = \frac{ds'}{\zeta'} \quad \frac{dw}{\zeta^2} = \frac{dw'}{\zeta'^2} \quad \frac{dv}{\zeta^3} = \frac{dv'}{\zeta'^3}$$

We say that two figures are **congruent** when one is the transform of the other under a generalized transformation (1).

We call the integral

$$\int \frac{ds}{\zeta}$$

taken over an arc, the L of that arc. We call the integral

$$\int \frac{dw}{\zeta^2}$$

over a plane or curved region, the S of that region, and finally we call the integral

$$\int \frac{dv}{\zeta^3}$$

taken over a solid, the V of that solid.

It follows from the equations (6) that two congruent arcs have the same L, that two congruent surfaces have the same S, and that two congruent solids have the same V.

Now suppose that we take away the usual meaning of the words **line** and **plane** and let **lines** and **planes** be circles and spheres which cut the $\xi\eta$ plane orthogonally. Suppose also that we take away the usual meanings of **length**, **area**, and **volume** and let them mean L, S and V. Finally, suppose that we retain the usual meanings of **circle**, **sphere**, and **angle**. We then recognize that, under this interpretation, all the theorems of Lobachevsky's nonEuclidean geometry, i.e., the geometry which does not admit Euclid's parallel postulate, are perfectly exact. We see also that we have a connection between the theory of linear transformations and non–Euclidean geometry. It is the same connection that Klein has used to find all the groups of finite order contained in the linear group.

II. Discontinuous groups

I propose to construct the discontinuous groups generated by finitely many linear transformations with complex coefficients. But, in order to go further, it is necessary to make a distinction which does not arise in the theory of Fuchsian groups, though it is of fundamental importance here. This distinction has been very clearly made by Klein (**Mathematische Annalen**, **XXI**, p.176).

By an **infinitesimal transformation** we mean any linear transformation for which the absolute values of $\alpha - 1$, β, γ, $\delta - 1$ are infinitely small.

If a group contains an infinitesimal transformation, i.e., if we can find in it a transformation for which the four absolute values just mentioned are smaller than any given quantity without being zero, then the group is continuous.

But, among the groups which do not satisfy this condition, i.e., among the discontinuous groups, it is important to distinguish two classes which we call **properly** and **improperly** discontinuous groups.

A group will be improperly discontinuous in the neighborhood of a point z if, in a domain surrounding the point z, we can find infinitely many transforms of this point by transformations in the group, no matter how small the domain D may be.

In the contrary case the group will be properly discontinuous.

For example, if we are dealing with a group of transformations applied to a real quantity z, then we can take the domain D to be the segment of the line between the point $z - \epsilon$ and the point $z + \epsilon$; if we are dealing with transformations applied to a complex quantity z, or a point z of the plane, then the domain D will be a circle with center at the point z; if we are dealing with a point P of space, then D will be a sphere with center P, etc.

To clarify the preceding with a simple example, consider the group of transformations

$$\left(z, \frac{\alpha z+\beta}{\gamma z+\delta}\right)$$

where α, β, γ, δ are real integers such that $\alpha\delta - \beta\gamma = 1$. It is clear that this group contains no infinitesimal transformation, hence it is discontinuous. We also see that if z is real, we can choose α, β, γ, δ so that

$$\left|z - \frac{\alpha z+\beta}{\gamma z+\delta}\right|$$

is as small as we please, without being zero; whereas this is impossible if z is imaginary. Thus the group is improperly discontinuous for z real and properly discontinuous for z imaginary.

We do not have to make this distinction in the case of Fuchsian groups, because it happens that each discontinuous group of real transformations is properly discontinuous whenever z is imaginary. I am going to prove this here, since the proof I give will permit an easier understanding of the more general case dealt with in this memoir.

Let G be an arbitrary group of real transformations, let z be any imaginary point, and suppose that each domain D surrounding the point z contains infinitely many transforms of this point by transformations in G. I then claim that the group contains infinitesimal transformations.

I shall say that a quantity which depends on a variable ϵ is of order ϵ when it vanishes with ϵ, and that a transformation is of order ϵ if $\alpha - 1$, $\delta - 1$, β and γ vanish with ϵ. Then:

1° Each transformation Σ which changes z to $z + \zeta$, where ζ is an infinitely small quantity of order ϵ, can be regarded as the resultant of an elliptic transformation S which has z as a fixed point and a hyperbolic transformation s which is infinitesimal of order ϵ. We write

$$\Sigma = Ss$$

2° If s and s_1 are two transformations of order ϵ, then their resultant ss_1 will also be infinitesimal and of order ϵ.

3° If s is a transformation of order ϵ and S is a finite transformation, then the resultant of S inverse, s and S, which we write $S^{-1}sS$*, will be infinitesimal and of order ϵ. (*As is customary, we use the notation S^{-1} for the inverse transformation of S, SS_1 for the resultant of S and S_1, S^m for the resultant of m transformations S in succession.)

4° If S and S_1 are two elliptic transformations with z as fixed point, then they are permutable, i.e.,

$$SS_1 = S_1S \qquad S_1 = S^{-1}S_1S \tag{1}$$

Now, by hypothesis, we have an infinity of transformations in the group which change z into $z + \zeta$, $z + \zeta_1$, ..., where ζ, ζ_1, ... are quantities of order ϵ. Let Σ and Σ_1 be two of them. We have

$$\Sigma = Ss \qquad \Sigma_1 = S_1s_1$$

where S and S_1 both have z as a fixed point, and s and s_1 are infinitesimal.

The transformation

$$\Sigma\ \Sigma_1\Sigma^{-1}\Sigma_1^{-1}$$

belongs to the group, and it is not the identity (z,z) because in general

Σ and Σ_1 do not have the same fixed point.

I claim that it will be infinitesimal. In fact it can be written

$$SsS_1s_1s^{-1}S^{-1}s_1^{-1}S_1^{-1}$$

or, by virtue of the relation (1),

$$SsS^{-1}S_1Ss_1s^{-1}S^{-1}s_1^{-1}S_1^{-1}$$

But, applying principles 2^0 and 3^0, we see in succession that SsS^{-1}, s_1s^{-1}, $S(s_1s^{-1})S^{-1}$, $(Ss_1s^{-1}S^{-1})s_1^{-1}$, $S_1(Ss_1s^{-1}S^{-1}s_1^{-1})S_1^{-1}$, and finally

$$(SsS^{-1})(S_1Ss_1s^{-1}S^{-1}s_1^{-1}S_1^{-1})$$

are infinitesimal of order ϵ.

Thus the group contains an infinitesimal transformation; it is continuous.

But if a group of real transformations cannot be improperly discontinuous in the neighborhood of an imaginary point z, it can on the contrary be improperly discontinuous in the neighborhood of a real point z. In fact, the Fuchsian groups of the first, second, and sixth families, i.e., those whose generating polygons have no sides of the second kind, are improperly discontinuous in the neighborhood of all real points z. On the other hand, Fuchsian groups of the other families are properly discontinuous in the neighborhood of real points belonging to a side of the second kind in the generating polygon or one of its transforms; they are not improperly discontinuous except in the neighborhood of singular points.

We now return to the study of groups generated by imaginary transformations, or rather, the more general transformations defined in the preceding paragraph.

I claim that such a group cannot be improperly discontinuous in the neighborhood of a point P situated outside the $\xi\eta$ plane. Suppose in fact that we have such a point P outside the $\xi\eta$ plane, and a domain D containing this point P. If D, no matter how small it is, contains infinitely many transforms of the point P by transformations in the group, then I claim that the group will be continuous, in fact:

1° Let Σ be a transformation which changes P to P', where the distance PP' is infinitesimal of order ϵ, then I claim that we can write

$$\Sigma = Ss$$

where S is an elliptic transformation whose fixed circle passes through P and s is an infinitesimal hyperbolic transformation of order ϵ. In fact, the transformation Σ^{-1} will change P' into P. There will be an infinitesimal hyperbolic transformation which changes P' into P, because the points P' and P are infinitely close. I call it s^{-1} and I put

$$\Sigma^{-1} = s^{-1}S^{-1} \quad \text{whence} \quad \Sigma = Ss$$

The transformation S will not change the point P, hence, by the principles of the preceding paragraph, it will be an elliptic transformation whose fixed circle passes through P.

Q.E.D.

2° and 3°: The previous principles 2° and 3° continue to hold here; this is evident.

4° If S and S_1 are two elliptic transformations with the same fixed circle, then we have

$$SS_1 = S_1S \qquad S_1 = S^{-1}S_1S$$

5° If S_1 is an elliptic transformation with fixed circle C_1 and if C'_1 is a circle orthogonal to the $\xi\eta$ plane which meets C_1 at a point P above the $\xi\eta$ plane at an infinitely small angle, then we can put

$$S_1 = S'_1\sigma$$

where S'_1 is an elliptic transformation with C'_1 as fixed circle and σ is an infinitesimal elliptic transformation.

Having established this, we take up the hypothesis that the group contains infinitely many transformations Σ, Σ_1, Σ_2, ... which change P into infinitely near points. We have

$$\Sigma = Ss \quad \Sigma_1 = S_1s_1 \quad \Sigma_2 = S_2s_2, \ldots$$

where s, s_1, ... are infinitesimal while S, S_1, S_2, ... are elliptic transformations whose fixed circles pass through P. We then have an infinity of elliptic transformations whose fixed circles pass through P. It necessarily follows that we can find two transformations S and S_1 whose fixed circles C and C_1 meet at an angle which is zero or infinitesimally small. We can then put

$$S_1 = S'_1\sigma$$

where σ is infinitesimal or the identity and S'_1 has the same fixed circle C as S, so that we have

$$S'_1 = S^{-1}S'_1S$$

Consider the two transformations S and S_1 and the corresponding transformations Σ, Σ_1. The transformation

$$\Sigma\,\Sigma_1\Sigma^{-1}\Sigma_1^{-1}$$

will belong to the group; I claim that it is infinitesimal. In fact it can be written

$$SsS_1s_1s^{-1}S^{-1}s_1^{-1}S_1^{-1}$$

or

$$SsS'_1\sigma s_1s^{-1}S^{-1}s_1^{-1}\sigma^{-1}S'_1{}^{-1}$$

or finally

$$T = SsS^{-1}S'_1S\sigma s_1s^{-1}S^{-1}s_1^{-1}\sigma^{-1}S'_1{}^{-1}$$

But, by virtue of	3°,	$T_1 = SsS^{-1}$	will be infinitesimal
" " "	2°,	$T_2 = \sigma s_1 s^{-1}$	" " "
" " "	3°,	$T_3 = ST_2S^{-1}$	" " "
" " "	2°,	$T_4 = T_3s_1^{-1}\sigma^{-1}$	" " "
" " "	3°,	$T_5 = S'_1T_4S'_1{}^{-1}$	" " "
Finally, "	2°,	$T = T_1T_5$	" " "

Q.E.D

Thus, if we consider a discontinuous group, it will be properly discontinuous at each point P outside the $\xi\eta$ plane. It follows that the half space above this plane will be divided into an infinity of regions R_0, R_1, ..., R_i, ... each of which corresponds to a transformation in the group; the region R_i for example, to the transformation S_i which changes R_0 into R_i. It is exactly the same thing as we have observed with Fuchsian groups.

But what we now propose is to find the groups of imaginary transformations which are properly discontinuous at the points of the

$\xi\eta$ plane themselves. These are the ones which can be used in the theory of functions.

Well, recall what we have observed for Fuchsian groups. These groups are all properly discontinuous for imaginary values of z, so that the half plane above the real axis is divided into infinitely many polygons R_0, R_1, ..., R_i, ... How do we tell whether the group remains properly discontinuous for real values of z? If the polygon R_0 has no sides of the second kind, i.e., if it is entirely above the real axis, then the group is improperly discontinuous for real values of z; it is properly discontinuous, on the contrary, if R_0 has a side of the second kind, i.e., if a whole side of this polygon belongs to the real axis.

It is the same thing here; if the region R_0 defined above is entirely above the $\xi\eta$ plane, or if it has no more than a point or a line in common with this plane, then the group is improperly discontinuous for the points of the $\xi\eta$ plane; it is properly discontinuous, on the contrary, if a portion of the surface of the region R_0 belongs to this plane.

III. Generating polygons

Consider a group of imaginary transformations which is properly discontinuous at the points of the $\xi\eta$ plane.

We say that is a **Kleinian group**. Such a group subdivides a part of the $\xi\eta$ plane into an infinity of regions R_0, R_1, ..., R_i, ... so that each region R_i corresponds to a transformation S_i in the group which changes R_0 into R_i. Consider the region R_0 and any other region R_1 adjacent to R_0; they are separated by a portion of the perimeter of R_0 which serves as their common frontier and which I call a **side** of R_0. Now such a side can be a closed curve or an arc bounded by two points which are **vertices** of R_0. Let C be any side of R_0, then there will always be another side C of R_0 such that one of the transformations in the group changes C into C′. The sides C and C' will be called **equivalent**.

The subdivision of the plane, or of a part of the plane, into an infinity of regions R can be made in infinitely many ways (cf. Sections IV and IX of the memoir on Fuchsian groups). We commence by giving several definitions. The region R_0 is called a **generating polygon of the group**; I say that two regions R and R'' are piecewise equivalent* with respect to a group G when we can pass from one to the other in the following way. (*Translator's Note: Poincaré simply says "equivalent". However, since this translation used "equivalent" in the stricter sense of "congruent via a motion in G", it seems desirable to describe the more general, piecewise, notion as such.) We decompose R into a certain number of parts $r'_1, r'_2, \ldots, r'_p$. Let r''_i be the transform of r'_i by a transformation S_i in the group G. The partial regions r''_i must then form the new region R''. It is clear that if R_0 is a generating polygon for the group G, then so is any region piecewise equivalent to R_0. Suppose, in particular, that we remove from R_0 any portion P_0, and adjoin to R_0 the transform P'_0 of P_0 by any transformation in the group. We then obtain a region $R'_0 = R_0 + P'_0 - P_0$ piecewise equivalent to R_0, and consequently we can replace the generating polygon R_0 by the polygon R'_0.

We can exploit this circumstance to simplify the region R_0. In fact, let C be any side of R_0 and let C' be its equivalent. If C is a closed side, then so too is C'; if, on the contrary, C is an open side ab, then C' will also be an open side $a'b'$. In the first case we let K be an arbitrary circle; in the second case let K be a circular arc bounded by the two vertices a and b of the side C. Now let K' be the transform of K under the transformation in the group which changes C into C'. In the first case K' will be, like K, a closed circle; in the second case K' will be a circular arc bounded by the two vertices a' and b' of the side C'. We let P_0 be the portion of the plane between K and C; and let P'_0 be the portion between K' and C'. P'_0 will be the transform of P_0 under one of the transformations in the group. We can then replace the region R_0 by $R'_0 = R_0 - P_0 + P'_0$. In the new region R'_0 the sides C and C' are replaced by the sides K and K' which are arcs of a circle.

It is worth noting that the region R'_0 is not completely

defined by this, since the circle K is not absolutely determined by the conditions we have imposed on it.

Operating in the same way on each side of R_0, we finally succeed in replacing R_0 by an analogous region whose sides are all circles or circular arcs.

Here we can encounter the same difficulty that we have observed in Section IV of the memoir on Fuchsian groups. It can happen that the region P_0, as we have defined it, is not completely contained in R_0. Under these conditions we arrive at a region R'_0 which is concave in the sense of the Section cited, and we overcome the difficulty as in that Section.

Thus we can always assume that the region R_0 is a polygon bounded by circles and circular arcs, though it can happen that this polygon is not simply connected, but of a higher order of connectivity.

We now go back to the case of Fuchsian groups, to those of the third family, for example. The polygon R_0, as we have defined it in the memoir in **Acta Mathematica, I** (Theory of Fuchsian Groups), is simply connected and bounded by sides of the first and second kind; but if we adjoin to R_0 its reflection R'_0 in the real axis, then the total region $R_0 + R'_0$, which is the analogue of the region R_0 studied here, will be multiply connected.

We are led, as in Section V of the memoir on Fuchsian groups, to distribute the vertices of R_0 is cycles; but since there is nothing here analogous to sides of the second kind, we have only **closed** cycles and not open ones. Let A_0 be any vertex of R_0 and let $A_1, A_2, \ldots, A_{n-1}$ be the vertices which belong to the same cycle as A_0, written in the order they are obtained by applying the rule of the Section cited. We describe an infinitely small contour around A_0 and suppose that in following this contour we successively encounter the polygons $R_0, R_1, R_2, \ldots, R_{n-1}, R_n, \ldots$ If $i < n$ then the vertex A_0, viewed as a part of R_i, will be homologous to A_i; viewed as a part of R_n it will be homologous to A_0. It follows that the transformation which changes R_0 into R_n has A_0 as a fixed point. Moreover, it can be elliptic, parabolic or hyperbolic, but not loxodromic. This fourth hypothesis must be rejected. (The third hypothesis must also be rejected. For the same reason as in the case of a Fuchsian group, the polygon R_0 cannot admit hyperbolic cycles. See the note at the end of Section V of the memoir on Fuchsian groups. N.E.N.)

In fact, let $K = \rho e^{i\omega}$ be the multiplier of our loxodromic transformation, and let p be an integer so large that $p\omega > 2\pi$. Let Σ_0 denote the union of the polygons R_0, R_1, ..., R_{n-1}, and let Σ_1 denote the union of the polygons R_n, R_{n+1}, ..., R_{2n-1}, i.e., the transform of Σ_0 by our loxodromic transformation. Then let Σ_2 be the transform of Σ_1, let Σ_3 be that of Σ_2, etc. It is easy to see that Σ_p partially covers one of the regions Σ_0, Σ_1, ..., Σ_{p-1}, which is absurd because the group is assumed to be properly discontinuous.

It remains to examine the first three hypotheses.

In the first case we say that the cycle is elliptic*; the sum of the angles of the cycle must be an aliquot part of 2π. (*The cycles that we call **elliptic**, **parabolic** and **hyperbolic** here are respective analogues of the cycles of the first category, third subcategory and fourth subcategory in the memoir on Fuchsian groups.)

In the second case the cycle will be parabolic; all the angles of the cycle will be zero.

In the third case we have a hyperbolic cycle; all the angles of the cycle will again be zero. In the next paragraph we shall see that we can always replace the polygon R_0 by a piecewise equivalent which admits no hyperbolic cycle.* (*We have seen in Sections IX and XI of the memoir on Fuchsian groups, and in Section II of the memoir on Fuchsian functions, that each group of the second order in the second, fourth, sixth or seventh family is identical with a group of the third or fifth family, or with a group of the first order in the sixth or seventh family. In other words, if the generating polygon of a Fuchsian group admits a cycle of the fourth subcategory then we can always, by the procedure of Section IX, replace it by another polygon not admitting any cycle of this subcategory. It is the same here.)

I now claim that we can always assume that there is a cycle consisting of only a single vertex. In fact, take the polygon R_0, vertex A_0, and the vertices A_1, A_2, ..., A_{n-1} which belong to the same cycle. Also consider the infinitely small contour that we have drawn around A_0, and the polygons R_1, R_2, ... encountered successively when following that contour. Let S_i be the transformation which changes R_0 into R_i. We decompose the polygon R_0 into n partial polygons $r_0, r_1, r_2, \ldots, r_{n-1}$ in such a way that r_i includes the vertex A_i but none of the others among $A_0, A_1, \ldots, A_{n-1}$. Now let r'_i be the transform of R_i by S_i. Then r'_0 is no different from r_0, and the partial regions $r'_0, r'_1, \ldots, r'_{n-1}$ make up a polygon R'_0 piecewise equivalent to R_0, which can therefore serve as generating polygon for our group. But R'_0 includes the vertex A_0 in such a way that the cycle containing A_0 consists of only a single vertex.

IV. Generating polyhedra

Now consider the half space situated above the $\xi\eta$ plane, and the subdivision of that portion of space into an infinity of regions $P_0, P_1, \ldots, P_i, \ldots$ such that the transformation S_i in our group changes P_0 into P_i.

Consider the region P_0 and any region P_1 adjacent to P_0; they are separated by a portion of the surface of P_0 which I call a **face** of P_0; I call this a face of first kind. If a portion of the $\xi\eta$ plane forms part of the surface of P_0, this will be a face of the second kind. These terms are entirely analogous to those we have used in the theory of Fuchsian groups. Two adjacent faces of P_0 are separated by an **edge**. Edges will be of the first kind if they separate two faces of the first kind, and of the second kind if they separate a face of the first kind and a face of the second kind. The faces of the first kind will be equivalent in pairs, so that each is changed into its equivalent by a transformation in the group. It follows that two equivalent faces are congruent.

Just as the faces are partitioned into pairs, so are the edges partitioned into cycles, like those of the vertices of generating polygons. Let A_0 be any edge of the first kind; here is how we find the edges in its cycle. Let F_0 be one of the two faces separated by the edge A_0, F'_0 its equivalent, so that the faces F_0 and F'_0 are congruent. Let A_1 be the edge of F'_0 homologous to A_0; it separates F'_0 from another face F_1. We operate on the face F_1 and the edge A_1 as we did on F_0 and A_0. We obtain in this way a sequence of faces F_0, F_1, F_2, F_3, ... and a sequence of edges A_0, A_1, A_2, A_3, ..., and we stop when we return to the edge A_0. All the edges so obtained belong to the same cycle.

Let A_0, A_1, ..., A_{n-1} be these edges, which I assume to be n in number. Suppose that we describe an infinitely small contour around the edge A_0; in doing so we successively traverse q regions P_0, P_1, ..., P_{q-1}. If $i < n$ the edge A_0, considered as part of P_i, will be homologous to A_i; considered as part of P_n, it will be homologous to A_0. Thus the transformation which changes P_0 into P_n does not change A_0. It follows that:

1° The number $\frac{q}{n}$ is an integer;

2° The transformation which changes P_0 into P_n is elliptic, with multiplier $e^{2i\pi \frac{n}{q}}$;

3° **The edge A_0 is a symmetric circle with respect to the $\xi\eta$ plane**;

4° The sum of the dihedral angles of the different edges in the cycle is an aliquot part of 2π.

Thus all the edges of the first kind are circles whose planes are perpendicular to the $\xi\eta$ plane and whose centers are in that plane.

The extremities of the edges are called **vertices**. Among the vertices we distinguish:

1° Those outside the $\xi\eta$ plane, at which three or more edges of the first kind meet;

2° Those on the $\xi\eta$ plane, at which one or more edges of the first kind meet two or more edges of the second kind;

3° It is also necessary to adjoin **isolated vertices**. If two faces are tangential, the point of contact can be regarded as a vertex though it nevertheless does not belong to any edge of the first kind.

As in the preceding paragraph, we can clearly replace the region P_0 by any other **piecewise equivalent** region P'_0. I claim that we can always arrange this in such a way that P'_0 is a polyhedron bounded by spheres or portions of spheres. All the faces of the first kind will be spheres or portions of spheres with their centers in the $\xi\eta$ plane; all the edges will be circles or circular arcs.

Suppose, in fact, that F is any face of the first kind, F′ is its equivalent, and that S is the transformation which changes F into F′. Several cases are possible:

1° Either, the face F is bounded by an edge of the second kind and this edge is a closed curve. We then let F_1 be any hemisphere with its center in the $\xi\eta$ plane, and let Q_0 be the region bounded by F and F_1; the transformation S then changes F_1 into another hemisphere F'_1, and Q_0 into Q'_0, the region bounded by F′ and F'_1. The regions P_0 and $P'_0 = P_0 - Q_0 + Q'_0$ will then be piecewise equivalent;

2° Or, the face F is bounded by two edges, one of which is of the first kind, with two vertices. We argue in the same way, except that we are obliged to make the sphere F_1 pass through the edge of the first kind, which is a circle, as we have seen above;

3° Or, the face F has several edges of the first kind lying on the same sphere Σ with its center in the $\xi\eta$ plane. In this case the sphere F_1 must be the sphere Σ;

4° Or, finally, the different edges of the first kind on the face F do not lie on the same sphere with its center on the $\xi\eta$ plane. In this case we can partition the face F into several others f, f′, f″, ... Suppose, for example, that the partial face f is contiguous only with the face f′, that f′ is contiguous only with f and f″, f″ only with f′ and f‴, etc. I assume that the partial face f is separated from the face f′ by a line α' whose extremities are β' and γ', that the face f′ is separated from the face f″ by a line α' whose extremities are β'' and γ'', etc. I suppose that all the edges of the first kind contained in f, and the vertices β' and γ', are on the same sphere f_1 with its center in the $\xi\eta$ plane; that the edges of the first kind in f′, and the vertices β', γ', β'', γ'' are on the same sphere f'_1 with its center in the $\xi\eta$ plane, and so on. We consider the portion of f_1 bounded by the edges of the first kind in f and the intersection of f_1 with f'_1; the portion of f'_1 bounded by the edges of the first kind in f' and the intersections of f'_1 with f_1 and f''_1, etc. These portions of spherical surfaces will form a new face F_1 from which we can argue as above.

In all cases we can replace the region P_0 by a piecewise equivalent, replacing the faces F and F' by the faces F_1 and F'_1 which are portions of spheres with their centers in the $\xi\eta$ plane. Operating in the same way on all the faces of the first kind on the region P_0, we replace this region by a piecewise equivalent, all of whose faces of the first kind will be portions of such spheres.

In summary, we can always assume that our region P_0 is a polyhedron, all of whose faces of the first kind are portions of spheres with their centers in the $\xi\eta$ plane. We call it a **generating polyhedron of the group**.

One easily sees that such a polyhedron can have an **isolated vertex** outside the $\xi\eta$ plane.

The faces of the second kind on the generating polyhedron P_0 will be polygons bounded by circular arcs, and these arcs will be the outlines of the faces of the first kind on the $\xi\eta$ plane. These polygons can be regarded as the generating polygons of a Kleinian group.

Suppose that our polyhedron P_0 has n faces of the second kind, F_0^2, F_0^2, ..., F_0^n; then the homologous polyhedron P_i will also have n faces of the second kind, F_i^1, F_i^2, ..., F_i^n. If we except certain singular points and lines, then each point of the $\xi\eta$ plane will belong to one of the faces F_i^k of one of the polyhedra P_i, and moreover, it can belong to only one, since each point of space belongs to at most one polyhedron P_i.

Thus the group is properly discontinuous, even at the points of the $\xi\eta$ plane, provided always that we except the singular points and lines mentioned above.

The group considered is therefore Kleinian.

The $\xi\eta$ plane will be partitioned into n regions R^1, R^2, ..., R^n; the region R^k itself will be subdivided into an infinity of polygons

$$F_0^k,\ F_1^k,\ F_2^k,\ \dots,\ F_i^k,\ \dots$$

such that the transformation S_i changes F_0^k into F_i^k.

Thus we can take the generating polygon of the group to be any face of the second kind on the generating polyhedron, i.e., one of the polygons whose sides are circles with the same centers and radii as the spheres which form the faces of the first kind on this polyhedron.

The converse is not true. Consider an arbitrary Kleinian group, and let R_0 be one of the polygons which can be chosen as its generating polygon. Construct the spheres with the same centers and radii as the circular arcs which serve as sides of this polygon, and consider the polyhedron bounded by these spheres. In general, it will not be a generating polyhedron for the group. In fact, in a generating polyhedron, two equivalent faces must be congruent. But, consider an arbitrary side bc of R_0, the adjacent sides ab and cd, the equivalent side b′c′, and the sides a′b′ and c′d′ adjacent to it. Construct the spheres S_1, S_2, S_3, S_4, S_5,. S_6 corresponding respectively to these six sides. Corresponding to each side of R_0 there is a face of P_0. Corresponding to the side bc we have a portion of the surface of S_1 bounded by the intersections of that sphere with S_2 and S_3; corresponding to the side b′c′ we have the portion of surface of S_4 which is bounded by the intersections of that sphere with S_5 and S_6. These must be equivalent faces of P_0, however, these two faces will not be congruent in general.

For the polyhedron P_0 to be the generating polyhedron of a group, it is necessary and sufficient that equivalent faces be congruent, and here is the necessary and sufficient condition for this:

Again, let bc be a side of R_0, let ab and cd be the adjacent sides, let b′c′ be the equivalent side, and let a′b′ and c′d′ be the sides adjacent to it. We prolong the arcs which are these six sides to complete circles. Let b_1 be the second intersection point of the circles ab and bc, and similarly let c_1, b'_1, c'_1 be the respective intersections of bc and cd, a′b′ and b′c′, b′c′ and c′d′. The cross ratio of the four points b, c, b_1, c_1 must equal that of the four points b′, c′, b'_1, c'_1.

Among the polygons piecewise equivalent to R_0, which are infinitely numerous and each able to be chosen as generating polygon, there is always one which satisfies this condition, because each Kleinian group is properly discontinuous for points outside the $\xi\eta$ plane and consequently has a generating polyhedron. We always assume that the polygon R_0 has been chosen to satisfy this condition.

Then there is a face of P_0 corresponding to each side of R_0, and equivalent faces corresponding to equivalent sides. Corresponding to each vertex of R_0 there is an elliptic cycle corresponding to an edge of P_0, and corresponding to different vertices in the same cycle we have the different edges in the same cycle; corresponding to a vertex of R_0 in a parabolic or hyperbolic cycle we have an isolated vertex of P_0. Thus we see that the isolated vertices of P_0 are partitioned into cycles.

I now claim that we can always assume that R_0, and consequently P_0, admits no hyperbolic cycle. In fact, suppose that P_0 has no isolated vertex belonging to a hyperbolic cycle; I claim that we can replace this polyhedron by a piecewise equivalent which admits no hyperbolic cycle. In fact, from what we have seen at the end of the preceding paragraph, we can always assume that this hyperbolic cycle consists of a single vertex A.

The vertex A is an isolated vertex of P_0, i.e., it is a point of contact between two faces of the first kind on P_0, tangential to each other, which I shall call F and F′. These two faces are equivalent, because if not the cycle of A would contain other vertices. Thus one of the transformations in the group, which I call S, will change F into F′, and it will be hyperbolic by hypothesis. The face F will either have no boundary edge, or else one edge, or else more than one. I first claim that we can always assume that the third case does not occur. In fact, if it does, we trace on the face F a semicircle, whose plane must be perpendicular to the $\xi\eta$ plane, and which must be small enough to be entirely contained in the face F; this is always possible. This semicircle partitions the face F into two others, F_1 and F_2, with F_1, say, containing the vertex A. The face F′ congruent to F is similarly divided into two others F'_1, F'_2 congruent to F_1, F_2 respectively. The vertex A is then part of two faces F_1 and F'_1, each of which is bounded by only a single edge.

Suppose then that F has at most one edge. We construct a sphere Φ slightly different from F. If the face F has one edge, then we require the sphere Φ to pass through that edge. The hyperbolic transformation S will change Φ into another sphere Φ', and we can always choose Φ in such a way that these two spheres do not cut or touch. To do this it suffices to make Φ differ sufficiently little from F and, if A and B are the two fixed points of the transformation S, to make one outside Φ and one inside. Let p_0 be the portion of space between Φ and F, let p'_0 be the portion between Φ' and F'. The transformation S will change p_0 into p'_0. Thus the polyhedron $P'_0 = P_0 - p_0 + p'_0$ is piecewise equivalent to P_0; hence it can serve as a generating polyhedron. However, it does not possess the hyperbolic vertex A, because the spheres Φ and Φ' do not meet.

Thus we can always suppose that our polygon R_0 and our polyhedron P_0 have no hyperbolic cycles; this is what we shall do from now on.* (*This proof is not exact. The polyhedron P'_0 is not piecewise equivalent to the polyhedron P_0, but to a more complicated domain formed by:

1° the polyhedron P_0;

2° a segment of the sphere, with diameter AB, which is orthogonal to the $\xi\eta$ plane;

3° a polyhedron inside this sphere which has A as a point of contact with the polyhedron P_0. The proof is faulty because the domain p_0 contains **an infinity** of points mutually homologous under the hyperbolic transformation and the principles of Section III evidently do not apply in this case. We conclude that only elliptic and parabolic cycles exist. N.E.N.)

V. Existence of Kleinian groups

Suppose that a generating polyhedron satisfies the conditions stated in the preceding paragraph: its equivalent faces are congruent, its edges of the first kind are partitioned into elliptic cycles, in such a way that the sum of the dihedral angles corresponding to edges in the same cycle is an aliquot part of 2π. Then the group corresponding to this polyhedron is completely defined. It remains to show that this group is discontinuous, i.e., that the transforms of P_0 fill the whole half space above the $\xi\eta$ plane without overlapping.

The proof is entirely the same as in the case of Fuchsian groups.

In fact, let A be any point inside P_0, and let B be a point above the $\xi\eta$ plane. We join A and B by an arc AMB not meeting this plane. This arc will leave the polyhedron P_0 by a face of the first kind, and we construct the polyhedron P_1 adjacent to P_0 along this face; the arc AMB will leave P_1 by a certain face, we construct the polyhedron P_2 adjacent to P_1 along this face, and so on.

Here is what we have to show:

1° That after a finite number of operations we arrive at a polyhedron P_n containing B;

2° That if the point B coincides with the point A, so that the arc AMB becomes a closed contour AMA, then the polyhedron P_n coincides with P_0.

The first point is established as in the theory of Fuchsian groups.

The L of the arc AMB will be a finite length L_1. I claim that this arc cannot encounter more than a finite number of polyhedra P, or, what amounts to the same thing, a finite number of faces of the first kind belonging to these polyhedra. In fact, one easily establishes the following lemmas:

I. **One can find an integer ν so large that any ν polyhedra P and any ν faces F have no common point other than a parabolic vertex.**

II. **If ν faces have no parabolic vertex, and hence no vertex at all, in common, and if an arc traverses these ν faces, then the L of that arc exceeds a certain bound λ.**

III. **Each arc which does not meet the $\xi\eta$ plane cannot traverse more than a finite number of faces F with a common**

parabolic vertex.

(See Sections I and VI of the memoir on Fuchsian groups.)

It follows that when the arc AMB has traversed $\frac{\nu L}{\lambda}$ faces F, it cannot traverse any more faces except some with a common parabolic vertex; then, by virtue of Lemma III, it will traverse only a finite number.

Now that the first point is proved, the second is established without difficulty. In fact, we see immediately that it suffices to prove it for an infinitely small contour. But the theorem is evident for such a contour, whether or not it loops around an edge of the first kind because, by hypothesis, this edge is part of a cycle whose dihedral angle sum is an aliquot part of 2π.

For the details of the proof I refer to Section VI of the memoir on Fuchsian groups.

Thus the polyhedra P_i do not overlap; if P_0 has a face R_0 of the second kind whose transforms are the faces R_i of the polyhedra P_i, then these faces R_i will not overlap. Thus **our generating polygon and its transforms do not overlap**.

This is the point which I wanted to establish but to reach it I had to resort to an artifice which I have not been able to dispense with in the general case: I have had to pass from the plane to space and from the polygons R to the polyhedra P. But if this detour is necessary in the most general case, one can do without it in a great number of particular cases; this is what we shall see later.

VI. Classification

Among the Kleinian groups, some attract particular attention because of their importance from the point of view of applications. There are the ones which include the Fuchsian groups as a particular case, inasmuch as we can pass from such a Kleinian group to a Fuchsian group by varying certain parameters **in a continuous manner**. These are the groups of the first species.

Those of the second species are the ones which do not enjoy this property.

We can also classify the Kleinian groups into genera. We define the genus of the generating polygon R_0 as in Section VII of the memoir on Fuchsian groups, and the genus of a group will be that of its generating polygon.

Now here is the analogue of the classification into families.

We first classify the generating polyhedra by the number of their faces of the second kind. This in fact is a very important point, because if a polyhedron P_0 has n faces of the second kind, then the $\xi\eta$ plane is divided into n parts, and each of these parts into an infinity of polygons R, in such a way that each transformation in the group corresponds to one and only one polygon R.

We then classify the polyhedra with the same number of faces into a second kind of family, according as they do or do not have elliptic cycles, or parabolic cycles.

We shall give this classification in detail for most important groups, which are those of the first species, retaining the same numbers for the families as in the theory of Fuchsian groups.

1° Polyhedra with two faces of the second kind.

First family. With elliptic cycles.

Second family. With parabolic cycles.

Sixth family. With elliptic and parabolic cycles.

2° Polyhedra with one face of the second kind.

Third family. Polyhedra whose faces of the first kind are all complete hemispheres, not cutting or touching each other, and without any elliptic or parabolic cycle.

Fourth family. With parabolic cycles.

Fifth family. With elliptic cycles.

Seventh family. With elliptic and parabolic cycles.

VII. Third family

Here is the manner of generation of the groups of the third family.

Consider 2n circles which neither cut nor touch; I assume, to fix ideas, that each of these circles is outside all others. The polygon R_0 will be the portion of the plane which is outside all these circles. I call the circles C_1, C_2, ..., C_n; C'_1, C'_2, ..., C'_n, where the circles C_i and C'_i are equivalent. Let S_i be a transformation which changes C_i into C'_i, in such a way that the outside of C_i becomes the inside of C'_i. S_i is then a hyperbolic or loxodromic transformation, with one fixed point inside C_i and the other inside C'_i.

The group generated by these transformations S_i is then a Kleinian group of the third family.

To show that this group is discontinuous, it is not necessary to take the roundabout route of Section V. In fact, we construct the polygon R_i adjacent to R_0 along C'_i, i.e., the transform of R_0 by the transformation S_i; it will lie entirely inside C'_i. The polygons R_0 and R_i then make up the part of the plane outside 4n − 2 circles (which are outside each other) which serve as the sides of these two polygons. Let C_k be one of these circles; if we construct the polygon R_k adjacent to R_0 or R_i along C_k, then this polygon will be entirely inside C_k, and so on. We easily see that, when this process is continued, the polygons so constructed cannot overlap, and consequently that the group is discontinuous.

Now what are the conditions imposed on the transformations S_i? These n transformations must be such that we can find n circles C_1, C_2, ..., C_n such that each of them and their transforms C'_1, C'_2, ..., C'_n is outside all the others. This amounts to some inequality conditions; thus the group generated by the n transformations is discontinuous, provided that the coefficients of the transformations satisfy certain inequalities.

Among the groups of the third family, there are some which

deserve particular mention. Assume that the polygon R_0 is symmetric with respect to a certain circle C_{n+1}, in such a way that the equivalent circles C_i and C'_i are symmetric to each other, and so that all transformations in the group can be obtained as combinations of inversions with respect to the circles $C_1, C_2, \ldots, C_n, C_{n+1}$. We then say that the group is symmetrical.

VIII. Second family

Suppose that 2n circles $C_1, C_2, \ldots, C_{2n}$ are situated so that:

1° they are all outside each other;

2° the circle C_i touches the circles C_{i-1} and C_{i+2} from outside;

3° the circle C_{2n} touches the circles C_{2n-1} and C_1 from outside.

We let A_i denote the point of contact of the circles C_{i-1} and C_i, and let A_1 be the point of contact of the circles C_{2n} and C_1. The plane becomes divided into three parts:

1° the polygon R_0 outside the circles C but inside the figure formed by these circles; this will be our generating polygon;

2° the polygon R'_0 outside all the circles and also outside the figure formed by them;

3° finally, the interior of the different circles C.

If we form the generating polyhedron P_0, then this polyhedron will have 2n faces of the first kind, formed by the surfaces of the spheres with the same centers and radii as the circles C, two faces of the second kind which will be the polygons R_0 and R'_0, and 2n isolated vertices $A_1, A_2, \ldots, A_{2n}$. I shall assume that the faces C_i and C_{2n+1-i} are equivalent and that the polyhedron has

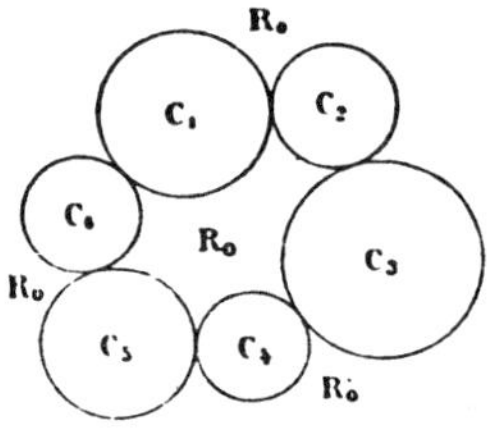

Figure 11

$n+1$ parabolic cycles, consisting respectively of the vertices A_1; A_2 and A_{2n}; ...; A_i and A_{2n+2-i}; A_n and A_{n+2}; A_{n+1}. This means that we have the relation

$$(A_2-A_1)(A_4-A_3)\ldots(A_{2n}-A_{2n-1}) = -(A_3-A_2)(A_5-A_4)\ldots(A_1-A_{2n})$$

between the vertices A.

Here, as in the preceding paragraph, the discontinuity of the group can be shown directly, without having to resort to the artifice of Section V. We see that the plane is decomposed into two domains D and D′; the first is covered by the polygon R_0 and its transforms, the second by the polygon R'_0 and its transforms. These two domains are separated by a line L, if one can call it a line.

Suppose that we have constructed a certain number of polygons R_0, R_1, ..., R_n, and that the corresponding polygons are R'_0, R'_1, ..., R'_n. We can be certain:

1° that each point of one of the polygons R is in the domain D;

2° that each point of one of the polygons R′ is in the domain D′;

3° that each vertex of one of the polygons R belongs to the line L.

The vertices of the different polygons R form **eine unendliche Punktmenge*** P, and to obtain the line L one has to adjoin to this **Punktmenge** its **erste Ableitung** P′. (*For the precise meaning of the various German expressions that I use, see G. Cantor: **Grundlagen einer allgemeinen Mannigfaltigskeitslehre**. Leipzig, Teubner, 1883. See also the French translation of this memoir: **Acta mathematica 2**, pp. 381–408.) We see that the line L is **eine perfecte und zusammenhängende Punktmenge**. This is the sense in which it is a line. However, we shall see that it does not enjoy all the properties that we usually attribute to lines.

We first investigate whether this line has a tangent. Here we must distinguish between the points of the **Punktmenge** P and those which belong to its **Ableitung** P′ without belonging to P. First consider a point of P; I claim that there is a tangent at this point.

We can first assume that this point is a vertex of R_0, since there is nothing to distinguish R_0 from the other polygons R_i; we can assume that this point is precisely A_1, since the only thing which distinguishes A_1 from the other vertices of R_0 is that the cycle containing A_1 contains no other vertex, and we have seen at the end of Section III that we can always assume that a given cycle consists of only a single vertex. Thus the group considered contains a certain parabolic transformation S which has A_1 as a fixed point. Let

$$\left(\frac{1}{z-A_1}, \frac{1}{z-A_1}+h\right)$$

be this transformation. Let N be a point such that

$$\arg(N-A_1) = -\arg h$$

We join A_1N and I claim that A_1N will be tangent to our line L. Here is what I mean by this. Suppose that ρ and ω are the polar coordinates of a point of L when A_1 is taken as the polar and A_1N is the polar axis; then I claim that when ρ tends to zero, ω tends to zero, so that A_1N will be the limit of a secant A_1B_1 of the line L as the point B_1 approaches indefinitely close to A_1.

To prove this I construct two circles K and K′ which touch externally at A_1 and which are both tangent to A_1N. I shall choose the circle K in such a way that it cuts the sides A_1A_2 and A_2A_{2n} of the polygon R_0 and does not cut any other side of this polygon. Let r_0 be the part of R_0 which is inside K and let r'_0 be the part of R'_0 which is inside K′. It is clear that the successive transforms of r_0 by the positive or negative powers of the transformation S fill the whole circle K, so that this circle is entirely contained in the domain D. Similarly, the circle K′ lies entirely inside the domain D′. It follows that the line L lies entirely in the part of the plane outside the two circles K and K′. This suffices to prove the theorem claimed.

Nevertheless, the line A_1N does not enjoy the same properties as tangents to ordinary lines. One easily sees, in fact, that if we join two points B and C of the **Punktmenge** P by a line BC, and let B and

C tend simultaneously to the point A_1, then the limit of the line BC is not in general the tangent A_1N. Consider two points B_0 and C_0 of the **Punktmenge** P, chosen in such a way that the circle $A_1B_0C_0$ is not tangent to A_1N. Consider the successive transforms

$$B_1C_1,\ B_2C_2,\ \ldots,\ B_nC_n$$

of B_0C_0 under the positive powers of S. When n increases indefinitely, B_n and C_n approach A_1, and the angle between B_nC_n and A_1N tends towards a finite limit. Moreover, I have every reason to believe that L has no tangent at points which are not in P*. (*Translator's Note: This was proved by Fricke and Klein, **Vorlesungen über die Theorie der automorphen Funktionen, Vol. 1**, Teubner, Leipzig, 1897, pp. 415–428.)

I now claim that the line L has no osculating circle. I claim that if we take a circle k tangent to the line A_1N at A_1 and passing through a point B_1 of L, then this circle does not tend to a determinate limit as the point B_1 approaches the point A_1. In fact, take two circles k′ and k″, both tangent to the line A_1N at A_1 and passing through two points, B'_0 and B''_0 respectively, of the line L. Let B'_1, B'_2, ..., B'_n be the successive transforms of B'_0 by powers of S. They will all be on k′ and will approach infinitely close to A_1 as n becomes infinite.

Thus if the osculating circle exists, it must be the circle k′. But the same is equally true of the circle k″. Hence the osculating circle does not exist at all.

I have said enough, I believe, to explain the way in which L differs from an analytic line.

IX. Symmetrical groups

Consider a polygon Π_0 bounded by n + 1 circular arcs

$$A_1A_2,\ A_2A_3,\ \ldots,\ A_{n-1}A_n,\ A_nA_{n+1},\ A_{n+1}A_1,$$

meeting at the n + 1 points A_1, A_2, ..., A_{n+1} which are the vertices of this polygon. We construct the polygons Π symmetric to Π_0 in its various sides, then the polygons symmetric to the n + 1 polygons Π in their various sides, and so on. If the polygons constructed in this way do not overlap, then we have a Kleinian group.

Let Π'_0 be the polygon symmetric to Π_0 in the side $A_{n+1}A_1$; then the polygon $\Pi_0 + \Pi'_0 = R_0$ will be generating polygon of the group. It has 2n sides; namely the n sides

$$A_1A_2,\ A_2A_3,\ \ldots,\ A_nA_{n+1}$$

of Π_0, and the sides in Π'_0 symmetric to these. Moreover, the two symmetric sides will be equivalent.

The first condition obviously necessary for the group to be discontinuous is that all the angles A_1, A_2, ..., A_n, A_{n+1} of the polygon Π_0 be zero or aliquot parts of Π. Thus they are all right angles or sharp points.

We assume that this condition is satisfied and construct the generating polyhedron P_0 of the group. To do this, we construct the spheres Σ_1, Σ_2, ..., Σ_{n+1} which have the same centers and radii as the circular arcs

$$A_1A_2,\ A_2A_3,\ \ldots,\ A_{n+1}A_1.$$

These spheres bound a certain polyhedron K_0. We then construct the polyhedron K'_0 symmetric to K_0 in the sphere Σ_{n+1}. The polyhedron

$$P_0 = K_0 + K'_0$$

will then be the generating polyhedron of the group.

The spheres Σ_{n+1} and Σ_1 cut along a circle C_1 which cuts the $\xi\eta$ plane at two points A_1 and B_1, the first of which is a vertex of Π_0. Similarly, the spheres Σ_{i-1} and Σ_i cut along a circle C_i which cuts the $\xi\eta$ plane at two points A_i and B_i, the first of which is a vertex of Π_0.

Having done this, we can make various hypotheses.

1° We can assume that the spheres Σ have no intersections other than the circles $C_1, C_2, \ldots, C_{n+1}$. This is what happens, for example, in the case of Figure 12. In this case the polyhedron K_0 has two faces of the second kind, Π_0 and Λ_0. The face Λ_0 in Figure 12 is the portion of the plane outside the polygonal contour $B_1B_2B_3 \ldots B_{n+1}$. The polyhedron K_0 has in addition n + 1 faces of the first kind which are portions of the spheres $\Sigma_1, \Sigma_2, \ldots, \Sigma_{n+1}$ bounded respectively by the circles C_1 and C_2, C_2 and C_3, ..., C_{n+1} and C_1, and n + 1 edges of the first kind which are these circles themselves. The polyhedron $P_0 = K_0 + K'_0$ has the same two faces of the second kind, 2n faces of the first kind, and 2n edges of the first kind.

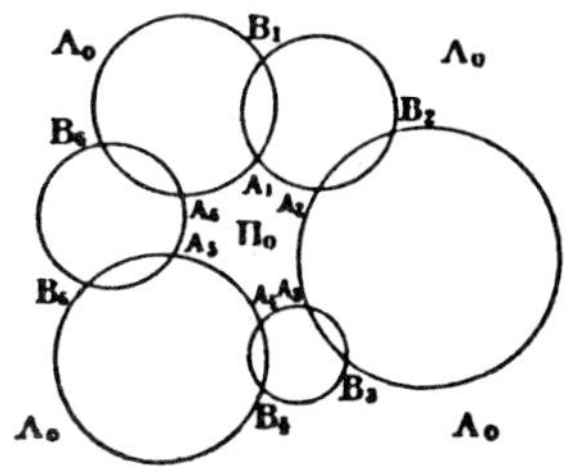

Figure 12

The conditions for discontinuity of the group are satisfied and we have a Kleinian group. Moreover, it is of the first species, since by continuously deforming the polygon Π_0, we can pass to the case where the sides of this polygon are orthogonal to a circle, i.e., to the case of a Fuchsian group. The plane is divided into two domains D and D′, the first filled by the polygon Π and its transforms, the second by the polygon Λ_0 and its transforms.

2° We can also assume that the spheres Σ have intersections other than the circles $C_1, C_2, \ldots, C_{n+1}$; for example, that they cut along other circles $k_1, k_2, \ldots, k_p$, but that these circles remain entirely outside the polyhedron K_0, so that they are not edges of this polyhedron. However, it is easy to see that this supposition is incompatible with our hypothesis that the angles of Π_0 are either right angles or sharp points.

3° We can assume that the polyhedron K_0 has edges which include, in addition to the circles C, one or more of the circles k, and that the corresponding dihedral angles are not aliquot parts of Π. It is then clear that the group is not discontinuous.

4° Finally, we arrive at the possibility that K has one or more of the circles k as edges, and that the corresponding dihedral angles are aliquot parts of Π. In this case the group is discontinuous, but we cannot pass to the case of Fuchsian groups by continuous deformation of the polygon Π_0. Thus the group is of the second species.

Take for example the case of Figure 13.

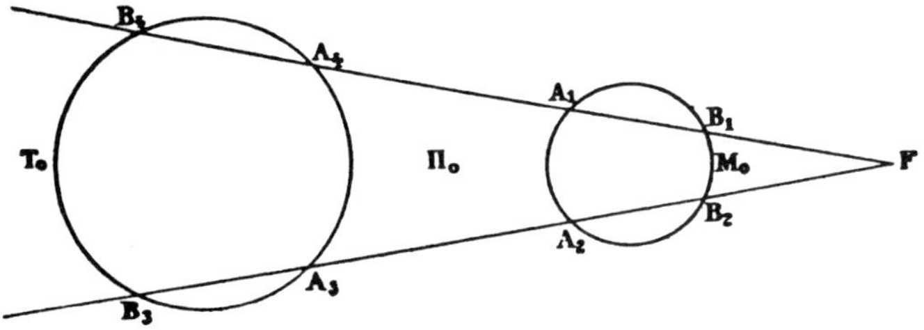

Figure 13

The polygon Π_0 is a quadrilateral $A_1A_2A_3A_4$ whose opposite sides A_1A_4 and A_2A_3 are straight lines whose prolongations cut at F. I suppose in addition that the angles A_1, A_2, A_3, A_4 and F are aliquot parts of 2π. The polyhedron K_0 then has dihedral angles equal to aliquot parts of π. It has three faces of the second kind, Π_0, M_0 and T_0, and four faces of the first kind which are respectively parts of the spheres A_1A_2 and A_3A_4, and the planes A_1A_4 and A_2A_3. The group G considered is therefore discontinuous.

If we take the polyhedron K'_0 symmetric to K_0 in the plane A_2A_3, then the union of these polyhedra will be P_0 and its faces of the second kind will be $\Pi_0 + \Pi'_0$, $M_0 + M'_0$ and $T_0 + T'_0$, where Π'_0, M'_0 and T'_0 are the polygons symmetric to Π_0, M_0 and T_0. The $\xi\eta$ plane is divided into three domains D, D′ and D″ covered respectively by the transforms of $\Pi_0 + \Pi'_0$, by those of

$M_0 + M'_0$, and by those of $T_0 + T'_0$.

We first study the domain D′. Suppose that we construct the triangles symmetric to M_0 in its three sides, then the triangles symmetric to these in their various sides, and so on. All the triangles thus constructed will cover a certain circle H which has center F and which cuts the circle $A_1A_2B_1B_2$ orthogonally. This circle H will then be a part of the domain D′, but only a part. In fact, the circle H is covered by the transforms of $M_0 + M'_0$ under certain transformations in the group G. These transformations are obtained by combining, in all possible ways, the three inversions with respect to the circles FB_1, FB_2 and B_1B_2, in such a way that total number combined is even. These transformations form a group g which is Fuchsian and which is a subgroup of the Kleinian group G. To obtain the other transforms of the quadrilateral $M_0 + M'_0$ and consequently the other parts of the domain D′, it is necessary to take the reflections of the circle H in the circle A_3A_4 and its transforms. In this way one obtains an infinity of circles H_1, H_2, ... whose union is the domain D′. This domain is therefore not a single piece.

It is the same for D″. In fact, we prove it similarly by constructing the triangles symmetric to T_0 in its three sides, then the triangles symmetric to these in their sides, and so on, obtaining all the transforms of $T_0 + T'_0$ under the transformations of a Fuchsian subgroup g_1 of the Kleinian group G. These transforms cover the portion of the plane outside a certain circle J, and hence this portion, together with the portions inside the various circles J_1, J_2, ... symmetric to J in the circle A_1A_2 and its transforms, make up the domain D″.

On the other hand, the domain D is a single piece. In fact, if we construct the polygons symmetric to Π_0 in the sides of that quadrilateral, then the polygons symmetric to these in their various sides, and so on, then we evidently obtain all the transforms of $\Pi_0 + \Pi'_0$, and consequently all the domain D. But the figure we have constructed from these polygons alongside one another is a single piece, hence this is true of D.

This domain D, moreover, is bounded by the circles H, H_1, H_2, ..., J, J_1, J_2, ...; hence **it is not simply connected**. This circumstance makes a direct proof of the discontinuity of the group almost impossible, and necessitates use of our artifice.

The existence of these domains bounded by an infinite number of circles was first pointed out by Klein.

X. First family

In Section XI of the memoir on Fuchsian groups we have considered a hexagon ABCDEF whose equivalent sides are AB and CB, CD and ED, EF and AF, and whose angles B, D, F and A + C + E are aliquot parts of 2π. We deform this hexagon in such a way that the angles continue to satisfy this condition, though the sides do not remain orthogonal to the same circle, and investigate the conditions under which it remains the generating polygon of a Kleinian group.

Consider then, the group generated by our deformed hexagon ABCDEF. It is evidently generated by three elliptic transformations S_1, S_2, S_3 which have the respective fixed points B and B′, D and D′, F and F′, and multipliers $e^{i\beta}$, $e^{i\delta}$, $e^{i\phi}$, where β, δ and ϕ denote the angles B, D and F.

S_1 changes BA into BC

S_2 changes DC into DE

S_3 changes FE into FA

The combination

$$S_1S_2S_3 = S_4$$

is also an elliptic transformation with fixed points A and A′ and multiplier $e^{-i\alpha}$, where α denotes the angle $A + C + E$.

The combinations

$$S_2S_3S_1 = S_5 \quad \text{and} \quad S_3S_1S_2 = S_6$$

also have multiplier $e^{-i\alpha}$ and the respective fixed points C and C', E and E'.

We now see what relations must hold between these fixed points and these multipliers.

Let z_1 be the transform of z by S_1, let z_2 be that of z_1 by S_2, let z_3 be that of z_2 by S_3, so that z_3 is the transform of z by S_4 and we have the relations

$$\frac{z_1-B}{z_1-B'} = e^{i\beta}\frac{z-B}{z-B'}$$

$$\frac{z_2-D}{z_2-D'} = e^{i\delta}\frac{z_1-D}{z_1-D'}$$

$$\frac{z_3-F}{z_3-F'} = e^{i\phi}\frac{z_2-F}{z_2-F'} \tag{1}$$

$$\frac{z_3-A}{z_3-A'} = e^{-i\alpha}\frac{z-A}{z-A'}$$

Differentiating the relations (1), we find

$$
(2)\qquad
\begin{array}{ll}
\dfrac{dz_1}{(z_1-B')^2} = e^{i\beta}\,\dfrac{dz}{(z-B')^2}, & \dfrac{dz_1}{(z_1-B)^2} = e^{-i\beta}\,\dfrac{dz}{(z-B)^2} \\[2ex]
\dfrac{dz_2}{(z_2-D')^2} = e^{i\delta}\,\dfrac{dz_1}{(z_1-D')^2}, & \dfrac{dz_2}{(z_2-D)^2} = e^{-i\delta}\,\dfrac{dz_1}{(z_1-D)^2} \\[2ex]
\dfrac{dz_3}{(z_3-F')^2} = e^{i\phi}\,\dfrac{dz_2}{(z_2-F')^2}, & \dfrac{dz_3}{(z_3-F)^2} = e^{-i\phi}\,\dfrac{dz_2}{(z_2-F)^2} \\[2ex]
\dfrac{dz_3}{(z_3-A')^2} = e^{-i\alpha}\,\dfrac{dz}{(z-A')^2}, & \dfrac{dz_3}{(z_3-A)^2} = e^{i\alpha}\,\dfrac{dz}{(z-A)^2}
\end{array}
$$

Among the relations (2) we consider the first three in the second column and the last in the first column, and put $z = A$, whence $z_1 = C$, $z_2 = E$, $z_3 = A$; then, combining the resulting relations so as to eliminate dz, dz_1, dz_2, dz_3, we obtain

$$
\frac{(A-B)^2(C-D)^2(E-F)^2}{(C-B)^2(E-D)^2(A-F)^2} = e^{i(\alpha-\beta-\delta-\phi)}
$$

or

$$
(3)\qquad \frac{(A-B)(C-D)(E-F)}{(B-C)(D-E)(F-A)} = \pm\, e^{i\frac{\alpha-\beta-\delta-\phi}{2}}
$$

An easy discussion shows that it is the $-$ sign which fits.

In the case where the four angles α, β, δ and ϕ are zero, i.e., in the case where the group reduces to the second family, the relation (3) becomes $(A-B)(C-D)(E-F) = -(B-C)(D-E)(F-A)$.

Suppose then that we are given, in addition to the angles α, β, δ, ϕ six fixed points A, B, C, D, E, F which satisfy the relation (3). The three fixed points B', D', F' are then determined by the equations

$$\frac{C-B}{C-B'} = e^{i\beta} \frac{A-B}{A-B'}, \quad \frac{E-D}{E-D'} = e^{i\delta} \frac{C-D}{C-D'}, \quad \frac{A-F}{A-F'} = e^{i\phi} \frac{E-F}{E-F'}$$

and the three fixed points A′, C′, E′ by the relations

$$\frac{C'-B}{C'-B'} = e^{i\beta} \frac{A'-B}{A'-B'}, \quad \frac{E'-D}{E'-D'} = e^{i\delta} \frac{C'-D}{C'-D'}, \quad \frac{A'-F}{A'-F'} = e^{i\phi} \frac{E'-F}{E'-F'}$$

In general, the four points A, B, A′, B′ will not end up on the same circle. Likewise with the points B, B′, C, C′ and C, C′, D, D′, etc. It follows that, if we construct the generating polyhedron P_0 of the group, the face of the second kind on this polyhedron will in general not be the hexagon ABCDEF, but a dodecagon, as we shall see.

In fact, here is how we can construct the polyhedron P_0:

We consider six circles K_1, K_2, ..., K_6 passing respectively through A and A′, B and B′, C and C′, D and D′, E and E′, F and F′. The first two cut at H_1 and H'_1, the next two at H_2 and H'_2, the last two at H_3 and H'_3. Now let K'_1 and K'_2 be the transforms of K_1 and K_2 by S_1; the first passes through C and C′, the second through B and B′ , and they cut at J_1 and J'_1.

Similarly, let K'_3 and K'_4 be the transforms of K_3 and K_4 by S_2, and let K'_5 and K'_6 be the transforms of K_5 and K_6 by S_3. These four circles pass respectively through E and E′, D and D′, A and A′, F and F′; the first two cut at J_2 and J'_2, and the last two cut at J_3 and J'_3.

I assume that these different circles have no points of intersection other than those I have enumerated, and that the relative positions of these different circles and points are as shown in Figure 4.

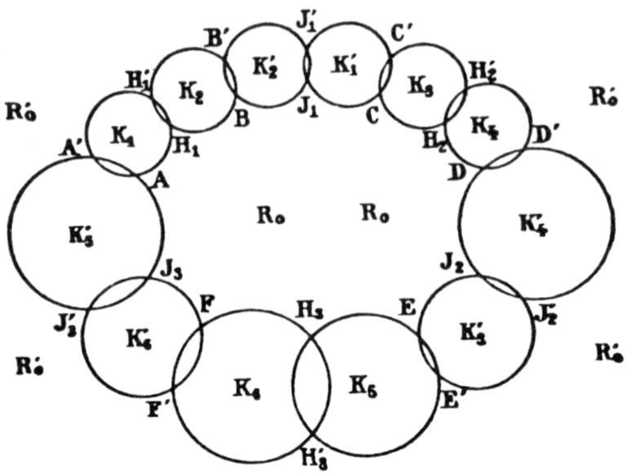

Figure 14

In this case, we construct the spheres Σ_i and Σ'_i with the same centers and radii as the circles K_i and K'_i. Next consider the polyhedron P_0 formed by the portion of space outside these twelve spheres. This polyhedron has two faces of the second kind, R_0 and R'_0, which are the respective portions of the plane inside and outside the ring formed by the twelve circles K_i and K'_i; it has twelve faces of the first kind formed by portions of the spheres Σ_i and Σ'_i; these twelve faces are equivalent in pairs, with Σ'_i equivalent to Σ_i. The polyhedron P_0 has twelve edges of the first kind, formed by the intersections in pairs of the spheres Σ_i and Σ'_i; these edges fall into seven cycles as indicated in the following table:

Cycle Number	Edges making up the cycle	Dihedral angle sum of the cycle
1	AA', CC', EE'	α
2	BB'	β
3	DD'	δ
4	FF'	ϕ
5	$H_1H'_1, J_1J'_1$	2π
6	$H_2H'_2, J_2J'_2$	2π
7	$H_3H'_3, J_3J'_3$	2π

The first four cycles correspond to the elliptic transformations S_4, S_1, S_2 and S_3, the last three to the identity transformation. Thus if the circles K_i and K'_i are in the relative positions indicated by Figure 4, then the group considered, whose generating polyhedron is P_0, will be discontinuous.

Our group will therefore be Kleinian, provided only that the fixed points A, B, C, D, E, F satisfy the relation (3) and that one has the inequalities which express the relative positions of the twelve circles K_i and K'_i shown in Figure 4.

The plane becomes divided into two domains bounded by a line L. Here is how the line L can be generated: we consider the set of fixed points of all hyperbolic and loxodromic transformations in the group. These fixed points form a **Punktmenge** P; if we adjoin its **erste Ableitung** P′, then we have the line L.

XI. Kleinian functions

Let G be an arbitrary Kleinian group and let

$$\left(z, \frac{\alpha_i z+\beta_i}{\gamma_i z+\delta_i}\right)$$

be the different transformations in the group. As in the theory of Fuchsian functions, we form the following series

$$\Theta(z) = \Sigma\, H\left(\frac{\alpha_i z+\beta_i}{\gamma_i z+\delta_i}\right)(\gamma_i z+\delta_i)^{-2m} \tag{1}$$

where the algorithm H(z) represents a rational function of z, each infinity of which coincides with a singular point of the group, and m denotes an integer greater than 1.

This series is convergent. In Section I of the memoir on Fuchsian functions we have given two proofs of the convergence of this series. The first of these proofs remains valid in the present case, though the second does not.

Thus, corresponding to each Kleinian group, there is an infinity of theta–Kleinian functions $\Theta(z)$ and Kleinian functions F(z) enjoying the properties

$$\Theta\left(\frac{\alpha_i z+\beta_i}{\gamma_i z+\delta_i}\right) = \Theta(z)(\gamma_i z+\delta_i)^{2m}$$

$$F\left(\frac{\alpha_i z+\beta_i}{\gamma_i z+\delta_i}\right) = F(z)$$

These functions enjoy the same properties as Fuchsian and theta–Fuchsian functions. Consequently, all Kleinian functions can be expressed rationally in terms of two particular ones x and y, between which there is an algebraic relation

$$f(x,y) = 0 \tag{2}$$

whose genus is that of the group G. When this genus is zero, all the Kleinian functions can be expressed rationally in terms of one particular one, which I call x.

If I put

$$\nu_1 = \sqrt{\frac{dz}{dx}}, \qquad \nu_2 = z\sqrt{\frac{dz}{dx}}$$

then ν_1 and ν_2 are two integrals of the linear equation

$$\frac{d^2\nu}{dx^2} = \phi(x,y)\nu \tag{3}$$

which reduces to

$$\frac{d^2\nu}{dx^2} = \phi(x)\nu \tag{3'}$$

in the case where the genus is zero (cf. Section IV of the memoir on

Fuchsian functions); in these equations, ϕ denotes a rational function.

We now examine some particular cases, and first recall equation (3′) from Sections V and VII of the memoir on Fuchsian functions:

$$\frac{d^2\nu}{dx^2} = \frac{P(x)}{Q^2(x)}\,\nu$$

where

$$Q(x) = (x-a_1)(x-a_2)\ldots(x-a_n)$$

The polynomial P(x) is of degree 2n − 2 and I assume that it satisfies the following n + 1 conditions

$$P(a_1) = -\,Q'^2(a_i)\left[\frac{1}{4} - \frac{1}{4\beta_i^2}\right]$$

$$\text{coefficient of } x^{2n-2} \text{ in } P(x) = -\left[\frac{1}{4} - \frac{1}{4\beta_{n+1}^2}\right] \tag{4}$$

The numbers β_1, β_2, …, β_{n+1} are positive integers which can become infinite.

We have seen that if the polynomial P(x) satisfies (in addition to the n + 1 conditions (4) which are **complex**, and consequently equivalent to 2n + 2 real conditions) 2n − 4 real and transcendental conditions, then the variable x is a Fuchsian function of the ratio of the integrals.

It follows from the preceding theory that if the polynomial satisfies, not only the conditions (4), but also certain **inequalities**, then the variable x will be a Kleinian function of the ratio of the integrals. Suppose, in fact, that the conditions (4) are satisfied; we let z be the ratio of the integrals of equation (3); when x returns to its initial value, after describing a closed contour C_i, z will be changed into

$\frac{\alpha_i z+\beta_i}{\gamma_i z+\delta_i}$, and the transformations

$$\left(z, \frac{\alpha_i z+\beta_i}{\gamma_i z+\delta_i}\right)$$

will form a group G.

If this group is Kleinian, then x will be a Kleinian function of z. But this group G, as one easily sees, is generated by n elliptic or parabolic transformations S_1, S_2, ..., S_n with the respective multipliers

$$e^{\frac{2i\pi}{\beta_1}}, e^{\frac{2i\pi}{\beta_2}}, e^{\frac{2i\pi}{\beta_3}}, \ldots, e^{\frac{2i\pi}{\beta_n}}$$

The combination

$$S_1 S_2 \ldots S_n$$

will also be an elliptic or parabolic transformation with multiplier

$$e^{-\frac{2i\pi}{\beta_{n+1}}}$$

But we have seen in the preceding paragraph that for such a group to be discontinuous it is sufficient that its coefficients satisfy certain inequalities. It therefore suffices to impose certain inequalities on the coefficients of P(x) in order to make x a uniform function of z.

Now recall equation (3) of Section VI of the memoir on Fuchsian functions

$$(3) \qquad \frac{d^2\nu}{dx^2} = \phi(x,y)\nu, \qquad \psi(x,y) = 0$$

I let c_i, d_i denote the analytic points at which we simultaneously have

$$\frac{d\psi}{dy} = 0, \qquad \frac{d\psi}{dx} \gtrless 0$$

I let a_i, b_i denote the analytic points different from c_i, d_i at which the function ϕ becomes infinite. I suppose that the function ϕ satisfies the following conditions:

$$\lim (y-d_i)^2 \left[\frac{dx}{dy}\right]^2 \phi(x,y) = -\frac{3}{4} \quad (\text{for } x = c_i,\ y = d_i)$$

(5)

$$\lim (x-a_i)\phi(x,y) = \frac{1-\beta_i^{\,2}}{4\beta_i^{\,2}} \quad (\text{for } x = a_i,\ y = b_i)$$

Here the numbers β_i are again positive integers which can become infinite.

We have seen that if the function ϕ satisfies in addition certain transcendental conditions, then x is a Fuchsian function of the ratio z of the variables. Similarly, if this function ϕ satisfies, not only equations (5), but also certain inequalities, then x will be a Kleinian function of z. The proof is entirely analogous to the preceding one.

Thus if x in equation (3) is to express a Kleinian function of z in the first, second or sixth family, it is sufficient that certain **algebraic equations** and certain **inequalities** hold. We do not have to impose a **transcendental equation**. It is not the same for the functions in the other families. For example, if we want x in equation (3) to be a Kleinian function, in the third family, of the ratio z of the integrals, then it is necessary to impose certain transcendental equations.

Let us go back, in fact, to equation (3) of Section VIII of the memoir on Fuchsian functions:

$$\frac{d^2\nu}{dx^2} = \phi(x,y)\nu, \qquad \psi(x,y) = 0$$

Supposing that the function ϕ satisfies the conditions stated on page 236 of that section (lines 11 onward). Let n be the genus of the relation $\psi = 0$; then the Kleinian group of our equation (3) must be of genus n, and consequently it will depend on $3n - 3$ complex parameters, i.e., precisely as many parameters as there are moduli in the relation $\psi = 0$. Thus when we are given this relation $\psi = 0$, the Kleinian group will be completely determined, and likewise the function ϕ. It follows that this function is subject, independently of the algebraic conditions on page 236 of the memoir on Fuchsian functions, to n transcendental conditions, because these algebraic conditions will not suffice to determine it.

XII. Historical

It was Schottky who first noted the discontinuity of certain Kleinian groups (**Journal für die reine und angewandte Mathematik, 83,** p.346), namely the symmetrical groups of the third family. Later, Klein went deeply into the theory of these groups in various notes inserted in the **Mathematische Annalen** (Vols. 19, 20 and 21) and in a memoir also in Volume 21 of the same annals, entitled **Ueber Riemannsche Functionentheorie**.

I have myself, in two notes I had the honor of presenting to the Académie des Sciences de Paris on 27 June and 11 July 1881 (see **Comptes rendu**, Vols. 92 and 93), succinctly announced the greater part of the results given in the present memoir.

Extract from an unedited memoir of Henri Poincaré

on Fuchsian functions

Acta Mathematica, 39(1923), 58-93

The Academy of Sciences of Paris proposed, as subject for the 1880 prize in mathematical sciences the following question: "To perfect, in some important point, the theory of linear differential equations of a single variable." The memoir before us was entered in the competition; it is very interesting because it permits us to follow Poincaré's ideas on Fuchsian functions from their origins.

The memoir consists of two distinct parts. The first part, which we do not reproduce, contains the researches on the irregular integrals of linear differential equations which Poincaré developed with more details in two memoirs published respectively in the **American Journal of Mathematics**, **Vol. 7**(1885), 203—258, and in **Acta Mathematica**, **Vol. 8**(1886), 295—344. As for the second part, Poincaré says that it contains the reflections inspired by his reading of a memoir of L. Fuchs. Poincaré had received the latter memoir at the beginning of May 1880, and his own memoir arrived at the Academy on May 28, 1880. We see the extreme rapidity with which Poincaré worked.

Letting f(x) and ϕ(x) denote two integrals of a linear second order differential equation and setting

$$\frac{f(x)}{\phi(x)} = z$$

Fuchs remarked that, under certain conditions, x is a meromorphic function of z. Poincaré showed that the conditions stated by Fuchs are not sufficient. Poincaré is concerned mainly with the case where the integration of the equation can be carried out by means of doubly periodic functions and the case where the differential equation admits no more than two singular points at a finite distance. It is from a deep study of these two cases that Poincaré is led to create the general theory of functions which he named after Fuchs.

N. E. Nörlund

In the latest issue of **Borchardt's Journal, Vol. 89**(1880), 151–169, Fuchs has published a memoir whose contents are summarized in a letter by Hermite in **Comptes rendus, Vol. 90**(1880), 678–680. This memoir concerns equations of second order. I suppose that the differential equation is reduced to the canonical form

$$\frac{d^2y}{dx^2} = Qy$$

Fuchs shows that, **under certain conditions**, if f(x) and $\phi(x)$ are two integrals of the given equation then:

1° If we put

$$\frac{f(x)}{\phi(x)} = z$$

x is a meromorphic function of z;

2° If we put

$$\int_0^x f(x)dx + \int_0^{x_1} f(x_1)dx_1 = u_1$$

$$\int_0^x \phi(x)dx + \int_0^{x_1} \phi(x_1)dx_1 = u_2$$

each rational symmetric function of x and x_1, is a meromorphic function of u_1 and u_2.

The latter result permits him to define functions analogous to Abelian functions, but here I am concerned with the first, which permits the definition of functions analogous to doubly periodic functions.

For this first result to be true, the conditions of Fuchs are not necessary and sufficient; what is lacking, if x is to be a meromorphic function, is that at all the singular points, including the point ∞, the difference between the roots of the determinant equation must be an aliquot part of unity.

In fact, let

$$z = a$$

be any value of z not corresponding to a singular point of the given equation. We have

$$f(x) - z\phi(x) = 0$$

hence if we expand x in powers of $z - a$ we see that the two expressions

$$f(x) - a\phi(x)$$

$$f'(x) - a\phi'(x)$$

do not simultaneously vanish. But since we have assumed that the value $z = a$ corresponds to a value of x which is not a singular point of the given equation, these two expressions cannot vanish for that value of x unless

$$f(x) - a\phi(x)$$

is identically zero, i.e., unless $f(x)$ and $\phi(x)$ are linearly dependent, which is contrary to my assumption.

Now suppose that a corresponds to a singular point $x = b$ situated at a finite distance: we then have functions $f_1(x)$ and $f_2(x)$ of x which are holomorphic and $\neq 0$ for $x = b$; α, β, γ, δ are the constants and ρ_1 and ρ_2 the roots of the determinant equation

$$\alpha(x-b)^{\rho_1}f_1(x) + \beta(x-b)^{\rho_2}f_2(x) +$$

$$+ z[\gamma(x-b)^{\rho_1}f_1(x) + \delta(x-b)^{\rho_2}f_2(x)] = 0$$

or, if real part of ρ_1 > real part of ρ_2,

$$(x-b)^{\rho_1-\rho_2}f_1(x)(\alpha+\gamma z) + f_2(x)(\beta+\delta z) = 0$$

For $z = a$ we must have $x = b$, i.e.,

$$\beta + \delta a = 0$$

Moreover, we have

$$\alpha + \gamma a \neq 0$$

otherwise

$$\frac{\alpha}{\beta} = \frac{\gamma}{\delta}$$

contrary to hypothesis, and then

$$(x-b)^{\rho_1-\rho_2} = -\frac{f_2(x)}{f_1(x)}\frac{\delta(z-a)}{\alpha + \gamma a + \gamma(z-a)}$$

or

$$(\rho_1-\rho_2)\log(x-b) = \log(z-a) + \log\frac{-\delta f_2(x)}{f_1(x)[\alpha+\gamma a+\gamma(z-a)]}$$

or

$$\frac{\log(z-a)}{\log(x-b)} = \rho_1 - \rho_2 - \frac{\log \dfrac{-\delta f_2(x)}{f_1(x)[\alpha+\gamma a+\gamma(z-a)]}}{\log(x-b)}$$

or, for z = a, x = b

$$\lim \frac{\log(z-a)}{\log(x-b)} = \rho_1 - \rho_2$$

But if x is a holomorphic function of z, for z = a we have

$$x - b = A_n(z-a)^n + A_{n+1}(z-a)^{n+1} + \ldots$$

or

$$\lim \frac{\log(z-a)}{\log(x-b)} = \frac{1}{n}$$

hence

$$\rho_1 - \rho_2 = \frac{1}{n}$$

Thus **the difference between the roots of the determinant equation must be an aliquot part of unity**.

Conversely, if

$$\rho_1 - \rho_2 = \frac{1}{n}$$

we have

$$f(x,z) = (x-b)f_1^n(x)(\alpha+\gamma z)^n + (-1)^{n-1}f_1^n(x)(\beta+\delta z)^n = 0$$

and we extract x as a holomorphic function of z, since for z = a

$$\frac{dF}{dx} = f_1^n(x)(\alpha+\gamma a)^n$$

The reasoning is the same if $x = \infty$ for z = a.

Consequence: for x to be a meromorphic function of z whenever z takes a corresponding value, whether it is a finite value of x which is non–singular, or a finite value of x which is a singular point, or an infinite value of x;

it is necessary and sufficient that for all singular points, including the point ∞, $\rho_1 - \rho_2$ be an aliquot part of unity.

These conditions are then necessary for x to be a meromorphic function of z over the whole plane.

Are they sufficient? They would be if we were able to obtain all the values of z by letting x describe a **finite** number of **finite** contours on the sphere. This is what Fuchs seems to have assumed without proof.

If this is the case, then if x describes any contour in the plane not crossing any of the cuts (which we can make between the singular points) more than a finite number of times, z will take all possible values, and the function x of z will be meromorphic not only over the whole plane **but over the whole sphere**, and consequently **rational.**

It follows that the equation (1) will admit an algebraic integral, which happens sometimes, but not always, as Fuchs himself shows.

Thus by describing a certain contour (around more than one singular point) an infinite number of times, we arrive at a certain singular value of z which cannot be obtained by describing the finite contours a finite number of times.

If there are no more than one or two of these singular values of z on the sphere, then there is no difficulty.

In fact, if α and β are the two singular values we put

$$z = \frac{\beta e^{it} + \alpha}{e^{it} + 1}$$

Then z cannot be equal to α or β for any finite value of t, thus for all finite values of t, x is a meromorphic function of z and consequently of t. Hence x is a monodromic* function of t over the whole plane and consequently over the whole sphere. (*Translator's Note: I.e., single valued.) Change of variable has therefore reduced us to the case where x is meromorphic over the whole plane, except only that x is a meromorphic function of t and not of z; for this to be true of z also, it is necessary that x, as a function of t, have period 2π, which we cannot expect **a priori**.

If there are more than two singular values on the sphere, such a device is not applicable. Whatever change of variable is made, at least one singular point at a finite distance remains, and if x is to be a monodromic function over the whole plane it is necessary that it be monodromic in the neighborhood of this singular point. However, **the proof of Fuchs does not apply to such points.**

This objection does not occur in analogous proofs that we encounter in the theory of elliptic or Abelian functions. For example, suppose that

$$\int_0^x \frac{dx}{\sqrt{(1-x^2)(1-k^2x^2)}} = z$$

One easily proves that x is meromorphic in z for all values of z that we can obtain by letting x describe a finite contour a finite number of times on the sphere. We can conclude that x is monodromic over the whole plane, and consequently over the whole sphere, since z tends to ∞ when we let x describe an arbitrary contour an infinite number of times.

Nothing like this occurs in the theory of linear differential equations.

I believe I have shown that the proof of Fuchs is insufficient. We now consider the question again from another point of view.

The differential equation can always be put in the form

(1)
$$\frac{d^2y}{dx^2} = Qy$$

where Q is a function of x.

Putting $\frac{dx}{dz} = t$, we find the following differential equations between the functions x, y, z, t:

$$\frac{dx}{dz} = y^2, \quad \frac{dy}{dz} = ty^2, \quad \frac{dt}{dz} = Qy^3$$

For x to be a meromorphic function of z, it is necessary and sufficient that **whenever** z **is finite, all** the relations between x and z derived from these differential equations be of the form

$$x = \text{monodromic function of } z$$

or

$$z = \text{constant.}$$

But x, y, t are meromorphic in z except when
1° $Q = \infty$;
2° $x = \infty$;
3° $y = \infty$;
4° $z = \infty$.

Fuchs has examined the first two exceptions; it remains to examine the other two. Suppose then that $y = \infty$. How can y become infinite? Suppose that x describes a certain contour C an infinite number of times, that when x describes this contour once, the two integrals f and ϕ of equation (1) are changed respectively into αf and $\beta\phi$, and that

$$y = \lambda f + \mu\phi$$

When x describes the contour C, m times, y will change into

$$\lambda\alpha^m f + \mu\beta^m\phi$$

but it follows from the particular form of equation (1) that

$$\alpha\beta = 1$$

Then, unless $|\alpha| = 1$

$$\lim_{m\to\infty} y = \infty$$

Suppose than that $y = \infty$. We put $y = \frac{1}{\eta}$ and the differential equations become

$$\frac{dx}{\eta} = \frac{d\eta}{-\eta^3 t} = \frac{dt}{Q} = \frac{dz}{\eta^3}$$

whose integrals, if $Q \neq 0$, $t \neq 0$, reduce to

$$\eta = 0, \quad x = \text{const.}, \quad z = \text{const.}$$

The relation between x and z here reduces to z = const., and there are no difficulties.

Suppose then that Q = 0 and put

$$\eta = \eta_1 z^{1/2}, \quad t = t_1 z^{-1/2}$$

giving

$$\frac{dz}{\eta_1{}^3 z} = \frac{dz}{\eta_1} = \frac{d\eta_1}{-\eta_1{}^3 t_1 - 1/2\eta_1{}^4} = \frac{dt_2}{Q + 1/2 t_1 \eta_1{}^3}$$

It remains to show that x remains holomorphic in z when

$$\eta_1 = Q = 0, \quad t_1 \neq \infty$$

and this is what Fuchs has not done.

It is necessary after this to examine the following cases:

$t = \infty$

$y = Q = \infty$, $y = t = q = \infty$, $y = x = \infty$

$y = t = x = \infty$

These considerations show, I think, the insufficiency of Fuchs' proof and the need for a deeper study of the question.

Consider first an example cited by Fuchs, namely equation (1) **Borchardt's Journal**, Part 2, **Vol. 89**, p.168.

This equation has three singular points:

$$a_1,\ a_2 \text{ and } \infty$$

The difference between the roots of the fundamental determinant equation is

$$1/3 \text{ for } a_1,$$

$$1/6 \text{ for } a_2,$$

$$1/2 \text{ for } \infty .$$

We trace two cuts on the x sphere, one going from a_2 to a_1, the other from a_2 to infinity.

Let $z = \frac{f(x)}{\phi(x)}$, where $f(x)$ and $\phi(x)$ are two integrals of equation (1) which can always be chosen so that z changes into $-z$ when x turns around infinity; i.e., when it describes a closed contour crossing the second cut.

When x crosses the first cut, so as to turn around a_1, z changes into z', where z' is related to z by an equation of the form

$$\frac{z'-\alpha}{z'-\beta} = e^{\frac{2i\pi}{3}} \frac{z-\alpha}{z-\beta}$$

Then when x turns around a_2 in such a way as to cross the two cuts in succession, z is changed into z'', where

$$\frac{z''-\alpha}{z''-\beta} = e^{\frac{2i\pi}{3}} \frac{z+\alpha}{z+\beta} \tag{2}$$

But since the roots of the determinant equation for a_2 have difference 1/6, z must be related to z'' by an equation of the form

(3) $$\frac{z''-\gamma}{z''-\delta} = e^{\frac{\pi i}{3}} \frac{z-\gamma}{z-\delta}$$

Identifying equations (2) and (3), we find by easy algebraic calculation that

$$\alpha \text{ or } \beta = 0, \quad \text{or } \alpha = 0$$

and

$$\gamma = 0$$

If we then put

$$z = 1/t$$

x will be a function of t which does not change when we change

$$t \text{ into } -t$$

$$t \text{ into } \beta + e^{\frac{2i\pi}{3}}(t-\beta)$$

$$t \text{ into } \delta + e^{\frac{i\pi}{3}}(t-\delta)$$

whence we conclude that this function does not change when we change

$$t \text{ into } t + 4\beta\left[1 - e^{\frac{2i\pi}{3}}\right]$$

or

$$t \text{ into } t + 4\beta\left[t - e^{\frac{2i\pi}{3}}\right]$$

Moreover, in making an infinite number of changes, of

$$t \text{ into } -t$$

or

$$t \text{ into } \beta + e^{\frac{2 i \pi}{3}} (t-\beta)$$

or

$$t \text{ into } \delta + e^{\frac{i \pi}{3}} (t-\delta)$$

we either make t tend to infinity or else remain in a cycle consisting of a finite number of values of t. It follows that we have

$$t = \infty$$

as the only singular point. But x cannot cease to be monodromic in t except for values of t which correspond to singular points; x is therefore monodromic for all finite values of x, and hence meromorphic over the whole plane.

Consequently, x **is a doubly periodic function of** t.

The figure gives an idea of the properties of this function.

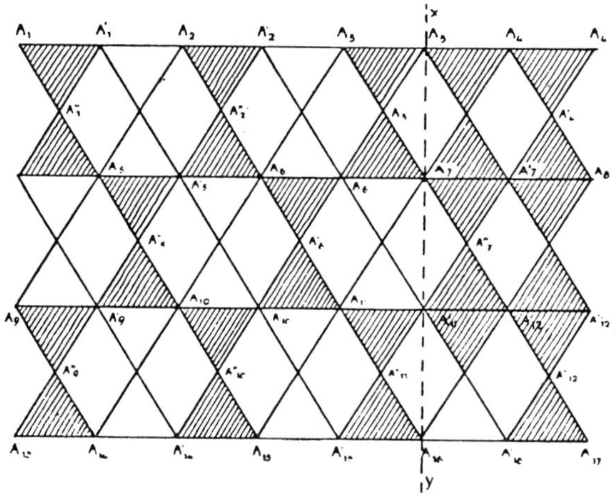

Figure 1.

I ask you to neglect the part to the right of the line XY, where the hatching is misplaced. In this figure the period parallelograms are

$$A_1A_2A_5A_6,\ A_2A_3A_6A_7,\ A_5A_6A_{10}A_{11},\ \text{etc.}$$

The parallelograms are the lozenges formed by two equilateral triangles. We decompose each of them into two equilateral triangles, each 1/8 of the surface of the parallelogram, and hatched, and a regular hexagon which remains white.

The function x does not change when t turns:

1° 180° around the vertex of a hatched triangle;

2° 120° around the center of one of these triangles;

3° or 60° around the center of one of the white regular hexagons.

When we know x as a function of t, equation (1) is easily integrated; in fact we have the integrals

$$\sqrt{\frac{dx}{dt}},\quad t\sqrt{\frac{dx}{dt}}$$

for it. But if x is a doubly periodic function of t, x will be **connected with $\frac{dx}{dt}$ by an algebraic equation. Thus one of the integrals will be algebraic in** x. In fact, if we form equation (1) we find

$$\frac{d^2y}{dx^2} = \left[-\frac{2}{9(x-a_1)^2} + \frac{5}{36(x-a_1)(x-a_2)} + \frac{35}{144(x-a_2)^2}\right]y$$

the integrals of which are evidently

$$y_1 = (x-a_1)^{\frac{1}{3}}(x-a_2)^{\frac{5}{12}}$$

and

$$y_2 = (x-a_1)^{\frac{1}{3}}(x-a_2)^{\frac{5}{12}} \int (x-a_1)^{-\frac{2}{3}}(x-a_2)^{-\frac{5}{6}}dx$$

We then have

$$y_2/y_1 = t = \int (x-a_1)^{-\frac{2}{3}}(x-a_2)^{-\frac{5}{6}}dx$$

whence we extract x as a doubly periodic function of t.

Remark. Equation (1) therefore admits no more than one algebraic integral and it does admit one; in fact it belongs to a very numerous class of differential equations which each have a single algebraic integral.

Let

$$\frac{d^2y}{dx^2} = y\left[\Sigma \frac{A_i}{(x-a_i)^2} + \Sigma \frac{2B_{ik}}{(x-a_i)(x-a_k)}\right] \tag{4}$$

This equation will admit an algebraic integral provided that we have

$$B_{ik} = \left[\frac{1}{2} \pm \sqrt{\frac{1}{4} + A_i}\right]\left[\frac{1}{2} \pm \sqrt{\frac{1}{4} + A_k}\right]$$

The second integral is found by a simple quadrature.

The preceding example shows that in certain cases the theorem of Fuchs is exact, and that x is a doubly periodic function of z.

We shall investigate how this comes about; we propose to find the cases in which x is a function of z capable of being reduced to doubly periodic functions.

In what amounts to the same thing, we investigate the cases in which x is a function of z which has no more than **one** or **two** singular points on the z sphere.

Supposing that the theorem of Fuchs is true, i.e., that x is a monodromic function of z, it will be a function of z which is reproduced when z is changed into z′, where

$$z' = \frac{az+b}{a'z+b'} \tag{5}$$

(and this will hold for an infinity of systems of values of a, b, a′, b′).

But the relation (5) between z and z′ can always be put in the form:

$$\frac{z'-\alpha}{z'-\beta} = \lambda \frac{z-\alpha}{z-\beta} \tag{6}$$

or in the form

$$\frac{1}{z'-\alpha} = \frac{1}{z-\alpha} + \lambda \tag{7}$$

First case

Suppose that x does not change when z changes into z'_0 or z'_1, or z'_2, ..., and that **all** the quantities z'_0, z'_1, z'_2, ... are connected to z by relations which can be put in the form (6) with

$$\lambda = e^{2i\pi h}$$

where h is rational. In that case, the number of quantities z'_0, z'_1, z'_2, ... is necessarily bounded and x will be a rational function of x. Equation (1) will be algebraically integrable.

Second case.

x does not change when we change z into z′, where

$$\frac{z'-\alpha}{z'-\beta} = \lambda \frac{z-\alpha}{z-\beta} \qquad |\lambda| \neq 1$$

We put

$$\frac{z-\alpha}{z-\beta} = t$$

and x will be a monodromic function of t which does not change when we change t into λt.

There are then two singular points

$$t = 0, \quad t = \infty$$

An infinity can no longer be avoided, since the points $t = 0$, $t = \infty$ are the only ones reproduced when we change t into λt. When x turns around one of the singular points of equation (1), t will change into t', where

$$\frac{t'-a}{t'-b} = \lambda_1 \frac{t-a}{t-b} \tag{8}$$

Here

$$\lambda_1 = e^{\frac{2i\pi}{n}}$$

where n is an integer. If we are to have no more than a finite number of singular points, then it is necessary that the transformation (8) reproduce the system of points

$$t = 0, \quad t = \infty$$

This can happen in two ways:

1° If the transformation (8) reproduces the point $t = 0$, and likewise reproduces the point $t = \infty$.

For this, the transformation (8) must be

$$t' = te^{\frac{2i\pi}{n}}$$

2° If the transformation (8) changes $t = 0$ into $t = \infty$, and $t = \infty$ into $t = 0$.

Such a change cannot occur unless transformation (8) changes t into t' and t' into t, i.e., unless

$$\lambda_1 = -1$$

If we put

$$\lambda_1 = -1, \quad t = 0, \quad t' = \infty$$

in equation (8) we get

$$a/b = -1$$

whence

$$\frac{t'-a}{t'+a} = -\frac{t-a}{t+a}$$

or

$$tt' - a^2 = 0$$

or

$$t' = \frac{a^2}{t}$$

Thus if we are to have no more than a finite number of singular points it is necessary that, when x turns around one of the singular points of equation (1), t change into

$$te^{\frac{2i\pi}{n}} \quad \text{or} \quad \frac{a^2}{t}$$

We therefore have a monodromic function of t which does not change when we change

$$t \text{ into } \lambda t$$

or t into

$$te^{\frac{2i\pi}{n_1}}, \quad te^{\frac{2i\pi}{n_2}}, \ldots, te^{\frac{2i\pi}{n_h}}$$

or t into

$$\frac{K_1}{t}, \frac{K_2}{t}, \ldots, \frac{K_p}{t}$$

and consequently when we multiply t by

$$\frac{K_1}{K_2}, \frac{K_1}{K_3}, \ldots, \frac{K_1}{K_p}$$

or again by

$$e^{\frac{2i\pi}{m}}$$

where m is the least common multiple of $n_1, n_2, \ldots, n_k$.

If we now put

$$t = e^u$$

then x will be a monodromic function of u which will not change when we change

$$u \text{ into } u + \log \lambda$$

$$u \text{ into } u + \frac{2i\pi}{m}$$

or

$$u \text{ into } u + \log K_1 - \log K_2 \quad \text{or } u + \log K_1 - \log K_3, \ldots$$

or into

$$u + \log K_1 - \log K_p.$$

This function therefore admits a certain number of periods, and it is necessary that these periods be compatible, i.e., that there exist commensurable quantities

$$\alpha_2, \alpha_3, \ldots, \alpha_p$$

such that

$$\alpha_2 \log \lambda + \log K_1 - \log K_2, \; \alpha_3 \log \lambda + \log K_1 - \log K_3, \ldots,$$

$$\ldots, \alpha_p \log \lambda + \log K_1 - \log K_p$$

are commensurable with $2i\pi$.

We can always assume

$$\lambda = \frac{K_1}{K_2}$$

since we can take λ to be **any one** of the quantities by which we can multiply t without changing x. It is then necessary to find commensurable quantities $\alpha_3, \ldots, \alpha_p$ such that

$$\alpha_3 \log \frac{K_2}{K_1} + \log \frac{K_3}{K_1}, \; \alpha_4 \log \frac{K_2}{K_1} + \log \frac{K_4}{K_1}, \ldots,$$

$$\ldots, \alpha_p \log \frac{K_2}{K_1} + \log \frac{K_p}{K_1}$$

are commensurable with $2i\pi$. So we have

Condition I. — 1° The logarithms of the absolute values of $\frac{K_2}{K_1}, \frac{K_3}{K_1}, \ldots$ must be commensurable

Condition II. — 2° The quantities

$$-\frac{\log \left|\frac{K_3}{K_1}\right|}{\log \left|\frac{K_2}{K_1}\right|} \log \frac{K_2}{K_1} + \log \frac{K_3}{K_1}, \ldots$$

must be commensurable with $2i\pi$.

If these conditions are satisfied, x will be a doubly periodic function of u.

To continue this discussion, we first remark that it can never happen that, when x turns around a singular point a of equation (1), t changes into

$$te^{\frac{2i\pi}{n}}$$

though it seems at first as though this would be possible.

In fact, if it were, then at x = a we would have t equal to 0 or ∞, i.e., a singular point, **which is absurd**.

Each time that x turns around a singular point of equation (1), t changes into $\frac{K_1}{t}$ or, $\frac{K_2}{t}$, ..., or $\frac{K_p}{t}$.

Hence at all singular points of equation (1), the difference between the roots of the fundamental determinant equation is equal to 1/2.

Problem. Can a doubly periodic function give rise to a linear differential equation of second order?

Letting h and k be the two periods of x as a doubly periodic function of u, we write

$$A \equiv 0 \qquad \text{mod}(h,k)$$

when we have

$$A = mh + nk$$

where m and n are integers.

We must have

$$\log K_1 - \log K_2 \equiv \log K_1 - \log K_3 \equiv \ldots \equiv \log K_1 - \log K_p \equiv 0 \quad \text{mod}(h,$$

$$2i\pi \equiv 0 \qquad \text{mod}(h,k)$$

For a given value of x, u can take infinitely many values. If

u_1 is one of them, then the others satisfy one of the congruences

$$u \equiv u_1,\ u \equiv \log K_1 - u_1,\ u \equiv \log K_2 - u_1,\ \ldots,\ u \equiv \log K_p - u_1$$

$$\mathrm{mod}(h,k)$$

i.e., one of the two congruences

$$u \equiv u_1,\ u \equiv \log K_1 - u_1 \quad \mathrm{mod}(h,k)$$

since we evidently have

$$\log K_1 - u_1 \equiv \log K_2 - u_1 \equiv \ldots \equiv \log K_p - u_1 \quad \mathrm{mod}(h,k)$$

In the period parallelogram there are no more than two values satisfying these congruences. Hence x is a doubly periodic function of u with two infinities. We suppose that these infinities are $-a$ and $+a$, i.e., that

$$\log K_1 = 0$$

as we always can, since this amounts to no more than multiplying t by a suitable factor.

We then put

$$x = \Lambda(u)$$

and we have

$$\frac{dx}{du} = 0$$

whenever

$$u \equiv 0 \quad \text{or} \quad u \equiv \frac{h}{2}$$

or

$$u \equiv \frac{k}{2} \quad \text{or} \quad u \equiv \frac{h+k}{2} \quad \mathrm{mod}(h,k)$$

Whenever $\frac{dx}{du} \neq 0$, u can be expanded in increasing powers of (x—a), if x is finite, or of 1/x if x is infinite.

Suppose on the contrary that u = 0 and let

$$\Lambda(0) = \alpha$$

when x turns around the point α, u is changed into —u, and since u = 0 for x = α we finally have

$$x - \alpha = A_2u^2 + A_3u^3 + \dots$$

where $A_2 \neq 0$, i.e., we have

$$u = \sqrt{x-\alpha}(B_0+B_1(x-\alpha)+B_2(x-\alpha)^2 + \dots)$$

where $B_0 \neq 0$.

Supposing likewise that

$$\Lambda(h/2) = \beta, \quad \Lambda(k/2) = \gamma, \quad \Lambda\left(\frac{h+k}{2}\right) = \delta$$

we have

$$u - h/2 = \sqrt{x-\beta}(B'_0+B'_1(x-\beta)+B'_2(x-\beta)^2+\dots)$$

$$u - k/2 = \sqrt{x-\gamma}(B''_0+B''_1(x-\gamma)+B''_2(x-\gamma)^2+\dots)$$

$$u - \frac{h+k}{2} = \sqrt{x-\delta}(B'''_0+B'''_1(x-\delta)+\dots)$$

where

$$B'_0 \neq 0, \quad B''_0 \neq 0, \quad B'''_0 \neq 0$$

Now let

$$y_1 = e^{-\frac{u}{2}}\sqrt{\frac{d\Lambda}{du}}$$

$$y_2 = e^{\frac{u}{2}}\sqrt{\frac{d\Lambda}{du}}$$

Corresponding to a given value of x there are infinitely many values of u; if u_0 is one of them then the others are

$$u_0 + mh + nk$$

$$- u_0 + mk + nk$$

where m and n are integers. We have

$$\Lambda(u_0 + mk + nk) = \Lambda(u_0)$$

$$\Lambda(- u_0 + mh + nk) = +\Lambda(u_0)$$

differentiation of which yields,

$$\Lambda'(u_0 + mh + nk) = \Lambda'(u_0)$$

$$-\Lambda'(- u_0 + mh + nk) = \Lambda'(u_0)$$

If we set $u = u_0$ in the formulae for y_1 and y_2, we find the values

$$y_{10} \quad \text{and} \quad y_{20}$$

for these functions. If we now set

$$u = u_0 + mh + nk$$

we find

$$y_1 = \pm y_{10}\, e^{-\frac{mh+nk}{2}}$$

$$y_2 = \pm y_{20}\, e^{+\frac{mh+nk}{2}}$$

Then setting

$$u = -u_0 + mh + nk$$

we get

$$y_1 = \pm \sqrt{-1}\, y_{20} e^{-\frac{mh+nk}{2}}$$

$$y_2 = \pm \sqrt{-1}\, y_{10}\, e^{\frac{mh+nk}{2}}$$

Thus y_1 and y_2 are functions of x which can take infinitely many values for each value of x, but if y_{10} and y_{20} are one pair of values for these functions, all others are of the form

$$y_1 = \alpha y_{10} + \beta y_{20}$$

$$y_2 = \alpha' y_{10} + \beta' y_{20}$$

moreover the determinant

$$\alpha\beta' - \alpha'\beta$$

is always equal to 1. This says that y_1 and y_2 satisfy an equation of the form

$$\frac{d^2y}{dx^2} = Uy$$

where U is a monodromic function of x over the whole plane.

To study the function U it is necessary to give x all possible values on the sphere and it suffices to give them a single time, so as to give u all values in the interior of a period parallelogram.

If we first give u a value such that

$$\frac{dx}{du} \neq 0$$

then it is clear that y_1 and y_2 can be expanded in powers of $x - a$, if x is finite. Moreover, neither y_1 nor y_2 is zero. This says that U is holomorphic in x if x is finite.

If we now put u = 0 we have

$$u = (x-\alpha)^{1/2}(B_0 + B_1(x-\alpha) + \ldots)$$

$$x - \alpha = A_2u^2 + A_3u^3 + \ldots$$

whence

$$\frac{dx}{du} = 2A_2u + 3A_3u^2 + \ldots$$

or

$$\frac{d\Lambda}{du} = \frac{dx}{du} = (x-\alpha)^{1/2}(C_0 + C_1(x-\alpha) + \ldots)$$

where

$$C_0 \neq 0$$

or

$$\sqrt{\frac{d\Lambda}{du}} = (x-\alpha)^{1/4}(D_0 + D_1(x-\alpha) + \ldots)$$

where

$$D_0 \neq 0$$

Moreover,

$$e^{-\frac{u}{2}} = (E_0 + E_1(x-\alpha)+\ldots) + (x-\alpha)^{1/2}(F_0 + F_1(x-\alpha)+\ldots)$$

where

$$E_0 \neq 0 \quad F_0 \neq 0$$

or finally

$$y_1 = (x-\alpha)^{1/4}(a_0+a_1(x-\alpha)+\ldots) + (x-\alpha)^{1/2}(x-\alpha)^{1/4}(b_0+b_1(x-\alpha)+\ldots)$$

and

$$y_2 = (x-\alpha)^{1/4}(a_0+a_1(x-\alpha)+\ldots) - (x-\alpha)^{1/2}(x-\alpha)^{1/4}(b_0+b_1(x-\alpha)+\ldots)$$

Here

$$a_0 \neq 0, \quad b_0 \neq 0$$

Thus for $x = \alpha$, U has a double infinity, hence the point $x = \alpha$ is a singular point of the equation

$$\frac{d^2y}{dx^2} = Uy$$

and the roots of the corresponding determinant equation are

$$\frac{1}{4} \quad \text{and} \quad \frac{3}{4}$$

We arrive at the same result by setting

$$u = \frac{h}{2}, \quad u = \frac{k}{2}, \quad u = \frac{h+k}{2}$$

and consequently

$$x = \beta, \quad x = \gamma, \quad x = \delta$$

Consequence. U **is a meromorphic function of** x **over the whole sphere; it is therefore a rational function.**

The equation

$$\frac{d^2y}{dx^2} = Uy$$

there has rational coefficients; it admits four singular points at a finite distance, and for these four singular points the roots of the determinant equation are

$$\frac{1}{4} \quad \text{and} \quad \frac{3}{4}$$

We conclude from this that:

We can always form a linear differential equation, with rational coefficients, integrable by means of a given doubly periodic function with two infinities.

If we choose this doubly periodic function (with periods h and k) so that

$$2i\pi \equiv 0 \quad \text{mod}(h,k)$$

then x will be a monodromic function of u admitting the period $2i\pi$; **it will therefore be a monodromic function of** t, **and consequently of** z.

Consequently, there is a case where the theorem of Fuchs is true.

If, on the contrary, we choose this doubly periodic function so that we do not have

$$2i\pi \equiv 0 \quad \text{mod}(h,k)$$

then x will be a monodromic function of u, but it will not admit the period $2i\pi$. **It will not be monodromic in** t **nor, consequently, in** z.

Thus there is a case where the theorem of Fuchs is false even

though the conditions set down by that geometer are satisfied.

We investigate how to form the linear differential equations which satisfy the preceding conditions.

These equations are written

$$\frac{1}{y}\frac{d^2y}{dx^2} = \frac{A_2}{(x-\alpha)^2} + \frac{B_2}{(x-\beta)^2} + \frac{C_2}{(x-\gamma)^2} + \frac{D_2}{(x-\delta)^2} +$$

$$+ \frac{A_1}{x-\alpha} + \frac{B_1}{x-\beta} + \frac{C_1}{x-\gamma} + \frac{D_1}{x-\delta}$$

If the roots of the determinant equation relative to the four singular points α, β, γ, δ are to be $\frac{1}{4}$ and $\frac{3}{4}$, then it is necessary that

(9) $$A_2 = B_2 = C_2 = D_2 = -\frac{3}{16}$$

We now put $x = \frac{1}{z}$ and study the equation in the neighborhood of $z = 0$; it becomes

$$z^4 \frac{d^2y}{dx^2} + 2z^3 \frac{dy}{dx} =$$

$$= \left[\frac{A_2 z^2}{(1-\alpha z)^2} + \frac{B_2 z^2}{(1-\beta z)^2} + \frac{C_2 z^2}{(1-\gamma z)^2} + \frac{D_2 z^2}{(1-\delta z)^2} + \right.$$

$$\left. + \frac{A_1 z}{1-\alpha z} + \frac{B_1 z}{1-\beta z} + \frac{C_1 z}{1-\gamma z} + \frac{D_1 z}{1-\delta z}\right] y$$

1° In the neighborhood of $z = 0$, the integrals of the equation must be regular; hence the coefficient of z in the expansion of the right hand must be zero, so

(10) $$A_1 + B_1 + C_1 + D_1 = 0$$

2° The roots of the determinant equation must be equal to 0 or -1, hence the coefficient of z^2 in the expansion of the right hand side must be zero, whence

$$(11) \qquad A_2 + B_2 + C_2 + D_2 + A_1\alpha + B_1\beta + C_1\gamma + D_1\delta = 0$$

3^0 The expansions of the integrals cannot contain logarithms, hence the coefficients of z^3 must also be zero, i.e., we have

$$(12) \qquad \begin{aligned} &2A_2\alpha + 2B_2\beta + 2C_2\gamma + 2D_2\delta + \\ &\qquad + A_1\alpha^2 + B_1\beta^2 + C_1\gamma^2 + D_1\delta^2 = 0 \end{aligned}$$

Thus the equation formed depends on five parameters. We originally had twelve parameters

$$\begin{matrix} \alpha & \beta & \gamma & \delta \\ A_2 & B_2 & C_2 & D_2 \\ A_1 & B_1 & C_1 & D_1 \end{matrix}$$

and we have found the seven equations (9), (10), (11), (12) between them.

Now consider an arbitrary doubly periodic function with two infinities. This function can be written $AZ(u-a) - AZ(u-b) + B = \Lambda(u)$. It depends on six parameters, namely

1^0 The two periods h and k;

2^0 The two infinities a and b;

3^0 The residue relative to the two infinities, i.e., A;

4^0 The constant B.

If two functions $\Lambda(u)$ do not differ in the periods h and k, nor in the quantities A and B, but only in the infinities a and b, and if $a - b$ has the same value for both functions also, then these functions give rise to the same differential equation.

On the other hand, if the two functions differ in any other way, then they cannot give rise to the same differential equation.

Thus the most general function $\Lambda(u)$ gives rise to a

differential equation depending on five, and only five, parameters.

Hence, for the equation

$$\frac{1}{y}\frac{d^2y}{dx^2} = \frac{A_2}{(x-\alpha)^2} + \frac{B_2}{(x-\beta)^2} + \frac{C_2}{(x-\gamma)^2} + \frac{D_2}{(x-\delta)^2} +$$

$$(13) \qquad + \frac{A_1}{x-\alpha} + \frac{B_1}{x-\beta} + \frac{C_1}{x-\gamma} + \frac{D_1}{x-\delta}$$

to be integrable by doubly periodic functions it is necessary and sufficient that the twelve parameters satisfy the relations (9), (10), (11) **and** (12).

With one more condition we can determine the periods of A(u) so that the theorem of Fuchs becomes true. However, this does not occur in general.

Third Case

x is a monodromic function of z which does not change when we change z to z′, where

$$\frac{1}{z'-\alpha} = \lambda + \frac{1}{z-\alpha}$$

We put

$$\frac{1}{z-\alpha} = t$$

and x will be a monodromic function of t, admitting the period λ. It will not have more than the one singular point

$$t = \infty$$

since if there were any more there would be infinitely many.

When x turns around one of the singular points of equation (1) t changes into t′, where

$$(14) \qquad \frac{t'-\gamma}{t'-\delta} = e^{\frac{2i\pi}{n}}\left(\frac{t-\gamma}{t-\delta}\right)$$

Thus x will be a function of t which is monodromic and unchanged when t changes into

$$t + \lambda$$

or into

$$\gamma_1 + e^{\frac{2i\pi}{n_1}}(t-\gamma_1),\ \gamma_2 + e^{\frac{2i\pi}{n_2}}(t-\gamma_2),\ \ldots,\ \gamma_p + e^{\frac{2i\pi}{n_p}}(t-\gamma_p)$$

It is easy to see, by combining these different transformations in all possible ways, that x admits a certain number of different periods.

It is therefore necessary for these periods to be compatible, and if they are, x is a doubly periodic function of t.

We could now discuss the compatibility of these periods. I do not do this, since in the case which concerns us **the different equation (1) always admits one algebraic integral and one other which can be found by quadrature, as I shall now show.**

Suppose that we have

$$x = \Lambda(t)$$

where $\Lambda(t)$ is a doubly periodic function of t. We then have

$$y_1 = \sqrt{\frac{d\Lambda}{dt}}$$

$$y_2 = t\sqrt{\frac{d\Lambda}{dt}}$$

for the two integrals of equation (1). It is clear that y_1 is connected with x by an algebraic relation, and that t, and consequently y_2, can be calculated as a function of x by quadrature.

Then, from what we have seen above, if the given linear

differential equation (1) can be written

$$\frac{1}{y}\frac{d^2y}{dx^2} = \Sigma \frac{A_i}{(x-a_i)^2} + \Sigma \frac{2B_{ik}}{(x-a_i)(x-a_k)}$$

(see p.317), then we must have the relations

$$B_{ik} = \left[\frac{1}{2} \pm \sqrt{\frac{1}{4} + A_i}\right]\left[\frac{1}{2} \pm \sqrt{\frac{1}{4} + A_k}\right] \tag{15}$$

We remark that if the conditions (15) are satisfied, and the conditions of Fuchs likewise, then the theorem of Fuchs will be true.

Example: the equation we have studied above from p.316 onwards.

Resumé. To summarize this long discussion:

For equation (1) to be integrable by means of a doubly periodic function, it is necessary and sufficient that either

1° The conditions (15) and also the conditions of Fuchs be satisfied, or else

2° It be of the form (13) and subject to the conditions (9), (10), (11), (12), from which it follows that the conditions of Fuchs are also satisfied.

In the first case there is always an algebraic integral and the theorem of Fuchs is always true. In the second case there is no algebraic integral, and the theorem of Fuchs is sometimes true and sometimes false.

Particular Case

We shall now make a special study of a very important particular case, which is that where we have no more than two singular points a_1 and a_2 at a finite distance. Equation (1) then reads

$$\frac{1}{y}\frac{d^2y}{dx^2} = \frac{A_1}{(x-a_1)^2} + \frac{2B}{(x-a_1)(x-a_2)} + \frac{A_2}{(x-a_2)^2}$$

Let ρ_1, ρ_2 and r be the differences between the roots of the determinant equations relative to

$$x = a_1, \quad x = a_2, \quad \text{and} \quad x = \infty$$

respectively. A_1, B and A_2 are completely determined as functions of ρ_1, ρ_2 and r.

If the conditions of Fuchs are satisfied we have

$$\rho_1 = \frac{1}{n_1}, \quad \rho_2 = \frac{1}{n_2}, \quad r = \frac{1}{p}$$

where n_1, n_2 and p are integers. **Suppose for the moment that they are not.**

As we have done up to now, let $\phi(x)$ and f(x) be two solutions of equation (1) and let

$$\frac{f(x)}{\phi(x)} = z$$

We have always been able to choose f(x) and $\phi(x)$ so that, when x turns around a_1, z changes into λz where

$$\lambda = e^{2i\pi\rho_1}$$

Then when x turns around a_2, z changes into z′, where

$$(16) \qquad \frac{z'-\alpha}{z'-\beta} = \mu \frac{z-\alpha}{z-\beta}, \quad \mu = e^{2i\pi\rho_2}$$

Also, we have always been able to choose f(x) and $\phi(x)$ so that $\alpha = 1$, for example (or so that $\beta = 1$, $\alpha = 0$). When x turns around ∞, z changes into z″, where

$$(17) \qquad \frac{z''-\gamma}{z''-\delta} = \nu \frac{z-\gamma}{z-\delta}, \quad \nu = e^{2i\pi r}$$

If x turns around a_1, then around a_2, this is the same as turning around infinity, hence z changes into z″. But z changes first into λz, when x turns around a_1, then when x turns around a_2, λz changes into z″, i.e., we have

$$\frac{z''-\alpha}{z''-\beta} = \mu \frac{\lambda z-\alpha}{\lambda z-\beta} \tag{18}$$

We identify equations (17) and (18). If equation (17) is expanded into the form

$$azz'' + bz + cz'' + d = 0$$

then we get

$$\nu^2 - \nu \frac{b^2+c^2-2ad}{ad-bc} + 1 = 0$$

But the expanded equation (18) is

$$zz''\lambda(1-\mu)+z\lambda(\beta\mu-\alpha)+z''(\alpha\mu-\beta)+\alpha\beta(1-\mu) = 0$$

We therefore have

$$\frac{a}{\lambda(1-\mu)} = \frac{b}{\lambda(\beta\mu-\alpha)} = \frac{c}{\alpha\mu-\beta} = \frac{d}{\alpha\beta(1-\mu)}$$

and consequently

$$(\nu^2+1)[\alpha\beta\lambda(1-\mu)^2 - \lambda(\beta\mu-\alpha)(\alpha\mu-\beta)] -$$

$$- \nu[\lambda^2(\beta\mu-\alpha)^2 + (\alpha\mu-\beta)^2 - 2\alpha\beta\lambda(1-\mu)^2] = 0$$

or

$$\alpha^2[(\nu^2+1)\lambda\mu - \nu(\lambda^2+\mu^2)] + \beta^2[(\nu^2+1)\lambda\mu - \nu(\lambda^2\mu^2+1)] -$$

$$- 2\alpha\beta[(\nu^2+1)\lambda\mu - \nu\lambda(1-\mu)^2 - \nu\mu(\lambda^2+1)] = 0$$

thus $\frac{\alpha}{\beta}$ is given by an equation of second degree. Forming the discriminant of this equation, we obtain

$$\Delta = [(\nu^2+1)\lambda\mu - \nu\lambda(1-\mu)^2 - \nu\mu(\lambda^2+1)]^2 -$$

$$- [(\nu^2+1)\lambda\mu - \nu(\lambda^2+\mu^2)][(\nu^2+1)\lambda\mu - \nu(\lambda^2\mu^2+1)]$$

or

$$\Delta = \nu(\nu^2+1)\lambda\mu(1+\lambda^2+\mu^2+\lambda^2\mu^2-2\lambda+4\lambda\mu-2\lambda\mu^2-2\mu\lambda^2-2\mu) +$$

$$+ 2\nu^2\lambda\mu(1+\lambda^2+\mu^2+\lambda^2\mu^2-2\lambda+4\lambda\mu-2\lambda\mu^2-2\mu\lambda^2-2\mu)$$

or finally

$$\Delta = \lambda\mu\nu(\nu+1)^2(\lambda-1)^2(\mu-1)^2$$

from which we obtain

$$\frac{\alpha}{\beta} = \frac{(\nu^2+1)\lambda\mu-\nu\lambda(1-\mu)^2-\nu\mu(\lambda^2+1)\pm(\nu+1)(\lambda-1)(\mu-1)\sqrt{\lambda\mu\nu}}{(\nu^2+1)\lambda\mu-\nu(\lambda^2+\mu^2)}$$

Which of the two roots must be chosen? The question is easy to decide. Suppose first that $\lambda = \lambda_0$, $\mu = \mu_0$, $\nu = \nu_0$; where λ_0, μ_0 and ν_0 are such that equation (1) admits an algebraic integral. Then the choice of the root is made without difficulty. Next we let λ, μ, ν vary continuously from the initial values λ_0, μ_0, ν_0 to any other values, and likewise let $\frac{\alpha}{\beta}$ vary continuously; then we will never be at a loss to know which of the two values of $\frac{\alpha}{\beta}$ is suitable.

Given two equations E and E$'$ of the form (1), suppose that for the first the differences between the roots of the determinant equations relative to a_1, a_2 and ∞ are respectively

$$\rho_1,\ \rho_2,\ r$$

and that for the second these differences are

$$\rho'_1,\ \rho'_2,\ r'$$

Suppose that the quantities $\rho_1 - \rho'_1$, $\rho_2 - \rho'_2$, $r - r'$ are integers, then λ, μ, ν will have the same values for the two equations E and E'. The second degree equation in $\frac{\alpha}{\beta}$ will be the same for the two differential equations E and E'. Must we choose the same root?

We remark that the two roots of the equation in $\frac{\alpha}{\beta}$ are permuted when λ, μ or ν describes a simple contour around the point 0. We therefore switch from one root to the other, or else, return to the same root, according as the number of simple contours described around the point 0, by either λ, μ or ν, is odd or even.

But when λ describes a simple contour around the point 0, ρ_1 changes into $\rho_1 + 1$ or $\rho_1 - 1$; when μ turns around 0, ρ_2 changes into $\rho_2 + 1$ or $\rho_2 - 1$; when ν turns around 0, r changes into $r + 1$ or $r - 1$.

Thus the value taken for $\frac{\alpha}{\beta}$ must be the same for the two equations E and E', or different, according as

$$\rho_1 - \rho'_1 + \rho_2 - \rho'_2 + r - r' \equiv 0 \mod 2$$

or

$$\rho_1 - \rho'_1 + \rho_2 - \rho'_2 + r - r' \equiv 1 \mod 2$$

Thus if $\rho_1 - \rho'_1$, $\rho_2 - \rho'_2$, $r - r'$ are integers, and if the sum of these integers is even, then we can choose two integrals of equation E,

$$\phi(x) \quad \text{and} \quad f(x)$$

and two integrals of equation E',

$$\phi'(x) \quad \text{and} \quad f'(x)$$

such that when x describes any closed contour the final values of $\phi'(x)$ and $f'(x)$ are expressible linearly in terms of their initial values **by the same formula** which expresses the final values of $\phi(x)$ and $f(x)$

linearly in terms of their initial values.

Remark. When equation (1) admits no more than two singular points at a finite distance, the preceding always gives the means of expressing the final values of the integrals of that equation linearly in terms of the initial values, by allowing x to describe an arbitrary closed contour, and consequently of recognizing whether these integrals are algebraic.

Discussion

Assuming that the conditions of Fuchs are satisfied, we have

$$\lambda = \cos 2\pi\rho_1 + i \sin 2\pi\rho_1, \quad \mu = \cos 2\pi\rho_2 + i \sin 2\pi\rho_2,$$

$$\nu = \cos 2\pi r + i \sin 2\pi r$$

whence

$$\frac{\alpha}{\beta} = \frac{\left(\nu + \frac{1}{\nu}\right) - \left(\mu + \frac{1}{\mu}\right) - \left(\lambda + \frac{1}{\lambda}\right) + 2}{\left(\nu + \frac{1}{\nu}\right) - \left(\frac{\lambda}{\mu} + \frac{\mu}{\lambda}\right)} \pm$$

$$\pm \frac{\left(\sqrt{\nu} + \frac{1}{\sqrt{\nu}}\right)\left(\sqrt{\lambda} - \frac{1}{\sqrt{\lambda}}\right)\left(\sqrt{\mu} - \frac{1}{\sqrt{\mu}}\right)}{\left(\nu + \frac{1}{\nu}\right) - \left(\frac{\lambda}{\mu} + \frac{\mu}{\lambda}\right)}$$

or

$$\frac{\alpha}{\beta} = \frac{\cos 2\pi r - \cos 2\pi\rho_1 - \cos 2\pi\rho_2 + 1}{\cos 2\pi r - \cos 2\pi(\rho_1 - \rho_2)} \mp$$

$$\mp \frac{4 \cos \pi r \sin \pi\rho_1 \sin \pi\rho_2}{\cos 2\pi r - \cos 2\pi(\rho_1 - \rho_2)}$$

This formula expresses $\frac{\alpha}{\beta}$ as a monodromic function of ρ_1, ρ_2 and r. Then, by reason of continuity, the appropriate root will either be **always** the one corresponding to the + sign, or else the one corresponding to the − sign. Taking for example an equation with an algebraic integral, one easily decides which of the two signs to choose.

Suppose then that

$$y = (x-a_1)^{\frac{1}{2} - \frac{\rho_1}{2}}(x-a_2)^{\frac{1}{2} - \frac{\rho_2}{2}}$$

This function satisfies the differential equation

$$\frac{1}{y}\frac{d^2y}{dx^2} = \frac{\left(\frac{1}{2} - \frac{\rho_1}{2}\right)\left(-\frac{1}{2} - \frac{\rho_1}{2}\right)}{(x-a_1)^2} + \frac{2\left(\frac{1}{2} - \frac{\rho_1}{2}\right)\left(\frac{1}{2} - \frac{\rho_2}{2}\right)}{(x-a_1)(x-a_2)} +$$

$$+ \frac{\left(\frac{1}{2} - \frac{\rho_2}{2}\right)\left(-\frac{1}{2} - \frac{\rho_2}{2}\right)}{(x-a_2)^2}$$

and it is clear that the difference between the roots of the determinant equation here is

ρ_1	for a_1
ρ_2	for a_2
$1 - \rho_1 - \rho_2$	for ∞

Then, putting $r = 1 - \rho_1 - \rho_2$ yields

$$\frac{\alpha}{\beta} = \frac{(1 - \cos 2\pi\rho_1)(1 - \cos 2\pi\rho_2) - \sin 2\pi\rho_1 \sin 2\pi\rho_2}{-2 \sin 2\pi\rho_1 \sin 2\pi\rho_2} \mp$$

$$\mp \frac{[-\sin 2\pi\rho_1 \sin 2\pi\rho_2 + (1 - \cos 2\pi\rho_1)(1 - \cos 2\pi\rho_2)]}{-2 \sin 2\pi\rho_1 \sin 2\pi\rho_2}$$

but it is clear that in the particular case we now have that $\frac{\alpha}{\beta} = 0$; hence whenever

$$r = 1 - \rho_1 - \rho_2$$

we must take the − sign, and then by reason of continuity we must take the − sign whatever the values of r, ρ_1 and ρ_2.

Thus for any r, ρ_1 and ρ_2 we have

$$\frac{\alpha}{\beta} = \frac{\cos 2\pi r - \cos 2\pi\rho_1 - \cos 2\pi\rho_2 +}{\cos 2\pi r - \cos 2\pi(\rho_1 - \rho_2)}$$

$$\frac{+ 1 - 4 \cos \pi r \sin \pi\rho_1 \sin \pi\rho_2}{\cos 2\pi r - \cos 2\pi(\rho_1 - \rho_2)}$$

If the conditions of Fuchs are satisfied, then this value is real. Supposing in particular that

$$\rho_2 = \frac{1}{2}$$

then

$$\frac{\alpha}{\beta} = \frac{\cos 2\pi r - \cos 2\pi\rho_1 + 2 - 4 \cos \pi r \sin \pi\rho_1}{\cos 2\pi r + \cos 2\pi\rho_1}$$

If we no longer suppose

$$\rho_2 = \frac{1}{2}$$

but return to the general expression, we see that it can be simplified and put in the following form

$$\frac{\alpha}{\beta} = \frac{\cos\frac{\pi}{2}(\rho_1+\rho_2-r)\cos\frac{\pi}{2}(\rho_1+\rho_2+r)}{\cos\frac{\pi}{2}(\rho_1-\rho_2+r)\cos\frac{\pi}{2}(\rho_1-\rho_2-r)}$$

or again,

$$\frac{\alpha}{\beta} = \frac{\cos\pi r + \cos\pi(\rho_1+\rho_2)}{\cos\pi r + \cos\pi(\rho_1-\rho_2)}$$

In general, let $\phi(x)$ and $f(x)$ be two integrals of an equation of second order. When x describes a certain closed contour, the final values of the integrals are given as linear functions of the initial values by the formulae

$$\alpha\phi(x) + \beta f(x)$$

$$\gamma\phi(x) + \delta f(x)$$

We do not know how to calculate α, β, γ, δ in general. But if the closed contour contains no more than one singular point we shall always be able to find the roots of the equation in ω

$$(\alpha-\omega)(\delta-\omega) - \beta\gamma = 0$$

We can no longer do this, in general, when the contour contains more than one singular point.

Nevertheless, we shall see that if there are no more than two singular points at a finite distance we can always make this calculation, whatever the contour may be; we can likewise, by suitable choice of the integrals $\phi(x)$ and $f(x)$, calculate the coefficients α, β, γ, δ themselves.

These circumstances will permit us to discuss more completely, in this particular case, the theorem of Fuchs.

Suppose, to fix ideas, that

$$\rho_1 = \frac{1}{4}, \quad \rho_2 = \frac{1}{2}, \quad r = \frac{1}{6}$$

then x does not change when we change

$$z \text{ into } iz$$

or

$$\frac{z-\alpha}{z-\beta} \text{ into } -\frac{z-\alpha}{z-\beta}$$

or

$$\frac{z-\gamma}{z-\delta} \text{ into } e^{\frac{2i\pi}{6}} \frac{z-\gamma}{z-\delta}$$

I shall call the first of these changes the operation L, the second the operation M, and the third the operation N, and I shall designate, for example, the compound operation which consists in making ℓ applications of operation L, then m of operation M, then n of operation N, then ℓ_1 of operation L again, by the notation

$$L^{\ell}M^{m}N^{n}L^{\ell_1}$$

We have

$$\frac{\alpha}{\beta} = \frac{\sqrt{3}-\sqrt{2}}{\sqrt{3}+\sqrt{2}}$$

hence $\frac{\alpha}{\beta}$ is positive, and we easily see that $\frac{\gamma}{\alpha}$ has argument $\frac{\pi}{4}$.

I now suppose that we make two cuts in the x plane, one from a_1 to a_2 and the other from a_2 to infinity. Suppose that a_1 and a_2 are real. Suppose that the integrals $\phi(x)$ and f(x) have been chosen so that

$$\frac{\phi(x)}{f(x)} = z$$

that $z = 0$ for $x = a_1$, and that $z = 1$ for $x = a_2$ or $\alpha = 1$.

All these assumptions can be realized.

When x follows the cut from a_1 to a_2 on a certain side, which we associate with α, then $\phi(x)$ and $f(x)$ will remain real and consequently z will be real, and it will vary along the straight line from 0 to α, i.e., from 0 to 1. When x follows the same cut on the other side, the argument of

$$\frac{\phi(x)}{f(x)} = z$$

an expression which contains a factor

$$(x-a_1)^{\frac{1}{4}}$$

will be constant and equal to $\frac{\pi}{2}$; thus z varies from 0 to $\alpha\sqrt{-1}$, i.e., from 0 to $\sqrt{-1}$. When x follows the cut from a_2 to infinity on the α side of that cut the argument of

$$\frac{z-\alpha}{z-\beta}$$

which is 0 when x follows the α side of the cut a_1, a_2, becomes $\frac{\pi}{2}$, because the expression

$$\frac{z-\alpha}{z-\beta}$$

contains $(x-a_2)^{\frac{1}{2}}$ as a factor, and in following the line $a_1a_2\infty$ we pass the point a_2 and see the argument of $x - a_2$ increase by π.

Thus when x describes this cut $a_2\infty$, z describes, on its plane, an arc of the circle with diameter $\alpha\beta$, the arc running from α to β.

We similarly show that when x describes the same cut on the other side, z follows an arc of the circle with diameter $\alpha\sqrt{-1}$

$\beta\sqrt{-1}$, the arc running from $\alpha\sqrt{-1}$ to γ.

It follows that if x **moves on its plane without crossing any cut, then** z **remains inside the mixtilinear quadrilateral** $\alpha 0 \alpha' \gamma$.

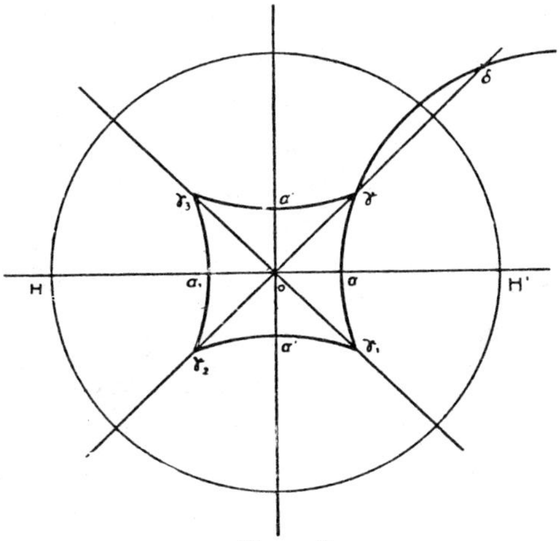

Figure 2.

In Figure 2, α represents the imaginary quantity α; α', α_1, α'_1 represent the quantities $\alpha\sqrt{-1}$, $-\alpha$, $-\alpha\sqrt{-1}$; γ, γ_1, γ_2, γ_3 represent γ, $-\gamma\sqrt{-1}$, $-\gamma$, $\gamma\sqrt{-1}$.

The circle $\gamma_1\alpha\gamma\delta$ is the circle on diameter $\alpha\beta$. If we now allow x to cross the cut a_1a_2, then we find the region to which z is confined by applying the operations

$$L,\ L^2,\ L^3$$

to the polygon $\alpha 0 \alpha' \gamma$, and we obtain the curvilinear quadrilateral

$$\gamma\gamma_1\gamma_2\gamma_3$$

the circular sides of which meet at angles of 60°.

With the point 0 as center we describe a circle HH$'$ cutting the circle $\alpha\gamma\delta$ orthogonally.

This circle is invariant under the operations L, M, N.

But the quadrilateral $\gamma\gamma_1\gamma_2\gamma_3$ is entirely inside this circle. If x describes any contour on its plane, then z remains inside the quadrilateral or else in one of the transforms of this quadrilateral by one of the operations composed from L, M, N. Now all these transforms are inside the circle HH′ because the transform of a point inside the circle is again inside the circle. **Thus, whatever contour** x **describes in its plane,** z **never goes outside the circle** HH′.

Another remark is that all the transforms of the circles $\gamma\gamma_1$, $\gamma_1\gamma_2$, $\gamma_2\gamma_3$ etc. by operations composed from L, M, N cut the circle HH′ orthogonally.

That being so, I suppose that we allow x to cross a large number of cuts; then the region described by z extends further and further. To give an account of it we must adjoin to the quadrilateral $\gamma\gamma_1\gamma_2\gamma_3$ a certain number of its transforms under the operations L, M, N repeated a certain number of times.

First, some definitions. By the conjugate of a point a I mean the point a_1 at the foot of the perpendicular from a to the polar of a with respect to the circle HH′.

By the operation S with respect to the point a I mean the operation which changes

$$\frac{z-a}{z-a_1} \quad \text{into} \quad e^{\frac{2i\pi}{6}} \frac{z-a}{z-a_1}$$

If a is the transform of γ under an operation composed from L, M, N then the operation S will be one of the operations composed from L, M, N.

I denote by Q the different curvilinear quadrilaterals which are transforms of the quadrilateral $\gamma\gamma_1\gamma_2\gamma_3$ under operations composed from L, M, N.

Consider an arbitrary curvilinear polygon P capable of being decomposed into a certain number of quadrilaterals Q. I apply, to the different quadrilaterals Q making up this polygon, the operations S, S^2, S^3, S^4, S^5 with respect to each of their vertices; except for quadrilaterals with a vertex on the perimeter of the polygon, these operations do no more than reproduce the quadrilaterals Q making up the polygon P. But when applied to quadrilaterals with a vertex on the perimeter of P, these operations lead to new quadrilaterals, which I **annex** to the polygon P to form a larger polygon P′. **I call this transformation of** P **into** P′ **the operation** T.

Thus if I apply the operation T to the quadrilateral $\gamma\gamma_1\gamma_2\gamma_3$ I obtain a first polygon P_1; I apply the operation T to this polygon and obtain a second polygon P_2, then a third, P_3, etc. The polygons P_m are the regions that z can traverse without having x cross an infinite number of cuts.

I claim the polygon P_1 has

angles equal to 60° or 120°. In fact, consider Figure 3, which roughly represents the decomposition of the polygon P_1 into quadrilaterals Q. (Translator's Note: Poincaré's figures 3, 4, 5 suggest that only the A angles are convex, whereas in fact the B angles are also, a fact which is crucial to the subsequent argument.) On this Figure the circular arcs

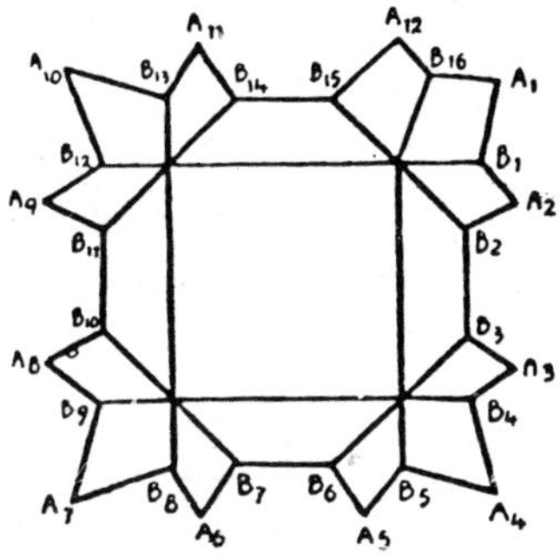

Figure 3.

have been replaced by striaght lines. Since the quadrilaterals Q have all their angles equal to 60°, we see that the angles of the polygon P_1

$$A_1, A_2, \ldots, A_{12} \text{ are } 60^o$$

and

$$B_1, B_2, \ldots, B_{16} \text{ are } 120^o$$

Q.E.D.

Moreover, we never have two 60° angles in succession. We pass to the polygon P_2.

On Figures 4 and 5, the angles marked A are 60°, the angles marked B are 120°.

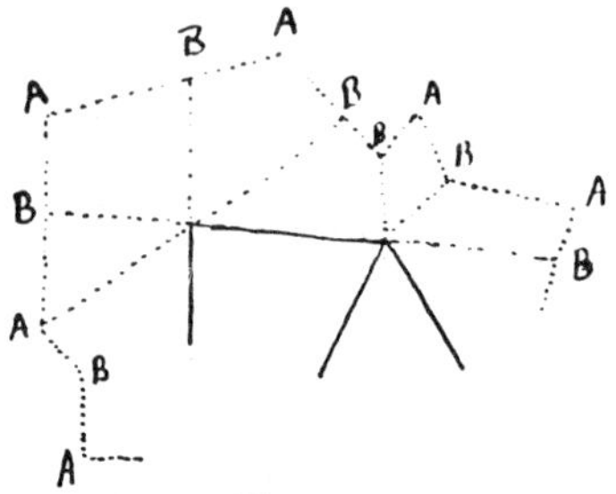

Figure 4.

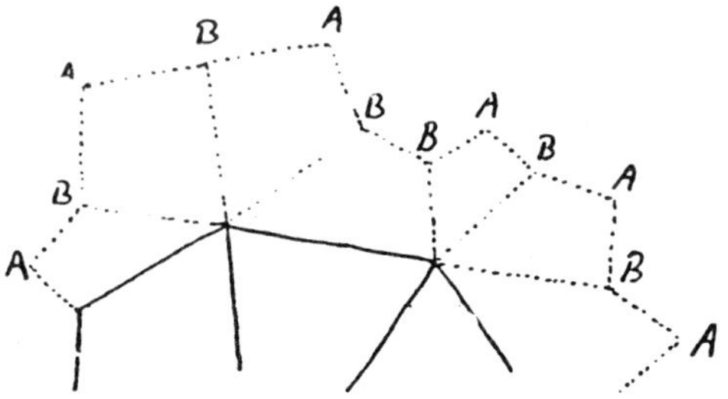

Figure 5.

In general, the quadrilaterals **annexed** to the polygon P_1 are divided into two categories:

1° those which have only one vertex in common with P_1 and three with P_2; at these three vertices we have one 60° angle and two 120° angles;

2° those which have two vertices in common with P_1 and two on P_2; the two corresponding angles of P_2 are 120°.

On Figures 4 and 5 the solid lines represent a portion of the polygon P_1 divided into quadrilaterals Q, and the dotted quadrilaterals are those which must be **annexed** to the polygon P_1 to form the polygon P_2. The Figure 4 shows a portion of the perimeter of P_1 where an angle of 120° succeeds an angle of 60°, Figure 5 shows a portion with two 120° angles in succession. We see by inspection of these figures that the angles of the polygon P_2 are again 60° or 120°.

In the same way, one shows that this is true of the angles of the polygon P_3, and in general, of the polygon P_m.

Consequence. The polygons P_m are curvilinear polygons whose sides are arcs of certain circles which cut the circle HH′ orthogonally, and whose angles are all projecting.

With these preliminaries established, we can now address ourselves to the following question:

We have seen that the function

$$x = \Theta(z)$$

does not exist when the absolute value of z is greater than OH; does it remain meromorphic when the absolute value of z is smaller than OH?

To resolve this question, we recall what has been said in Note 6*. (*Translator's Note: Placed at the end of the essay.) In that Note we considered certain regions R_m and I showed that the condition for x to remain monodromic in z is that this region R_m does not come to overlap itself. Also, we have seen that this region R_m can come to overlap itself in two different ways, either while leaving the rest of the sphere on the same side of its boundary, or else by forming a sort of ring which divides the portion of the sphere outside R_m into two distinct regions, so that we cannot go from one to the other without traversing R_m.

The proof of Fuchs shows that R_m cannot overlap itself in the first way. We therefore investigate whether it can overlap itself in the second way.

But here the regions R_m are represented by the polygons P_m, or at least the polygons P_m can play exactly the same role in the proof.

We consider the sphere which has the circle HH′ as a great circle, and **stereographically** project the figure in the plane of this great circle on to the sphere. The polygons P_m are projected into curvilinear spherical polygons H_m whose sides are small circles cutting HH′ orthogonally and consequently situated in planes perpendicular to that of this great circle, and whose angles are all projecting.

We project the figure again **orthogonally** onto the plane of this great circle HH′; the polygons H_m are then projected into rectilinear polygons K_m whose angles are all projecting.

But it is clear that a rectilinear polygon whose angles are all projecting cannot overlap itself in the second way.

Hence neither the polygons K_m, nor consequently the polygons P_m, can overlap themselves in the second way.

Hence x is meromorphic in z in the interior of the polygon P_m.

But as m tends to infinity the polygons P_m approach closer and closer to the circle HH'. If this were not so, then as m tended to infinity the polygon P_m would tend towards a certain contour P which would have to be a simple closed curve reproducible by all the operations composed from L, M, N, which is possible only for the circle HH'.

Hence x **is meromorphic in** z **inside the circle** HH'.

The function x, as we have seen, does not exist over the whole plane, so that we cannot say positively that it is an **analytic function** of z; however, it is a **perfectly determinate function** of that variable. This is how we must understand the theorem of Fuchs **in this case**, and moreover, this is sufficient for the conclusions this geometer draws from it.

We make some remarks on the function x.

First, it can be represented by a convergent series. In fact, for suitable choice of λ, $\frac{1}{x-\lambda}$ is a meromorphic function of z which remains finite along the whole perimeter of the quadrilateral $\gamma\gamma_1\gamma_2\gamma_3$, and consequently along the whole perimeter of the polygon P_m. Moreover, when m tends to infinity the perimeter of this polygon remains finite.

Thus the integral

$$\int \frac{d\zeta}{(x-\lambda)(z-\zeta)}$$

taken along the polygon P_m, remains finite when m tends to infinity.

But this integral is equal to a part of a convergent series in powers of z; the other part equals $-$ f(z) plus a series of terms in

$$\frac{A}{z-a}$$

which is easy to construct. In fact, let a be one of the infinities of the function $\frac{1}{x-\lambda}$; we shall have a term in

$$\frac{1}{z-a}$$

Suppose that x does not change when we change

$$z \text{ into } \frac{hz+k}{h'z+k'}$$

then we shall have another infinity

$$z = -\frac{k-ak'}{h-ah'}$$

with the residue

$$\frac{k'h-h'k}{(h-ah')^2}$$

giving us the term

$$\frac{k'h-h'k}{(h-ah')^2} \quad \frac{1}{z+\frac{k-ak'}{h-ah'}}$$

The sum of these terms, diminished by f(z) and multiplied by $2i\pi$, represents the integral considered. Of course, we must take only terms corresponding to infinities inside P_m.

We then have

$$f(z) = \Sigma \frac{A}{z-a} + \phi(z)$$

where $\phi(z)$ is holomorphic in z; when m tends to infinity $\phi(z)$, which equals the integral divided by $2i\pi$, tends to a finite limit, at the same time that $\Sigma \frac{A}{z-a}$ becomes an infinite series. This infinite series is therefore convergent.

Thus f(z) **can be put in the following form**:

$$\Sigma \frac{A}{z-a} + \phi(z)$$

where $\phi(z)$ **is holomorphic inside the circle** HH′ **and where** $\Sigma \frac{A}{z-a}$ **is a convergent series whose general term is easy to construct.**

One other remark: take for example a differential equation

$$\frac{d^2y}{dx^2} = y\left[\frac{A_1}{(x-a_1)^2} + \frac{2B+2}{(x-a_1)(x-a_2)} + \frac{A_2}{(x-a_2)^2}\right]$$

such that

$$\rho_1 = 1 + \frac{1}{4}, \quad \rho_2 = 1 + \frac{1}{2}, \quad r = 2 + \frac{1}{6}$$

In order to avoid confusing the variable x in this new equation with the old one, we shall call it x_1; but to continue, put

$$z = \frac{\phi_1(x_1)}{f_1(x_1)}$$

where $\phi_1(x_1)$ and $f_1(x_1)$ are suitably chosen integrals of the new equation; z will be a function of x which has infinitely many values, but we obtain all of them by applying to one of them the operations composed from L, M, N (see p.344), since x is a monodromic function of z which does not change when one of these operations is applied to the latter variable. **Thus x is monodromic in x_1, but here again x does not exist for all values of x_1; this function exists only for those values of x_1 for which z is inside the circle HH′.**

Concluding Remarks

The mode of discussion used here can always be applied when we have no more than two singular points. The difficulty increases with the number of these points. We see how we must approach the question if we have three singular points, for example.

Let a_1 and a_2 be two of these singular points, suppose that they are real and that the coefficients of the differential equation are likewise real. We join these two points by a rectilinear cut, and similarly join the other singular points by cuts.

When x traverses the cut a_1a_2 on a certain side, z remains real and varies from α to β; when we apply to this line segment the different operations which leave x invariant when applied to the variable z, then the successive transforms of this line segment will be an infinity of circular arcs.

For Fuchs' theorem to be true, it is necessary and sufficient that none of these circular arcs cut the line segment.

Note 6.

We can account for the inadequacy of Fuchs proof in the following way. I suppose that, on the x sphere, each singular point is joined to the point ∞ by a cut. If x describes a path which does not cross cuts more than m times, then the point representing z will describe a path confined to a certain region R_m of the sphere. When m increases, the region R_m extends further and further.

If R_m comes to overlap itself in this extension process, so that the point representing z can arrive at a certain position on the sphere in two different ways, then it is clear that x will not be a monodromic function of z; if, on the contrary, this does not happen, then x will be monodromic in z.

But we can conceive of two different ways in which the region R_m can overlap itself, as the following figures indicate.

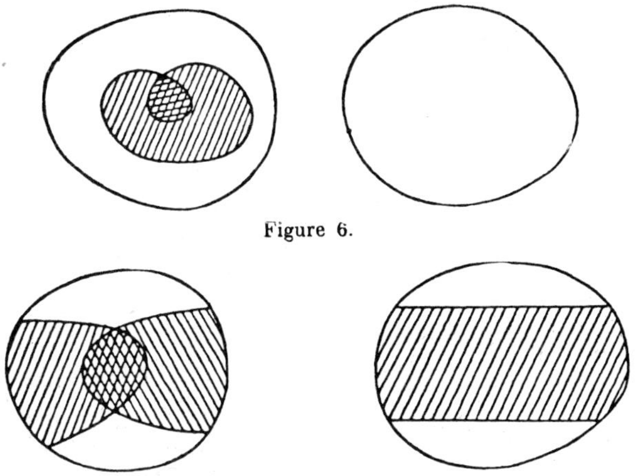

Figure 6.

In these figures*, the two circles represent the two hemispheres, the parts remaining white are the portions of the sphere not contained in the region R_m; the region R_m is hatched and we observe cross hatching in the portions of the sphere where the region R_m overlaps itself. (*Translator's Note: I have reversed the order of Poincaré's figures to agree with what he subsequently calls "the first way" and "the second way".)

Fuchs shows that the region R_m cannot overlap itself **in the first way**, because when z describes an infinitely small circle **in the interior of the region** R_m, then x returns to the same value.

However, he does not show that the region R_m **cannot overlap itself in the second way**.

ON THE GROUPS OF LINEAR EQUATIONS

Acta Mathematica 4(1884), 201–311

Received October 20, 1883; published February 9, 1884 N.E.N.

In three memoirs (Theory of Fuchsian Groups, On Fuchsian Functions, Memoir on Kleinian Groups), I have studied the discontinuous groups of linear transformations and the uniform functions which are invariant under these groups of transformations. To show how these functions, and others analogous to them, give the integrals of linear equations with algebraic coefficients, it is necessary to resolve two important problems:

1° Given a linear equation with algebraic coefficients, determine its group.

2° Given a linear second order equation dependent on certain arbitrary parameters, dispose these parameters in such a way that the equation becomes Fuchsian.

I. Fundamental invariants

We concern ourselves initially with the first of these problems. Consider an arbitrary equation with algebraic coefficients

$$(1)\qquad \frac{d^p v}{dx^p} + \phi_{p-1}(x,y)\frac{d^{p-1}v}{dx^{p-1}} + \phi_{p-2}(x,y)\frac{d^{p-2}v}{dx^{p-2}} + \dots$$
$$\dots + \phi_1(x,y)\frac{dv}{dx} + \phi_0(x,y)v = 0$$

In this equation the ϕ are rational functions of two variables x and y which are connected by an algebraic relation

$$(2)\qquad \psi(x,y) = 0$$

Suppose that we consider a fundamental system of integrals of the equation (1)

$$v_1, v_2, \ldots, v_p$$

and that x describes a closed contour C which returns y to the same value; the integrals v_1, v_2, ..., v_p will then take new values w_1, w_2, ..., w_p which are linear functions of their initial values, so that we have

$$w_k = \Sigma \alpha_{ik} v_i$$

In other words, the integrals v_1, v_2, ..., v_p undergo a linear transformation

$$S = (v_1, v_2, \ldots, v_p; w_1, w_2, \ldots, w_p)$$

which we can represent by the table of coefficients

$$\begin{matrix} \alpha_{11} & \alpha_{12} & \cdots & \alpha_{1p} \\ \alpha_{21} & \alpha_{22} & \cdots & \alpha_{2p} \\ \cdots & \cdots & \cdots & \cdots \\ \alpha_{p1} & \alpha_{p2} & \cdots & \alpha_{pp} \end{matrix}$$

If we proceed in this way for all possible contours C, then we obtain a set of linear transformations which form a group G. This is **the group of equation** (1).

We can always assume that the coefficient ϕ_{p-1} in equation (1) is identically zero, because we can always reduce to this case by a simple and well known transformation. We conclude that the determinant of the transformations S is equal to unity, and hence that

$$\Sigma \pm \alpha_{11}\alpha_{22} \ldots \alpha_{pp} = 1 \tag{3}$$

Knowledge of equation (1) and the contour C does not suffice to determine the transformation S. In fact, this transformation also depends on the choice of the system of fundamental integrals v_1, v_2, ..., v_p.

Suppose that we replace them by p other fundamental integrals

$$u_1, u_2, \dots, u_p$$

We will have

$$u_k = \Sigma \beta_{ik} v_i$$

so that the transformation

$$\sigma = (v_1, v_2, \dots, v_p; u_1, u_2, \dots, u_p)$$

will be linear. Then if we describe the contour C, and begin with the integrals u_1, u_2, ..., u_p, we obtain a linear transformation

$$\sigma^{-1} S \sigma$$

The set of transformations $\sigma^{-1}S\sigma$ form a group we can denote by $\sigma^{-1}G\sigma$, and which can be regarded as the group of equation (1) just as well as G, with the system of integrals u in place of the system of integrals v. We know that the group $\sigma^{-1}G\sigma$ is called the conjugate* of G by the linear transformation σ. (*Translator's Note: Poincaré says "transform".)

In order to have only a single group for equation (1), we do not consider the group G as distinct from its conjugates by the various linear transformations.

Having done that, we investigate the invariants of the transformation S, i.e., the functions of its coefficients which remain invariant when we replace S by $\sigma^{-1}S\sigma$. We form the equation

$$
(4) \qquad \begin{vmatrix} \alpha_{11} + t & \alpha_{12} & \cdots & \alpha_{1p} \\ \alpha_{21} & \alpha_{22} + t & \cdots & \alpha_{2p} \\ \cdots & \cdots & \cdots & \cdots \\ \alpha_{p1} & \alpha_{p2} & \cdots & \alpha_{pp} + t \end{vmatrix} = 0
$$

The roots of this equation, and hence also its coefficients, do not change when we replace S by $\sigma^{-1}S\sigma$. They are the invariants of S.

These invariants are $p - 1$ in number. Equation (4) in t is in fact of degree p, hence it has $p + 1$ coefficients, but the coefficient of t^p and the constant term equal 1, so $p - 1$ invariants remain. These invariants, which are the coefficients of equation (4), will be rational functions of the α.

Suppose that the integrals v_1, v_2, ..., v_p are defined in the following way:

At the initial point of the contour C (at $x = 0$, for example) v_1 is equal to 1 and its first $p - 1$ derivatives are zero; v_i is zero, together with its first $p - 1$ derivatives, with the exception of the $i - 1^{th}$, which equals 1. Now let $w_k{}^1$ be what the integral v_k becomes when x returns to the initial point after describing the contour C, and let $w_k{}^i$ be what its $i - 1^{th}$ derivative becomes. Then we have

$$
w_k = \Sigma\, w_k{}^i v_i; \quad w_k{}^i = \alpha_{ik}
$$

It follows that the invariants are rational functions of the $w_k{}^i$.

Suppose in particular that $p = 2$; equation (1) becomes

$$
\frac{d^2v}{dx^2} + v\phi_0(x,y) = 0
$$

and equation (4) can be written

$$\begin{vmatrix} w_1^1 + t & w_1^2 \\ w_2^1 & w_2^2 + t \end{vmatrix} = 0$$

or

$$t^2 + t\left(w_1^1 + w_2^2\right) + 1 = 0$$

Thus the transformation S has $w_1^1 + w_2^2$ as its single invariant. If k is its eigenvalue*, this invariant is precisely $k + \frac{1}{k}$. (*Translator's Note: Poincaré says "multiplier".)

If we know the invariants of all the transformations S, then the group G will be completely determined, because we do not regard it as distinct from its conjugates $\sigma^{-1}G\sigma$. But it will not be necessary to know all these invariants, it will suffice to know a certain number that we call **fundamental invariants**, all others being functions of them.

How do we obtain the fundamental invariants? Suppose that the group G is generated by n fundamental transformations. The coefficients of each transformation will be p^2 in number, but because of relation (3), only $p^2 - 1$ will remain independent. For the n transformations, we have $n(p^2 - 1)$ coefficients in all. But we do not consider the group G as distinct from its various conjugates. Hence it is necessary to subtract $p^2 - 1$ from the preceding number, and we arrive at the conclusion that $(n - 1)(p^2 - 1)$ conditions are necessary to determine the group G.

Thus if we are given any $(n - 1)(p^2 - 1)$ invariants (provided that they are indenpendent), then all the others will be functions of them. We could choose the fundamental invariants to be, for example, the invariants of $(n - 1)(p + 1)$ suitably chosen transformations.

We can always assume that equation (1) has rational coefficients. In fact, suppose that this is not the case and that equation (2) is an algebraic relation of degree m, so that there are m values of y

$$y_0, y_1, \ldots, y_{m-1}$$

corresponding to each value of x. In the x plane we can always draw $m - 1$ contours $C_1, C_2, \ldots, C_{m-1}$ such that, when x describes the contour C_i, y_0 changes into y_i. Now let

$$v_1^0, v_2^0, \ldots, v_p^0$$

be p fundamental integrals of equation (1) corresponding to the value y_0 of y. Suppose these integrals become

$$v_1^i, v_2^i, \ldots, v_p^i$$

when x describes the contour C_i. Then the mp functions

$$\begin{array}{cccc} v_1^0, & v_2^0, & \ldots, & v_p^0 \\ v_1^1, & v_2^1, & \ldots, & v_p^1 \\ \ldots\ldots & \ldots\ldots & \ldots & \ldots\ldots \\ v_1^{m-1}, & v_2^{m-1}, & \ldots, & v_p^{m-1} \end{array}$$

satisfy a linear equation (1′) of order mp whose coefficients are rational in x. Knowledge of the group of (1′) suffices to determine that of equation (1).

Suppose then that equation (1) has rational coefficients and that they have $n + 1$ singular points, including the point ∞, if it occurs, and not including apparent singular points. The group is then generated by n fundamental transformations, and we have $n + 1$ transformations whose invariants are immediately known, with the aid of the determinant equation, provided that the integrals are regular; they are the transformations to which we are led by making the variable describe a closed contour around a single singular point. It then remains to calculate the invariants of $(n - 1)p - 2$ transformations.

Let a be an arbitrary singular point and let

$$\lambda_1, \lambda_2, \ldots, \lambda_p$$

be the roots of the corresponding determinant equation.

In the neighborhood of the point a there are p particularly remarkable integrals. The kth of them equals $(x - a)^{\lambda_k}$ multiplied by a holomorphic function. Let

$$v_1, v_2, \ldots, v_p$$

be these p integrals, which I call the **canonical integrals** with respect to the point a. When the variable describes an infinitely small circle around the point a, these p integrals v undergo the linear transformation

$$\begin{matrix} e^{2\pi i\lambda_1} & 0 & \ldots & 0 \\ 0 & e^{2\pi i\lambda_2} & & 0 \\ 0 & 0 & \ldots & e^{2\pi i\lambda_p} \end{matrix}$$

which I call the **canonical transformation** relative to the point a.

Similarly, let

$$w_1, w_2, \ldots, w_p$$

be the canonical integrals with respect to a second singular point b.

We join the two points a and b by an arbitrary path amb. When the variable leaves the point a and describes the path amb, the canonical integrals $v_1, v_2, \ldots, v_p$ will become linear functions of the canonical integrals $w_1, w_2, \ldots, w_p$, in the neighborhood of the point b, so that we have

$$v_k = \Sigma\ \beta_{ik} w_i$$

I call the linear transformation

$$S = \begin{bmatrix} \beta_{11} & \beta_{21} & \cdots & \beta_{p1} \\ \beta_{12} & \beta_{22} & \cdots & \beta_{p2} \\ \cdots & \cdots & \cdots & \cdots \\ \beta_{1p} & \beta_{2p} & \cdots & \beta_{pp} \end{bmatrix}$$

the **auxiliary transformation relative to the path** amb. If we know this auxiliary transformation, then we also know a transformation in the group of equation (1); it will be that undergone by the canonical integrals $v_1, v_2, \ldots, v_p$ when the variable leaves the point a, describes the path amb up to the neighborhood of the point, then describes an infinitely small circle around the point b and finally returns to the neighborhood of the point a by the path bma.

Let Σ be the canonical transformation relative to the point b. When the variable describes the preceding contour, the integrals v will undergo the transformation $S^{-1}\Sigma S$.

Suppose that we join the n + 1 singular points together by n paths, in such a way that we can travel between any two of them by means of these n paths. Then when we know the auxiliary transformations relative to these n paths, the group of equation (1) will be completely determined.

In fact, let a be one of the singular points and let $b_1, b_2, \ldots, b_n$ be the n others. We join, for example, the point a to each point b_i by a line. Let $\Sigma, \Sigma_1, \Sigma_2, \ldots, \Sigma_n$ be the canonical transformations relative to the points a, $b_1, b_2, \ldots, b_n$ respectively; let $S_1, S_2, \ldots, S_n$ be the auxiliary transformations relative to the paths $ab_1, ab_2, \ldots, ab_n$. The group of equation (1) will be generated by the n + 1 transformations

$$\Sigma,\ S_1^{-1}\Sigma_1 S_1,\ S_2^{-1}\Sigma_2 S_2,\ \ldots,\ S_n^{-1}\Sigma_n S_n$$

between which there is the relation

$$\Sigma\, S_1^{-1}\Sigma_1 S_1 S_2^{-1}\Sigma_2 S_2 \ldots S_n^{-1}\Sigma_n S_n = 1$$

Now suppose that we distinguish two of the singular points, a and b, and let $c_1, c_2, \ldots, c_{n-1}$ be the $n - 1$ others. We join a and b by $n - 1$ different paths am_1b, am_2b, ..., $am_{n-1}b$ and let $S_1, S_2, \ldots, S_{n-1}$ be the corresponding auxiliary transformations. These $n - 1$ paths partition the plane into $n + 1$ regions; suppose that each of these regions contains exactly one singular point c_i. The group of equation (1) will be completely determined. In fact, if Σ is the canonical transformation relative to the point a, then the group will be generated by the n transformations

$$\Sigma,\ S_1^{-1}S_2,\ S_2^{-1}S_3,\ \ldots,\ S_{n-2}^{-1}S_{n-1},\ S_{n-1}^{-1}S_1$$

If the point c_i is contained in the region bounded by the two paths am_ib and $am_{i+1}b$, then we describe a contour inside this region and around the point c_i; when the variable describes this contour the canonical integrals relative to the point a undergo the transformation

$$S_i^{-1}S_{i+1}$$

We see that, to determine the group of the equation (1), it suffices to know either the fundamental invariants, or certain auxiliary transformations.

II. Numerical calculation of fundamental invariants

The fundamental invariants which completely define the group of a linear equation are evidently functions of the coefficients of that equation, whence the following problem naturally arises: to determine these invariants as functions of these coefficients. But in reality this problem is double: we can propose to numerically calculate these invariants when we have a given numerical equation, but it is of greater interest to study, from the point of view of function theory, the way in which these invariants depend on the coefficients as the latter vary. The methods appropriate for numerical calculation teach us nothing about the nature of these functions, while the most instructive formulae from the latter point of view lead to laborious calculations if we want to translate them into numbers.

From the point of view of numerical calculation, a great number of methods have been proposed, among which I would cite those of Fuchs (**Crelle's Journal, Vol. 75**) and Hamburger (**Crelle's Journal, Vol. 83**).

The method of Fuchs consists in distinguishing two of the $n + 1$ singular points, a and b, as at the end of the preceding section, then dividing the plane into $n - 1$ regions by $n + 1$ paths am_1b, am_2b, ..., $am_{n-1}b$ so that each of the regions contains exactly one of the $n - 1$ remaining singular points c_1, c_2, ..., c_{n-1}. The learned geometer of Heidelberg then gives expansions of the canonical integrals relative to the point a, and these expansions are valid in a certain region R_a. Similarly, the canonical integrals relative to the point b are susceptible to expansions valid in a region R_b. The two regions R_a and R_b have $n + 1$ common parts P_1, P_2, ..., P_{n-1}. Moreover, we can draw the path am_ib in such a way that it remains constantly inside one of the regions R_a, R_b or both, while traversing the region P_i. It then suffices to compare the two expansions of Fuchs for any value of x inside P_i (the region in which both expansions are valid simultaneously) in order to be able to calculate the coefficients of the auxiliary transformation relative to the path am_ib. In this way the group of equation (1) is entirely determined.

One can vary this method indefinitely; suppose in fact that we join any two singular points a and b by an arbitrary path amb and that we look for the auxiliary transformation relative to this path. Around the two points a and b we draw any two regions R_a and R_b such that:

1° the first contains a as its only singular point, and the second contains b as its only singular point;

2° they have a common part P;

3° the path amb remains constantly inside at least one of the regions R_a and R_b, and traverses the region P.

Suppose that two functions $f_a(x)$ (and $f_b(x)$) are such that, when x remains inside R_a (or R_b) the function f_a (or f_b) has absolute value less than 1, so that these two functions give respective conformal mappings of the circle of radius 1 and center 0 on the regions R_a and R_b. I assume also that

$$f_a(a) = 0, \quad f_b(b) = 0$$

The canonical integrals relative to the point a are expanded in powers of $f_a(x)$. This expansion, whose coefficients are calculated by recurrence, will be valid in all of the region R_a. Similarly, the canonical integrals relative to the point b will be expandable in powers of $f_b(x)$ in all of the region R_b. It will suffice to compare the two expansions at a point of the region P (where both are valid simultaneously) in order to calculate the coefficients of the auxiliary transformation sought.

We choose the regions R_a and R_b so that the functions f_a and f_b will be as simple as possible. We could take, for example, a rectangle, or a spindle region between two intersecting circular arcs, or the part of the plane between two parallel lines.

Here now is the method of Hamburger.

Let O be one of the singular points and let $a_1, a_2, \ldots, a_n$ be the n others, arranged in order of increasing absolute value; let $\rho_1, \rho_2, \ldots, \rho_n$ be their absolute values, and construct the two circles with center at the origin and radii ρ_i and ρ_{i+1}. Consider a closed contour C inside the annular region bounded by these two circles. Let A be a point of this contour C; suppose for example

$$A = \sqrt{\rho_i \rho_{i+1}}$$

Let $v_1, v_2, \ldots, v_p$ be a fundamental system of integrals in the neighborhood of the point A. Let these integrals become $w_1, w_2, \ldots, w_p$ when we return to the point A after traversing the contour C.

The integrals v can be expanded in increasing powers of

$$\log x - \log A$$

and if we have

$$\log \rho_{i+1} - \log A > 2\pi$$

the expansions are again valid when we take

$$\log x - \log A = 2i\pi$$

Then substituting $2i\pi$ in place of $\log x - \log A$ in the expansions of the integrals v and their derivatives, we obtain the values of the integrals w and their derivatives, which suffice to determine the transformation corresponding to the contour C in the group of equation (1).

The method of Hamburger can easily be extended to the case where

$$\log \rho_{i+1} - \log A < 2\pi$$

To abbreviate, we put

$$\log \rho_{i+1} - \log A = \frac{\pi}{2\alpha}$$

and then expand the integrals v in powers of

$$\frac{x^{i\alpha} - A^{i\alpha}}{x^{i\alpha} + A^{i\alpha}} = z$$

The expansion will again be convergent for

$$x = Ae^{2i\pi}$$

or

$$z = \frac{e^{-2\pi\alpha} - 1}{e^{-2\pi\alpha} + 1}$$

Replacing z by this value in the expansions of the integrals v and their derivatives, we obtain the values of the integrals w and their derivatives, and consequently the transformation which corresponds to the contour C.

An instant's reflection shows that this method can be varied indefinitely. The calculator's choice is guided by the more or less rapid convergence of the series employed. This convergence depends above all on the relative position of the n + 1 singular points. Hence it is this position which determines the choice of method.

However, none of the methods just proposed tells us anything about the properties of the fundamental invariants as functions of the coefficients, from the viewpoint of function theory. This is what we begin to study in the next section.

III. Properties of the fundamental invariants

We write equation (1) in the following form:

$$(1) \qquad \frac{d^p v}{dx^p} = \sum_{k=0}^{p=2} \sum_{i=1}^{n} \sum_{h} \frac{A_{hki}}{(x-a_i)^h} \frac{d^k v}{dx^k}$$

To write the equation in this way we decompose the coefficients of the various derivatives of v, which by hypothesis are rational functions of x, into simple elements. We have also assumed that these rational functions have no entire part, as is always permissible.

The fundamental invariants that we seek are functions of the a_i and the A_{hki}. First consider the a_i as constants and the A as the only variables. I claim that **the invariants will be entire functions of the** A.

To see this, it suffices to proceed as follows:

We write equation (1) in the following form:

$$\frac{d^p v}{dx^p} = \alpha \sum \phi_k \frac{d^k v}{dx^k} + \beta \sum \psi_k \frac{d^k v}{dx^k} \tag{1'}$$

The ϕ_k and ψ_k are rational functions of x which I suppose to be completely determined; α and β are parameters which I regard as variables. It suffices, I claim, to show that the invariants are entire functions of α and β.

Consider any point a of the plane and describe a circle K, centered on that point, so small that it contains no singular point. Let $v_1, v_2, \ldots, v_p$ be p integrals subject to the following conditions:

At $x = a$, $v_1 = 1$ and its first $p - 1$ derivatives are zero,

$v_i = 0$ and its first $p - 1$ derivatives are zero,

except the $i - 1^{th}$, which equals 1.

The p integrals $v_1, v_2, \ldots, v_p$ can be expanded in increasing powers of $x - a$ inside the circle K. The coefficient of $(x-a)^m$ is a polynomial of order $m - p + 1$ in α and β. We therefore have an expansion of these integrals in powers of $x - a$, α and β. It remains to see that this expansion is convergent.

If we regard v as a function of x, α and β, then equation (1) can be regarded as a partial differential equation and the theorem of Kowalevsky (**Crelle's Journal**, **Vol. 80**) applies to it, showing that v can be expanded in powers of $x - a$, $\alpha - \alpha_0$ and $\beta - \beta_0$, provided that x is inside the circle K and the absolute values of $\alpha - \alpha_0$ and $\beta - \beta_0$ are sufficiently small. Thus v is a holomorphic function of α and β in the neighborhood of an arbitrary point α_0, β_0; it is therefore an entire function of α and β and the expansion of v in powers of $x - a$, α and β is convergent for any α and β, provided x is inside the circle K.

Now it is necessary to show that if x is regarded as constant then v is again an entire function of α and β, **when** x **is also outside the circle** K. To completely define the function v in this case it is not sufficient to give the value of x, it is also necessary to know the path amx by which the variable arrives at this value from the point a.

Suppose, to fix ideas, that b is a point inside K, that K′ is a circle centered on b and not containing any singular point, that x is inside K′ and outside K, and finally that the path amx is entirely inside the figure formed by the two circles K and K′.

Let $u_1, u_2, \ldots, u_p$ be a fundamental system of integrals defined as follows: u_i must satisfy the same conditions for x = b as v_i does for x = a.

Let v_i^1 be the value of v_i for x = b, and v_i^{k+1} the value of its kth derivative. We have

$$v_i = \sum_{k=1}^{p} v_i^k u_k$$

But the v_i^k are entire functions of α and β because the point b is inside the circle K, and the u_k are also entire functions of α and β because the point x is inside the circle K′. Hence the v_i are also entire functions of α and β.

The argument will be the same if we have to consider several circles of convergence instead of just the two, K and K′.

Now suppose that the path amx is a closed contour C. The values of the integrals v_i and their derivatives are then nothing but what we called w_i^k in Section I. But the fundamental invariants are polynomials in the w_i^k, and hence entire functions of α and β.

Thus, when we regard the a_i as constants, the invariants are entire functions of the A_{hki} and consequently can be expanded in powers of these quantities A_{hki}. If we consider any coefficient in this expansion, it will evidently be a function of the a_i, and it remains to study the nature of this function. It is easy to see that these functions can be expressed in terms of successive quadratures. In fact, we write equation (1) in the form (1′) and consider the expansion of the integral v_i in powers of α and β

$$v_i = \sum v_{imn} \alpha^m \beta^n$$

I claim that we can obtain the coefficient v_{imn} by simple quadratures. Supposing that this is true for $v_{i,m-1,n}$ and $v_{i,m,n-1}$, it will also be true for v_{imn}, since we have identically

$$\frac{d^p v_{imn}}{dx^p} = \Sigma\, \phi_k \frac{d^k v_{i,m-1,n}}{dx^k} +$$

$$+ \Sigma\, \psi_k \frac{d^k v_{i,m,n-1}}{dx^k}$$

and if the pth derivative of v_{imn} is expressed by means of successive quadratures, then it is the same for the function v_{imn} itself.

Now let w_{imn}^k be the coefficient of $\alpha^m \beta^n$ in the expansion of w_i^k. This coefficient is expressible by quadratures, for the same reason, and it will be the same for the coefficients in the expansion of the invariants, since they are polynomials in the w_{imn}^k. Moreover, we can push the study of the expansion of the function v_i further. For this purpose we put

$$\Lambda(x,\alpha_1) = \int_0^x \frac{dx}{x-\alpha_1}$$

$$\Lambda(x,\alpha_1,\alpha_2) = \int_0^x \frac{dx\Lambda(x,\alpha_1)}{x-\alpha_2}$$

. . . .

$$\Lambda(x,\alpha_1,\alpha_2,\ \ldots,\ \alpha_{q-1},\alpha_q) = \int_0^x \frac{dx\Lambda(x,\alpha_1,\alpha_2,\ldots,\alpha_{q-1})}{x-\alpha_q}$$

We now remark that we can put equation (1) in the form of p

simultaneous first order equations. If the integrals are regular in the neighborhood of each of the singular points, then we can introduce p simultaneous variables $u_1 = v$, u_2, ..., u_p and replace equation (1) by the p simultaneous equations

$$\frac{du_i}{dx} = \Sigma\, \phi_{ik} u_k$$

The ϕ_{ik} will be rational functions of the following form:

(2) $$\phi_{ik} = \Sigma \frac{A_{i,k,\lambda}}{x - a_\lambda}$$

where the $A_{i,k,\lambda}$ are constants. Supposing, for example, that $p = 2$, then equation (1) will be

$$\frac{d^2 v}{dx^2} = \frac{P}{Q^2} v$$

Q will be the product $(x-a_1)(x-a_2)...(x-a_n)$ and P will be a polynomial in x of degree $2n - 2$. We can always find two integral polynomials A and C of degree $n - 1$ in x identically satisfying the relation

$$C + A^2 + A'Q - AQ' = P$$

We then put $v = u_1$ and introduce an auxiliary variable u_2. We can write

$$\frac{du_1}{dx} = \frac{A}{Q} u_1 + \frac{1}{Q} u_2$$

$$\frac{du_2}{dx} = \frac{C}{Q} u_1 + \frac{Q' - A}{Q} u_2$$

The coefficients are indeed of the form (2).

It is now easy to see that if we expand the integral v_i in powers of the $A_{ik\lambda}$, then the coefficient of any term will be a sum of functions such as Λ, when the parameters α_1, α_2, ..., α_q will be none other than the singular points $a_1, a_2, \ldots, a_n$ in an order in which each of them can be repeated a finite number of times.

IV. Inverse functions

We have until now studied the fundamental invariants as functions of the coefficients of equation (1), but if, on the contrary, we regard these coefficients as functions of the fundamental invariants, we are led to interesting transcendants which play the same role with respect to linear equations as the modular function does with respect to elliptic integrals.

One remark is important here: recall the equation

$$\frac{d^p v}{dx^p} = \Sigma\, \phi_k \frac{d^k v}{dx^k} \tag{1}$$

and consider the infinities of the coefficients ϕ_k; they are of two kinds: the first are singular points in the strict sense, and when the variable x describes a closed contour around one of these points the integrals undergo a linear transformation; the others are simple poles of the integrals, and if a is such a point then all the integrals can be put in the form

$$(x-a)^{\lambda} \phi(x)$$

where λ is a constant which is the same for all the integrals and $\phi(x)$ is holomorphic. It follows that when x describes a closed contour around the point a all the integrals are multiplied by the same factor, and their ratios retain their original values. These points are called **apparent singular points**. For an infinity of the coefficients ϕ_k to be an apparent singular point, $\frac{(p+2)(p+1)}{2}$ conditions must be satisfied.

Suppose then that an equation of the form (1) has n singular points apart from the point ∞, namely $a_1, a_2, \ldots, a_n$ and $b_1, b_2, \ldots,$

b_q.

I assume that the integrals are everywhere regular. Equation (1) can then be written

$$\frac{d^p v}{dx^p} = \sum_{k=0}^{p-2} \frac{P_k}{Q^{p-k}} \frac{d^k v}{dx^k}$$

We put

$$Q = (x-a_1)(x-a_2) \; \ldots \; (x-a_n)(x-b_1)(x-b_2) \; \ldots \; (x-b_q)$$

and P_k is a polynomial of order

$$(p-k)(n-q-1)$$

Equation (1) then contains

$$\frac{p(p+1)}{2} (n+q) - \frac{p(p-1)}{2}$$

parameters, namely

1° The $n + q$ infinities $a_1, a_2, \ldots, a_n, b_1, b_2, \ldots, b_q$;

2° The

$$\frac{(n+q-1)(p+2)(p-1)}{2} + p - 1$$

coefficients of the polynomials P_k.

But we have

$$q \frac{(p+2)(p-1)}{2}$$

conditions expressing the fact the b are apparent singular points. Hence there remain

$$n \frac{p(p+1)}{2} + q - \frac{p(p-1)}{2}$$

independent parameters.

We remark that we can always assume

$$a_1 = 0, \quad a_2 = 1$$

as well, for if this is not the case it can be brought about by a linear change of variable. It is therefore necessary to subtract two more parameters.

If we also assume that there are no apparent singular points, there finally remain

$$n\,\frac{p(p+1)}{2} - \frac{p(p-1)}{2} - 2 \tag{2}$$

parameters. But the group G of equation (1) is generated by n transformations, and we have seen in Section I that such a group G depends on

$$(n-1)(p^2-1) \tag{3}$$

fundamental invariants, when we do not distinguish it from its conjugates by different linear transformations.

If $p = 2$ the expressions (2) and (3) both reduce to $3n - 3$. The number of parameters of the equation is therefore equal to the number of parameters of the group, with the following consequence:

We can **in general** find a second order equation, without apparent singular points, which admits a given group.

I wish to point out that, for this equation to exist, it is not necessary for any relation to hold between the invariants of the group; if we have to impose certain conditions, they will be inequalities. Moreover, we can find infinitely many second order equations which have apparent singular points and which admit the same group.

Now suppose that $p > 2$ and of course $n > 1$. We easily see that the expression (2) is always smaller than (3), whence:

We cannot in general find an equation of order higher than second, without apparent singular points, which admits a given group.

Thus it is necessary in general, if we are to construct an

equation with a given group, to introduce apparent singular points.

It is to avoid these points, which greatly complicate the results and proofs, that I confine myself here to equations of second order. I therefore suppose that we have a second order equation without apparent singular points.

Consider the n + 1 singular points on the sphere

$$a_1 = 0,\ a_2 = 1,\ a_3,\ a_4,\ \ldots,\ a_n,\ a_{n+1} = \infty$$

and join them by n cuts

$$a_1a_2,\ a_2a_3,\ \ldots,\ a_na_{n+1}$$

Consider any two integrals of equation (1) and their ratio z; when x runs over the sphere without crossing the cuts, z will run over a certain region R. This region will be analogous to the generating polygons of Fuchsian and Kleinian groups, however it can overlap itself. It has 2n sides

$$\alpha_1\alpha_2,\ \alpha_2\alpha_3,\ \ldots,\ \alpha_n\alpha_{n+1}$$

and

$$\alpha_1\beta_2,\ \beta_2\beta_3,\ \ldots,\ \beta_n\alpha_{n+1}$$

corresponding to the n cuts

$$a_1a_2,\ a_2a_3,\ \ldots,\ a_na_{n+1}$$

The sides $\beta_i\beta_{i+1}$ and $\alpha_i\alpha_{i+1}$ are equivalent, and we pass from one to the other by a linear transformation S_1. The vertices α_1 and α_{n+1} then form a cycle. There are n + 1 other cycles formed by the respective α_i and β_i vertices. The sum of the angles α_i and β_i is therefore equal to

$$2\pi(1-2\lambda_i)$$

where λ_i is the smallest root of the determinant equation relative to the singular point a_i. Similarly, the angles α_i and α_{n+1} are equal to

$$2\pi(1-2\lambda_1) \quad \text{and} \quad 2\pi(1-2\lambda_{n+1})$$

When we know the value of

$$\alpha_1, \alpha_2, \ldots, \alpha_{n+1}, \beta_2, \beta_3, \ldots, \beta_n, \lambda_1, \lambda_2, \ldots, \lambda_{n+1} \tag{4}$$

the transformations S_i are completely determined and, since they are the generators of the group G, the fundamental invariants of this group are easily expressed as functions of the quantities (4). We therefore regard the coefficients of equation (1) as functions of the quantities (4).

The region R is not completely determined when the quantities (4) are given; in fact we can arbitrarily vary the form of the sides $\alpha_1\alpha_2$, $\alpha_2\alpha_3$, ..., $\alpha_n\alpha_{n+1}$, with corresponding variations in the equivalent sides $\alpha_1\beta_2$, $\beta_2\beta_3$, ..., $\beta_n\alpha_{n+1}$. Nevertheless, all the regions R obtained are **piecewise equivalent** in the sense of Section III of the Memoir on Kleinian Groups (**Acta Mathematica, Vol. 3**, p. 63, this volume p.271). Moreover, we can draw the cuts

$$a_1a_2, \ldots, a_na_{n+1}$$

so that the sides

$$\alpha_1\alpha_2, \ldots, \alpha_n\alpha_{n+1}$$

have whatever form we want.

It follows that if there are two equations (1) leading to the same system of values of the quantities (4), then we can always draw the cuts in such a way that the region R is the same for both equations.

Now suppose that the functions of z enjoy the following

properties:

1° they are uniform for z in the region R; it is clear that if the region R overlaps itself then two points z_0 and z_1 of that region can correspond to the same point of the plane; in this case the function can take two different values at the points z_0 and z_1;

2° they return the same values at two corresponding points on the perimeter of R;

3° they have no singularities other than poles (and logarithmic singular points in the case where one of the α and β angles is zero).

It is clear that all these functions can be expressed rationally in terms of one among them. (Cf. Schottky, **Crelle's Journal**, **.Vol. 83** and On Fuchsian Functions, **Acta Mathematica**, **Vol. 1**, p. 228, and this volume p.173).

If we now regard x as a function of z, then it will be one of the functions just mentioned, and it is clear that all the others are rational in x.

Now suppose that we have two equations (1) which lead to the same system of values of the quantities (4), and consequently to the same region R. Let x and x_1 be the corresponding variables which we regard as functions of z. By the preceding, x will be rational in x_1 and x_1 will be rational in x and consequently there will be a relation of the form

$$Axx_1 + Bx + Cx_1 + D = 0$$

Now let a_i and a_i^1 be the values of x and x_1 which correspond to the value $z = \alpha_i$. These will be the singular points of the two equations (1), and by hypothesis we have

$$a_1 = a_1^1 = 0, \quad a_2 = a_2^1 = 1, \quad a_{n+1} = a_{n+1}^1 = \infty$$

I.e., we have

$$x = x_1$$

for $z = \alpha_1$, $z = \alpha_2$, $z = \alpha_{n+1}$.

It follows that x will be indentically equal to x_1. The two equations (1) we have supposed to exist are therefore identical.

It follows that, when the quantities (4) are completely determined, so is equation (1), and the **coefficients of this equation are uniform functions of the quantities** (4).

Thus we are led to a new class of uniform functions of several variables. These functions are invariant when the quantities (4) are subjected to certain transformations I shall now say a few words about.

We recall the polygon R defined above and let S_k be the linear transformation which changes $\beta_k \beta_{k+1}$ into $\alpha_k \alpha_{k+1}$. For greater symmetry of notation we agree to put

$$\alpha_1 = \beta_1, \quad \alpha_2 = \beta_2, \quad \alpha_{i+n+1} = \alpha_i, \quad \beta_{i+n+1} = \beta_i$$

The following transformation:

$$(5) \qquad \begin{array}{l} \alpha_i, \ \beta_{k+1-i} S_k \\ \beta_i, \ \alpha_{k+1-i} \\ \lambda_i, \ \lambda_{k+1-i} \end{array}$$

applied to the quantities (4) leaves invariant the function we have been concerned with. Table (5) gives the new values of the quantities α, β and λ as functions of the old, since the coefficients of the linear transformation S_k are easily obtainable as functions of the α, β and λ.

V. Enunciation of the second problem

In the four preceding Sections we have shown how to resolve the following problem:

Find the group of a given equation, both from the point of view of numerical calculation, and from the point of view of the theory of functions.

Here is the second problem I have to resolve before going further.

We are given a second order equation

$$\frac{d^2v}{dx^2} = \phi(x,y)v \tag{1}$$

where ϕ is a rational function of two variables x and y connected by an algebraic relation

$$\psi(x,y) = 0 \tag{2}$$

We suppose that the function ϕ depends on a certain number of parameters and we demand that these parameters be disposed in such a way as to make x a Fuchsian function of the ratio of the integrals. We then say, for brevity, that equation (1) is **Fuchsian**.

In what follows we shall be concerned only with Fuchsian functions which do not exist outside the fundamental circle (the first, second and sixth families). Thus we systematically leave aside the Fuchsian functions which exist over the whole plane.

For equation (1) to be Fuchsian, it is necessary that it satisfy, on the one hand, certain algebraic conditions, on the other hand, a certain number of transcendental conditions.

We begin by stating the algebraic conditions.

The singular points of equation (1) are of two kinds:

1° The points where y ceases to be a holomorphic function of x and the points where x or y ceases to be finite;

2° The singular points in the strict sense.

We do not have to worry about the first kind, since we can put

$$x = \theta(x',y'), \qquad y = \theta_1(x',y')$$

where θ and θ_1 are rational functions, to arrange things so that, in the neighborhood of the point considered, the new variables x' and y' remain finite, with y' holomorphic in x'.

As for the singular points in the strict sense, to each of them there corresponds a determinant equation and **the difference between the roots of this equation must be zero or an aliquot part of unity**. These are the algebraic conditions in question and when they are satisfied I say that equation (1) is **normal**.

First consider two normal equations

$$\frac{d^2v}{dx^2} = \phi(x)v, \qquad \frac{d^2v}{dx_1^2} = \phi_1(x_1)v$$

where we assume that ϕ and ϕ_1 are rational functions of x and x_1. If we can pass from the first of these equations to the second by putting

$$x = \frac{ax_1 + d}{cx_1 + d}$$

then we do not consider the two equations as distinct.

Now consider a more general normal equation

$$\text{(1)} \qquad \frac{d^2v}{dx^2} = \phi(x,y)v$$

$$\text{(2)} \qquad \psi(x,y) = 0$$

and put

$$\text{(3)} \qquad x' = \theta(x,y), \qquad y' = \theta_1(x,y)$$

If we leave aside certain exceptional cases, we can solve (2) and (3) for x and y as rational functions of x′ and y′. I assume that we do not have these exceptional cases.

We then find

$$\text{(1}'\text{)} \qquad \frac{d^2v}{dx'^2} = \phi'(x',y')v$$

$$\text{(2}'\text{)} \qquad \psi'(x',y') = 0$$

I do not regard equations (1) and (1′) as distinct.

I say that two equations

$$(1) \qquad \frac{d^2v}{dx^2} = \phi(x,y)v$$

$$(1') \qquad \frac{d^2v}{dx^2} = \phi_1(x,y)v$$

$$(2) \qquad \psi(x,y) = 0$$

$$(2') \qquad \psi_1(x,y) = 0$$

belong to the same **type** if the two relations (2) and (2′) are identical and if corresponding singular points of equations (1) and (1′) have the same determinant equations.

It goes without saying that if an equation

$$(1'') \qquad \frac{d^2v}{dx'^2} = \phi_1'(x',y')v$$

$$(2'') \qquad \psi_1'(x',y') = 0$$

is regarded as the same as (1′) by virtue of the above conditions, then I also say that (1) and (1″) belong to the same type.

Here is the problem to be resolved:

1° **Recognize whether there is a Fuchsian equation among the equations of a given type.**

2° **Find this equation, if it exists.**

But first it is necessary to make a distinction, and to classify the types of normal equation into Fuchsian, elliptic and rational types.

Suppose that x is a Fuchsian function of the ratio of the integrals of equation (1); this function admits a generating polygon R_0 which has 2n sides of the first kind and 2n vertices of the first or second kind. The sum of the angles of this polygon must be smaller than

$$\pi(2n-2)$$

Let q be the genus of the relation (2); p the number of singular points of equation (1); a_1, a_2, ..., a_p these singular points; α_i the difference between the roots of the determinant equation relative to the point a_i.

If $p > 0$ the vertices of R_0 are partitioned into p cycles; we have

$$n = 2q + p - 1$$

and the sum of the angles will be equal to

$$2\pi \Sigma \alpha_i$$

We must then have the inequality

$$\Sigma \alpha_i < 2q + p - 2 \tag{4}$$

If $p = 0$ the vertices of R_0 will form a single cycle and the sum of the angles will be 2π. We have

$$n = 2q$$

whence the inequality

$$q > 1 \tag{4'}$$

If the inequalities (4) or (4') are satisfied, the type considered will be a **Fuchsian** type.

Now suppose that x is a doubly periodic function of the ratio of the integrals of (1), and we have the equations

$$\Sigma \alpha_i = 2q + p - 2 \tag{5}$$

or else

$$q = 1, \quad p = 0 \tag{5'}$$

The type is then **elliptic**.

Moreover, here is the enumeration of elliptic types:

$$q = 1 \quad p = 0$$
$$q = 0 \quad p = 3 \quad \alpha_1 = 1/3 \quad \alpha_2 = 1/3 \quad \alpha_3 = 1/3$$
$$q = 0 \quad p = 3 \quad \alpha_1 = 1/2 \quad \alpha_2 = 1/4 \quad \alpha_3 = 1/4$$
$$q = 0 \quad p = 3 \quad \alpha_1 = 1/2 \quad \alpha_2 = 1/3 \quad \alpha_3 = 1/6$$
$$q = 0 \quad p = 4 \quad \alpha_1 = \alpha_2 = \alpha_3 = \alpha_4 = 1/2$$

(If $q = 0$ we can have neither $p = 0$ nor $p = 1$. If we suppose $p = 2$, we are led to the equation

$$\frac{d^2v}{dx^2} = \frac{Av}{(x-a)^2}\ )$$

Finally, suppose that x is a rational function of z, and we have the inequalities

$$\Sigma\, \alpha_i > 2q + p - 2 \tag{6}$$

the hypothesis $p = q = 0$ must be rejected.

The type is then **rational**. Here is the enumeration of rational types, which is well known:

$$q = 0 \quad p = 3 \quad \alpha_1 = 1/2 \quad \alpha_2 = 1/2 \quad \alpha_3 = 1/n$$
$$q = 0 \quad p = 3 \quad \alpha_1 = 1/2 \quad \alpha_2 = 1/3 \quad \alpha_3 = 1/3$$
$$q = 0 \quad p = 3 \quad \alpha_1 = 1/2 \quad \alpha_2 = 1/3 \quad \alpha_3 = 1/4$$
$$q = 0 \quad p = 3 \quad \alpha_1 = 1/2 \quad \alpha_2 = 1/3 \quad \alpha_3 = 1/5$$

Thus the equations of the same type depend on a certain number P of arbitrary parameters, so that the type consists of ∞^P distinct equations.

These P parameters, being complex, correspond to 2P real parameters. But the number of these real parameters is precisely that of the transcendental conditions which must be satisfied by equation (1) (assuming it to belong to the given type) if x is to be a Fuchsian function of x (or else doubly periodic or rational, in the case of elliptic or rational type).

(cf. On Fuchsian functions, **Acta Mathematica**, **Vol. I**, p. 234, 257, 262 and 272, this volume p. 179, 210, 216, 228)

If it were a question of algebraic, rather than transcendental, conditions we could conclude that there was always a Fuchsian equation among those of a given type. But this conclusion is not legitimate and a special proof is necessary. This proof is the heart of the problem we are concerned with.

It is easy to understand its importance.

Suppose in fact that the equation

$$\frac{d^2 v}{dx^2} = \phi(x,y)v \tag{1}$$

$$\psi(x,y) = 0 \tag{2}$$

is Fuchsian. Let

$$(x=a_1, y=b_1),\ (x=a_2, y=b_2),\ \ldots,\ (x=a_p, y=b_p)$$

be the singular points of equation (1), and let $\frac{1}{k_i}$ be the difference between the roots of the determinant equation relative to the point (a_i, b_i). The number k_i will be a positive integer or infinity.

We consider a function of x and y with no singular points other than

$$(a_1, b_1),\ (a_2, b_2),\ \ldots,\ (a_p, b_p)$$

and such that the function returns to its original value when the analytic point (x,y) makes k_i turn around the singular point (a_i, b_i). This latter condition must be dropped when $k_i = \infty$.

This function will be uniform in x.

It will be the same for the integrals of the equation

$$\frac{d^q w}{dx^q} = \sum \phi_k(x,y) \frac{d^k w}{dx^k} \tag{7}$$

if the functions ϕ_k are rational in x and y, if there are no singular points other than

$$(a_1,b_1),\ (a_2,b_2),\ \ldots,\ (a_p,b_p)$$

and if all the roots of the determinant equation relative to the point (a_i,b_i) are multiples of $\frac{1}{k_i}$ (the latter condition being dropped when $k_i = \infty$).

It will be the same for every rational function of x and y and for a great number of algebraic functions.

Consequently, if we show that each Fuchsian type contains a Fuchsian equation we shall have:

1° Given any linear equation with algebraic coefficients, the variable and the integrals can be expressed as uniform functions of the same auxiliary variable z.

2° Given any algebraic curve, the coordinates x and y of any point of that curve can be expressed as uniform functions of the same auxiliary variable z.

VI. Subordination of types

We consider two normal equations

$$\frac{d^2 v}{dx^2} = \phi(x,y)v \tag{1}$$

$$\frac{d^2 v}{dx^2} = \phi_1(x,y)v \tag{1'}$$

The coefficients ϕ and ϕ_1 are rational functions of two variables x and y which are connected by an algebraic relation

(2) $$\psi(x,y) = 0$$

which **I assume to be the same for the two equations** (1) **and** (1′).

I assume that equation (1) has p singular points

$$(a_1,b_1),\ (a_2,b_2),\ \ldots,\ (a_p,b_p)$$

and that the difference between the roots of the determinant equation relative to (a_i,b_i) is $\frac{1}{k_i}$.

I assume that equation (1′) has the same singular points (a_1,b_1), (a_2,b_2), ..., (a_p,b_p) as equation (1), and in addition q others

$$(a_{p+1},b_{p+1}),\ (a_{p+2},b_{p+2}),\ \ldots,\ (a_{p+q},b_{p+q})$$

such that the difference between the roots of the determinant equation relative to (a_i,b_i) is $\frac{1}{N_i}$.

Finally, I assume that N_1 is divisible by k_1, N_2 by k_2, ..., N_p by k_p. If k_i is infinite, N_i must also be infinite. If $n_i = \infty$ the divisibility condition is considered to be satisfied, whatever k_i is.

I then say that the type of equation (1′) is subordinate to the type of equation (1).

Let a Fuchsian type T′ be subordinate to another Fuchsian type T. I assume that each of them contains a Fuchsian equation; the first the equation E′ and the second the equation E. Let z be the ratio of the integrals of equation E′, and t the ratio of the integrals of equation E. It is easy to see that t is a uniform function of z. It is evident that this function exists only when z is inside the fundamental circle. For its part, t cannot take any value outside this circle; on the other hand, t takes each value inside the circle infinitely often.

Again let (a_1,b_1), (a_2,b_2), ..., (a_p,b_p) be the singular points of type T and let $\frac{1}{k_i}$ be the difference between the roots of the determinant equation relative to the point (a_i,b_i). Let F be a function of the analytic point (x,y) with no singular point other than (a_1,b_1), (a_2,b_2), ..., (a_p,b_p) and such that it returns to its initial value after k_i

turns around the point (a_i, b_i). F could be, for example, an integral of the linear equation (7) is the preceding section. F will be a uniform function of t and consequently of z.

But suppose that we do not know whether the type T contains a Fuchsian function, so that we are unable to show the existence of t, whereas we know that T′ contains a Fuchsian equation and that the function z exists. It is easy to see that F is again a uniform function of z.

Thus to prove the following result: **the integrals of a linear equation with coefficients rational in** x **and** y **can be expressed, like** x **and** y, **as uniform functions of the same auxiliary variable**, it is not necessary to show that every Fuchsian type contains a Fuchsian equation. It suffices to show that, among the types subordinate to a given Fuchsian type, there is always one which contains a Fuchsian equation.

This is the importance of the notion of subordination of types.

In particular, consider an equation E with rational coefficients and let

$$a_1,\ a_2,\ \ldots,\ a_n$$

be its singular points. Let k_i be an integer such that all the roots of the determinant equation relative to the point a_i are multiples of $\frac{1}{k_i}$. If such an integer does not exist, we take $k_i = \infty$.

Instead of considering the type T with the singular points $a_1, a_2, \ldots, a_n$ and $\frac{1}{k_1}, \frac{1}{k_2}, \ldots, \frac{1}{k_n}$ as the differences of the roots of the determinant equations, we can consider a type T′ subordinate to T. We assume, for example, that T′ has p other singular points $b_1, b_2, \ldots, b_p$ in addition to the singular points a_1, a_2, a_n and that each of the p + n determinant equations has a double root.

Suppose that this type T′ contains a Fuchsian equation E′; then x will be a Fuchsian function of the ratio z of the integrals of that equation. The characteristic property of this Fuchsian function is that it cannot take any of the values $a_1, a_2, \ldots, a_n, b_1, b_2, \ldots, b_p$ and, if it exists, it is evident that the integrals of equation E will be uniform functions of z.

Thus it suffices to show that we can always find a Fuchsian function which does not take n given values $a_1, a_2, \ldots, a_n$, in order to see that all linear equations with rational coefficients can be integrated using only certain uniform functions, the study of which I reserve to a later memoir.

It is the same for equations with algebraic coefficients, since we know that integration of an equation with algebraic coefficients reduces to integration of a higher order equation with rational coefficients.

VII. Fundamental lemma

No Fuchsian type can contain more than one distinct Fuchsian equation.

Suppose in fact that a type T contains two Fuchsian equations E and E'; let z be the ratio of two integrals u_1 and u_2 of equation E; let t be the ratio of two integrals u'_1 and u'_2 of equation E'.

We know that z can only take values inside a certain fundamental circle. We can always choose the integrals u_1, u_2 and u'_1, u'_2 so that:

1° the fundamental circle for z, and likewise the fundamental circle for t, has center at 0 and radius 1;

2° the two variables z and t vanish at the same time, at x = a for example;

3° at any other value b of x the argument z equals that of t.

It is clear that z will be a uniform function of t, and conversely that t will be a uniform function of z. It then follows from the above hypotheses that $\frac{t}{z}$, considered either as a function of z or a function of t, does not exist outside the fundamental circle, and that inside this circle the ratio remains holomorphic and non–zero.

Consider, for example, $\frac{t}{z}$ as a function of z. We put

$$z = \xi + i\eta$$

$$u = \log \left| \frac{t}{z} \right|$$

and u will be a function of ξ and η, holomorphic inside the fundamental circle and satisfying the equation

(1) $$\frac{d^2u}{d\xi^2} + \frac{d^2u}{d\eta^2} = 0$$

Consider a circle C in the z plane which is concentric with the fundamental circle but of radius $r < 1$.

Along this circle, t has absolute value constantly less than one, and consequently

(2) $$u < \log \left|\frac{1}{r}\right|$$

But no holomorphic function satisfying equation (1) can take a value inside a contour which is greater than its maximum value along that contour. The inequality (2) therefore holds for all points inside the circle C.

But I can take the radius of this circle C as close to unity as I please. When I let r tend to unity, the right hand side of the inequality (2) tends to zero. It follows that u cannot take any positive value inside the fundamental circle, whence

$$\left|\frac{t}{z}\right| \geqslant 1$$

But there is no difference between t and z, so we can also write

$$\left|\frac{z}{t}\right| \geqslant 1$$

whence

$$|z| = |t|$$

whence

$$t = ze^{i\theta}$$

where θ is a constant.

But t and z have the same argument for $z = b$, by hypothesis, so we have

$$\theta = 0 \quad \text{and} \quad t = z$$

Thus the two equations E and E′ are not distinct.

Q.E.D.

This important lemma was first announced by me in a communication to the Academy of Sciences of Paris (**Comptes rendus, 93**, 17 October 1881, p. 582). It was later given by Klein (**Mathematische Annalen, 21**, p. 209). Moreover, Klein extended it to the case of functions which exist in a domain bounded by an infinity of circles (cf. Memoir on Kleinian Groups, Section IX, **Acta Mathematica, Vol. 3**, p. 83, this volume p.293). However, I will say nothing of this extension, since I do not need it for my purpose.

VIII. First glance at the method of continuity

Klein and I have been led independently to a method which permits a demonstration that each Fuchsian type contains a Fuchsian equation, and which can be called the **method of continuity**. We have made different applications of this method (see **Comptes rendus, 92**, pp. 1200 and 1486; Klein, **Mathematische Annalen, 19**, p. 565; **20**, p. 49, and **21**, p. 211; Poincare, **Comptes rendus, 94**, p. 1038). Here are the fundamental principles of this method.

First suppose that we have a real variable x and a real function y of that variable; I suppose that for each value of x, finite or infinite, we have exactly one value of y, and that this value is an analytic function of x (holomorphic or meromorphic); also, that y cannot take the same value twice. We can evidently conclude that y takes each possible value from $-\infty$ to $+\infty$ exactly once.

Suppose, on the contrary, that x can vary only between two limits a and b, that we know y is an analytic function of x at all values between a and b, but that we know nothing about it at these limits themselves, and that y cannot take the same value twice. When x tends to a, y tends to a certain limit A, and when x tends to b, y tends to B. Then we only know that y is either constantly increasing or constantly decreasing, and that it takes all intermediate values between A and B.

Instead of a single independent variable, consider n of them, $x_1, x_2, \ldots, x_n$, and n real functions of these n variables $y_1, y_2, \ldots, y_n$. I suppose that the y are analytic functions for all values of the x, and that these functions cannot take the same system of values twice. This is sufficient to ensure that the y can take all systems of possible values. In particular, let $n = 2$. We regard x_1 and x_2 as defining the position of a point on a sphere S, and y_1 and y_2 as defining the position of a point on a sphere S′. Each point of the sphere S corresponds to exactly one point of the sphere S′, and each point of s′ can correspond to no more than one point of S. This ensures that each point of S′ corresponds to a point of S.

It is again the same if S and S′, instead of being two spheres, are any two closed surfaces. But now suppose that S is not a closed surface, but an open one, with a kind of **boundary**, the **frontier**. Suppose that the coordinates of each point m′ or S′ are analytic functions of the coordinates of the corresponding point m of S, when the point m is not on the boundary of the surface, but that we know nothing when the point m is on the boundary of S. Then it will not be possible to conclude, from the above hypotheses, that each point of S′ corresponds to a point of S.

The same thing happens when S and S′ are regarded as surfaces in a space of more than three dimensions; they will be **Mannigfaltigkeiten** (manifolds) of more than two dimensions, as the Germans say.

We always assume that each point m of S corresponds to exactly one point m′ of S′, in such a way that the coordinates of m′ are analytic functions of those of m, at least when m is not at the boundary of the manifold S, in the case where this manifold has one. Suppose that no point of S′ corresponds to more than one point of S. If S is a **closed manifold**, we can be certain that each point of S′ corresponds to a point of S. If, on the contrary, S is an **open manifold** with a **boundary**, a **frontier**, then we can affirm nothing.

We now apply the preceding to the problem we are concerned with.

We say that two types

$$\frac{d^2v}{dx^2} = \phi(x,y)v \tag{1}$$

$$\frac{d^2v}{dx^2} = \phi_1(x,y)v \tag{1'}$$

$$\psi(x,y) = 0 \tag{2}$$

$$\psi_1(x,y) = 0 \tag{2'}$$

belong to the same class when the two relations (2) and (2′) are of the same genus, and when the two equations (1) and (1′) have the same number of singular points, which can be made to correspond in such a way that the determinant equations relative to two corresponding singular points are the same.

Having done this, we place each type in the same class C′ in correspondence with a point of a certain manifold S′.

Similarly, we say that two Fuchsian groups (of the first, second or sixth families) belong to the same class when their generating polygons have the same number of sides, and when the vertices are partitioned into the same number of cycles, which can be made to correspond in such a way that the angle sums of corresponding cycles are the same.

To each class C′ of types corresponds a class C of groups, in such a way that if a Fuchsian equation belongs to class C′, its group belongs to class C.

We make each group of class C correspond to a point of a manifold S.

Each point of S corresponds to exactly one point of S′; to

see that, it suffices to refer to the theory of Fuchsian functions which shows that each Fuchsian group corresponds to a Fuchsian equation.

Moreover, by virtue of the fundamental lemma, each point of S′ corresponds to at most one point of S.

It remains to see that each point of S′ corresponds to a point of S.

For this it is sufficient, by the preceding, that S be a **closed** manifold. **This is not at all evident** a priori.

In fact, among the groups of a given class, there are infinitely many which could be **limit groups**, corresponding to points on the **boundary** of the manifold S. These are the groups whose generating polygons have one or more infinitesimal sides. We see, in fact, that we can always construct a generating polygon with one side as small as we please. However, this is not enough to show that S is an open manifold since, by virtue of Section IX of Theory of Fuchsian Groups, the same group can be generated by an infinity of piecewise equivalent polygons, and it is possible that among these we can always choose one whose sides have lengths above a given bound.

Thus it is not evident that S is a closed manifold, and it is necessary to show, by a special discussion in each particular case, that each point of S′ corresponds to a point of S. This is what Klein has neglected to do. **It is a difficulty which cannot be overcome in a few lines.**

IX. Second lemma

Suppose that we vary our polygon R_0 continuously, so that the elements of this polygon are continuous functions of a certain parameter t. We now consider the Fuchsian equation which can be formed with the aid of R_0, and the type T to which this equation belongs. Let

$$(1) \qquad \frac{d^2 v}{dx^2} = \phi(x,y)v$$

$$(2) \qquad \psi(x,y) = 0 \quad \text{(of genus } p\text{)}$$

be this Fuchsian equation. The type T will be defined when we know the $3p - 3$ moduli of the relation (2) and the singular points of equation (1); these quantities are called the **parameters** of the type.

I have shown in Section I of the memoir On Fuchsian Functions that the functions Θ we can form with the aid of R_0 are continuous functions of t, and we can conclude that the parameters of type T are also continuous functions of t, which is most important from the point of view of applications of the method of continuity.

I want to push this further. I suppose that when t tends to zero, two or more sides decrease indefinitely. (Here we are talking about their geometric length and not their L.) I suppose, for example, that two equivalent sides, or 2q sides equivalent in pairs, become infinitely small, so that the sides of R_0 which remain finite are again equivalent in pairs. R_0 is then one of the limit polygons which we shall study in detail a little later.

Passing to the limit by setting $t = 0$, so that the 2q infinitesimal sides of R_0 vanish altogether, R_0 is reduced to another polygon R_0 having no more than $2n - 2q$ sides equivalent in pairs. We consider the group G generated by R_0. Corresponding to each pair of equivalent sides of R_0 there is a transformation in G which sends one of these two sides to the other, and the group G is generated precisely by the n transformations which correspond in this way to the different sides of R_0. Among these transformations we distinguish the $n - q$ which correspond to the $n - q$ pairs of sides of R_0 which remain finite. The group generated by these $n - q$ transformations is contained in G and will be called G''. Let G' be the group generated by R'_0; it will be isomorphic to G'', because to each **finite** side of R_0 there corresponds a side of R'_0. When t tends to zero, the transform of R_0 by any transformation in G'' tends to the transform of R'_0 by the corresponding transformation in G'.

We let A be the portion inside the fundamental circle which is occupied by the transforms of R_0 under transformations in G″, and let B be the portion occupied by the transforms of R_0 under transformations in G which do not belong to G″.

By hypothesis, the polygon R_0 is of the first family, or of the first order in the second or sixth family; in the limit, the polygon R'_0 will belong to the second or sixth family, but it can belong to the first or second order of these families, i.e., it may or may not admit hyperbolic cycles*. (*See note on p.280. N.E.N.)

In the first case, R_0 will be of the first category; in the second, of the second category.

In the first case, the transforms of R'_0 by transformations in G fill in the whole fundamental circle (cf. Theory of Fuchsian Groups, Section VI). Consequently, when t tends to zero, the total area of B will tend to zero, and at the same time the smallest distance from B to the origin will tend to 1, i.e., to the radius of the fundamental circle.

In the second case, conversely, the area of B does not tend to zero.

Suppose then that R_0 is of the first category. In this case R'_0 gives rise to a Fuchsian equation

$$(1') \qquad \frac{d^2v}{dx^2} = \phi'(x,y)v$$

$$(2') \qquad \psi'(x,y) = 0$$

which belongs to a type T′ less complicated than T. Suppose, to fix ideas, that q = 1. In this case R'_0 will have two sides less than R_0. It is then the case either that (1′) has one less singular point than (1), or else that relation (2′) has genus p − 1 instead of genus p, and that equation (1′) has two singular points more than equation (1). (See below, p.424)

Suppose for example, to fix ideas, that the first case occurs. I claim that the moduli of the relation (2) must tend to those of the relation (2′) and that the singular points of equation (1) must tend towards those of equation (1′), in such a way that, among the singular points of equation (1), there are two which tend toward the same singular point of equation (1′).

I shall first show that the theta–Fuchsian functions engendered by R_0 tend towards those engendered by R'_0.

Consider the series

$$\Theta(z,H) = \Sigma\, H\left[\frac{\alpha_i z+\beta_i}{\gamma_i z+\delta_i}\right](\gamma_i z+\delta_i)^{-2m}$$

$$\Theta'(z,H) = \Sigma\, H\left[\frac{\alpha'_i z+\beta'_i}{\gamma'_i z+\delta'_i}\right](\gamma'_i z+\delta'_i)^{-2m}$$

where H is the algorithm of an arbitrary rational function and

$$\left[z,\ \frac{\alpha_i z+\beta_i}{\gamma_i z+\delta_i}\right]$$

and

$$\left[z,\ \frac{\alpha'_i z+\beta'_i}{\gamma'_i z+\delta'_i}\right]$$

denote arbitrary transformations in the groups G and G′ respectively. I claim that

$$\lim_{t\to 0} \Theta(z) = \Theta'(z)$$

We can find a contour C around the point z which is so small that it is inside R_0 and R'_0 and it remains inside R_0 for all sufficiently small values of t. We let σ be the S of this contour, and let λ be the L of the longest arc of a circle orthogonal to the fundamental circle which can be drawn inside C. We let r be arbitrary, let A be the L of the line Oz and set

$$K = \frac{\pi}{4\sigma}\,(e^{2A} + e^{-2A} + 2)^m e^{2\lambda+2mr}$$

as in Section I of the memoir <u>On Fuchsian Functions</u> **(Acta**

Mathematica, I, p. 206, this volume p.145).

We let S_n and S'_n be the sums of terms of the series Θ and Θ' which correspond to points

$$\frac{\alpha_i z+\beta_i}{\gamma_i z+\delta_i}$$

and

$$\frac{\alpha'_i z+\beta'_i}{\gamma'_i z+\delta'_i}$$

inside the circle M with center at the origin and R equal to (n—1)r. Let

$$\Theta = S_n + R_n, \quad \Theta' = S'_n + R'_n$$

Let h be the greatest value taken by the absolute value of H at transforms of the point z; then we have

$$|R_n| < \frac{Khe^{2n(1-m)r}}{1-e^{2(1-m)r}}, \qquad |R'_n| < \frac{Khe^{2n(1-m)r}}{1-e^{2(1-m)r}}$$

We can then take n so large that

$$|R_n| < \frac{\epsilon}{3}, \qquad |R'_n| < \frac{\epsilon}{3}$$

where ϵ is a given quantity.

The number n is fixed from now on, and hence so is the circle M.

We can now take τ so small that, as t varies from τ to 0, no transform of z either enters or leaves the circle M, and so that the transforms of z by transformations in G which do not belong to G″ are all outside this circle.

Then, as t varies from τ to 0, S_n will be a continuous function of t, because it is the sum of a finite number of continuous functions of t. For t = 0, S_n reduces to S'_n. We can therefore take t so small that

$$|S'_n - S_n| < \frac{\epsilon}{3}$$

and we then have, for arbitrarily small ϵ

$$|\Theta - h'| < \epsilon$$

Q.E.D.

The Fuchsian functions x and y which appear in equation (2) can be chosen arbitrarily; we put

$$x = \frac{\Theta(z, H)}{\Theta(z, H_1)}, \quad y = \frac{\Theta(z, H_2)}{\Theta(z, H_1)}$$

where H, H_1 and H_2 are three arbitrarily chosen rational functions. If we likewise put

$$x_1 = \frac{\Theta'(z, H)}{\Theta'(z, H_1)}, \quad y_1 = \frac{\Theta'(z, H_2)}{\Theta'(z, H_1)}$$

then we obtain

$$x_1 = \lim_{t \to 0} x, \quad y_1 = \lim_{t \to 0} y$$

Thus we see that in the limit the relation between x and y reduces to that between x_1 and y_1, and consequently the moduli of relation (2) reduce to those of relation (2′).

Similarly, we have

$$\lim \left[\frac{1}{2} \frac{x''' x'}{(x')^4} - \frac{3}{4} \frac{(x'')^2}{(x')^4}\right] = \frac{1}{2} \frac{x'''_1 x'_1}{(x'_1)^4} - \frac{3}{4} \frac{(x''_1)^2}{(x'_1)^4}$$

where x', x'', ... denote the first, second, ... derivatives of x with respect to z.

It follows that

$$\lim \phi(x,y) = \phi'(x_1,y_1)$$

and consequently the singular points of equation (1) tend toward those of (1′).

Q.E.D.

All of this remains true when R_0 is a polygon of the second category.

In the case where the polygon R_0 is symmetric, the fact can be proved in an entirely different way.

Suppose for example that we have a curvilinear quadrilateral $\alpha\beta\gamma\delta$ whose sides are arcs of circles orthogonal to the fundamental circle, and whose angles are aliquot parts of π; I call the latter α, β, γ, δ. If we take the reflection of this quadrilateral in the side $\beta\gamma$ we obtain, by joining the two quadrilaterals, a hexagon which engenders a system of Fuchsian functions. But we can also engender the same functions by conformal mapping of the quadrilateral $\alpha\beta\gamma\delta$ onto a disc with center 0 and radius 1. If z is a point inside the quadrilateral and x is the corresponding point of the disc, then x is a Fuchsian function of z.

Now suppose that the side $\gamma\delta$ becomes infinitely small, until it vanishes, so that the quadrilateral reduces to a triangle $\alpha\beta\gamma$ whose vertex γ will be on the fundamental circle with the angle $\alpha\gamma\beta$ equal to zero. Now consider the quadrilateral $\alpha\beta\gamma\delta$ again; I suppose that the side $\gamma\delta$ is now very small, but not zero. We can draw an infinity of circles $\beta_i\gamma_i$ tangent to the circle $\beta\gamma$ at γ_i and cutting the circle $\alpha\beta$ at β_i in an angle equal to β. We draw one of these circles $\beta_1\gamma_1$ so that the triangle $\alpha\beta_1\gamma_1$ is entirely inside the quadrilateral, but also as large as possible. Then we draw another of these circles, $\beta_2\gamma_2$, so that the quadrilateral is entirely inside the triangle $\alpha\beta_2\gamma_2$, while this triangle is as small as possible.

Let z_0 be a point inside $\alpha\beta_1\gamma_1$. Suppose that we make conformal maps onto the disc of radius 1 and center 0, first of $\alpha\beta_1\gamma_1$, then $\alpha\beta\gamma\delta$, then $\alpha\beta_2\gamma_2$, in such a way that the point z_0 corresponds to the center 0. Let x_1, x and x_2 be the points of the disc which correspond to the point z in $\alpha\beta_1\gamma_1$, $\alpha\beta\gamma\delta$ and $\alpha\beta_2\gamma_2$ respectively. We have

$$|x_1| > |x| > |x_2|$$

When the side $\gamma\delta$ vanishes altogether, the two triangles $\alpha\beta_1\gamma_1$ and $\alpha\beta_2\gamma_2$ coincide and it is easy to show that we have

$$\lim |x_1| = \lim |x_2|$$

whence

$$\lim |x_1| = \lim |x|$$

$$\lim x_1 = \lim x$$

Q.E.D.

This proof, which I have only sketched, does not apply to the case where R_0 is not symmetrical.

X. Symmetric types

I shall first apply the method of continuity to a simple case. Let

$$\frac{d^2 y}{dx^2} = \phi(x)y$$

be an equation with $\phi(x)$ rational in x, with all singular points real, and such that the determinant equation relative to each of the latter has a double root. The type T to which this equation belongs will be called **symmetric**, because if this type contains a Fuchsian equation, the corresponding generating polygon is symmetric.

I claim that each symmetric type contains a Fuchsian equation.

If the number of singular points is 3 we can always, by a linear change of variable, suppose that these points are precisely 0, 1 and ∞. Then we know that there is a Fuchsian equation in the type T, since this equation is precisely that which defines the square of the modulus of an elliptic transcendant as a function of the ratio of the periods.

Now suppose that the type T admits four singular points. By a linear change of variable, we can always suppose that these four points are

$$0,\ 1,\ a \text{ and } \infty$$

where a is a positive real quantity greater than 1.

Now let α, β, γ, δ be four points on the fundamental circle, and I suppose that in traversing this circle we encountered them precisely in the order α, β, γ, δ. We describe four circles, orthogonal to the fundamental circle, and passing respectively through the points α and β, β and γ, γ and δ, δ and α. Thus we have a certain quadrilateral. We construct the quadrilateral symmetric to the first with respect to the circle $\delta\alpha$, for example. These two quadrilaterals form a polygon R_0 which generates a symmetric Fuchsian group of genus 0 and the second family which I call G.

We form the corresponding Fuchsian functions and in particular the function

$$x = f(z)$$

in terms of which all others are expressible rationally. To obtain a definition of the latter function we write

$$f(\alpha) = 0, \quad f(\beta) = 1, \quad f(\delta) = \infty$$

and we put

$$f(\gamma) = c$$

c will be real positive and greater than 1. Then z will be defined as a function of x as the ratio of two integrals of the Fuchsian equation

$$\frac{d^2 v}{dx^2} = \phi(x)v$$

where ϕ is rational in x, where the singular points are 0, 1, c, and ∞, for each of which the determinant equation has a double root.

To show that T contains a Fuchsian equation, it then suffices to show that we can choose α, β, γ, δ so that

$$c = a$$

Suppose that the points α, β, δ remain fixed and that we make the point γ describe the arc $\beta\delta$ of the fundamental circle, from β to δ.

Here is what will happen:

1° The theta Fuchsian series defined in Section I of the memoir **On Fuchsian Functions** will be continuous functions of γ, since they are uniformly convergent, as I have shown in that section.

2° It will then be the same for the Fuchsian functions themselves, and consequently for c.

3° By virtue of the fundamental lemma, c cannot take the same value twice.

4° As the point γ tends to the point β, the quadrilateral $\alpha\beta\gamma\delta$ reduces to a triangle $\alpha\beta\delta$, and the polygon R_0 reduces to the quadrilateral R'_0 formed by this triangle and its reflection in the circle $\alpha\delta$. The second lemma then tells us that c tends to unity.

5° For the same reason, when γ tends to δ, c tends to infinity.

In summary, when γ varies along the fundamental circle from β to δ, c will increase continuously from 1 to infinity. Hence c will take any value a between 1 and infinity.

Q.E.D.

Now suppose that a type T admits not four, but five singular points. By a linear change of variable we can always suppose that these five singular points are

$$0,\ 1,\ a,\ b,\ \infty$$

with the inequalities

(1) $$1 < a < b$$

where a and b are real.

If we regard a and b as the coordinates of a point in the plane, we see that each type T corresponds to a point of a plane region bounded by the two lines $a = 1$, $a = b$ and by the line $b = \infty$. This region is none other than the manifold S' defined in Section VIII. We see from our manner of considering these things that this manifold is **open**.

We now take five points α, β, γ, δ, ϵ on the fundamental circle, in precisely that order. It follows that we have

$$\arg \alpha > \arg \beta > \arg \gamma > \arg \delta > \arg \epsilon$$

We construct the pentagon $\alpha\beta\gamma\delta\epsilon$, then the symmetric pentagon with respect to the side $\alpha\epsilon$, and we have a polygon R_0 which generates a Fuchsian group G. Among the Fuchsian functions corresponding to this group, we let $x = f(z)$ be one in terms of which all the others are expressible rationally and which, moreover, is such that

$$f(\alpha) = 0, \quad f(\beta) = 1, \quad f(\epsilon) = \infty$$

The quantities

$$f(\gamma) = a, \quad f(\delta) = b$$

are real, positive and such that

$$1 < a < b$$

When $\gamma - \beta$ tends to zero, a tends to 1; when $\gamma - \delta$ tends to zero, a − b tends to zero; finally, when $\delta - \epsilon$ tends to zero, b increases indefinitely. Moreover, when α, β and ϵ remain fixed, and γ and δ vary, a and b are continuous functions of γ and δ.

If we consider the arguments of γ and δ as the coordinates of a point in the plane, these arguments are subject to the inequalities

$$\arg \beta < \arg \gamma < \arg \delta < \arg \epsilon$$

so the corresponding point will be situated in the interior of a triangle bounded by the three lines

$$\arg \beta = \arg \gamma, \quad \arg \gamma = \arg \delta, \quad \arg \delta = \arg \epsilon$$

This triangle is none other than the manifold S, which is open like the manifold S′.

Then each point of S corresponds to exactly one point of S′, each point of S′ corresponds to at most one point of S. Each point on the periphery of S corresponds to a point on the periphery of S′. It necessarily follows that each point of S′ corresponds to a point of S.

In other words, we can always choose the pentagon $\alpha\beta\gamma\delta\epsilon$ so that a and b take given values, i.e., each Fuchsian type with no more than five singular points contains a Fuchsian equation.

The proof is absolutely the same for a greater number of singular points, whence the conclusion:

Each symmetric Fuchsian type contains a Fuchsian equation.

We see that, in the particular case of symmetric types, application of the method of continuity does not present any difficulty.

We can draw an important conclusion from the preceding.

Consider a function y of x which is holomorphic except when x takes the n values $a_1, a_2, \ldots, a_n$; suppose also that these n values are real.

Consider the Fuchsian type which admits the n singular points $a_1, a_2, \ldots, a_n$. It contains a Fuchsian equation and, in this equation, the variable is a Fuchsian function of the ratio z of the integrals:

$$x = f(z)$$

This function does not exist outside the fundamental circle; if we assume $a_n = \infty$, then it is holomorphic inside this circle and it cannot take any of the values $a_1, a_2, \ldots, a_n$.

It follows that, inside the fundamental circle, y is holomorphic in z.

Thus, **if** y **is a function of** x **admitting only a finite number of singular points, which are all real, we can find a variable** z **such that** y **and** x, **expressed as functions of** z, **are non—existent outside the fundamental circle and holomorphic inside that circle.**

We can suppose in particular that y is an algebraic function, or the integral of a linear equation with algebraic coefficients, for which all the singular points are real.

XI. Generalization of the preceding theorem

We have not applied the method of continuity beyond the symmetric types, however, in order to commence the study of more complicated types, we have to extract as much as possible from those results. Let

$$\frac{d^2v}{dx^2} = \phi(x)v$$

be an equation such that $\phi(x)$ is rational in x, with singular points a_1, a_2, ..., a_n not all real, for each of which the determinant equation has a double root.

We let T be the type to which this equation belongs.

I shall show later that this type contains a Fuchsian equation, but this is not my present aim. I want to show that there is a type subordinate to T which contains a Fuchsian equation, in other words, that there is a Fuchsian function which cannot take any of the values a_1, a_2, ..., a_n inside the fundamental circle, and which also cannot take certain other values b_1, b_2, ..., b_k.

Thus we consider ourselves free to adjoin, to the table of quantities a, such quantities as are convenient. We can always suppose that if the table of numbers a contains an imaginary quantity, then it also contains its conjugate, since if the conjugate is not already present we can adjoin it.

I assume that the theorem is true when the table of quantities a contains no more than $q - 1$ pairs of imaginaries, and I shall show that it is then true when the table contains q such pairs. This is sufficient, since I have shown in the preceding section that the theorem is true when all the quantities a are real.

Suppose then that the table of quantities a_1, a_2, ..., a_n contains $n - 2q$ real quantities

(1) $$a_1, a_2, \ldots, a_{n-2q}$$

and 2q imaginary quantities conjugate in pairs

(2) $$a'_1, a'_2, \ldots, a'_{2q}$$

We put

(3) $$\phi(x) = (x-a'_1)(x-a'_2)\ldots(x-a'_{2q})$$

and $\phi(x)$ will be a polynomial with real coefficients. The equation

$$\phi'(x) = 0$$

will have at least one real root, and consequently at most $q - 1$ pairs of imaginary roots. Let

$$c_1, c_2, \ldots, c_{2q-1}$$

be the roots of this equation. We put

$$n - 2q = p, \quad 2q - 1 = m$$

Among the following quantities

$$(4) \qquad 0, \ \phi(a_1), \ \phi(a_2), \ \ldots, \ \phi(a_p); \ \phi(c_1), \ \phi(c_2), \ \ldots, \ \phi(c_m)$$

there will be at most $q - 1$ pairs of imaginary values.

We can then construct a Fuchsian function $F(z)$ which cannot take any of these values, since the theorem is assumed true for a system of values not containing more than $q - 1$ pairs of conjugate imaginaries.

We put

$$\phi(x) = F(z)$$

We see, first, that x is a uniform function of z; in fact, this will cease to hold only if we have

$$\phi'(x) = 0$$

whence

$$x = c_i, \quad F(z) = \phi(c_i)$$

which is impossible.

Moreover, x cannot take any of the values

$$a_1, a_2, \ldots, a_p$$

without giving

$$F(z) = \phi(a_i)$$

which is impossible, nor any of the values

$$a'_1, a'_2, \ldots, a'_{2q}$$

since these give

$$F(z) = 0$$

which is impossible.

The derivative of x with respect to z can never vanish without that of F with respect to z vanishing also, which never happens.

In addition, x is a Fuchsian function of z. In fact, let

$$F(z) = \phi(x) = y, \qquad v = \sqrt{\frac{dF}{dx}}$$

then, since F is a Fuchsian function, we have

$$\frac{d^2v}{dy^2} = \psi(y)v$$

where $\psi(y)$ is a rational function of y.

If we next put

$$w = \sqrt{\frac{dx}{dz}}$$

whence

$$v = w\sqrt{\phi'(x)}$$

we obtain

(5) $$\frac{d^2w}{dx^2} = \left\{[\phi'(x)]^2\psi[\phi(x)] + \frac{3}{4}\left[\frac{\phi''(x)}{\phi'(x)}\right]^2 - \frac{1}{2}\frac{\phi'''(x)}{\phi'(x)}\right\}w$$

where the coefficient of w is a rational function of x.

Thus x is a uniform function of z, i.e., of the ratio of the two integrals of equation (5). It is therefore a Fuchsian function.

Hence x is a Fuchsian function which cannot take any of the values (1) and (2), but which also cannot take any of the values such that

$$\phi(x) = \phi(a_i) \quad \text{or} \quad \phi(x) = \phi(c_i)$$

We call these values

$$b_1, b_2, \ldots, b_k$$

Equation (5), which is a Fuchsian equation, admits the singular points

$$a_1, a_2, \ldots, a_p, a'_1, a'_2, \ldots, a'_{2q}, b_1, b_2, \ldots, b_k$$

for each of which the determinant equation has a double root. It therefore belongs to a type subordinate to T.

Thus it is always possible to find a Fuchsian function $x = f(z)$ which cannot take n **given** values

$$a_1, a_2, \ldots, a_n$$

and which also cannot take k other non–assigned values

$$b_1, b_2, \ldots, b_k$$

Corresponding to this Fuchsian function is a Fuchsian equation whose singular points are the a and b and whose determinant equations all have a double root. This equation belongs to a type subordinate to T.

Now let y be a function of x which admits no singular points other than the a. If we put $x = f(z)$, then y will be a uniform function of z.

This is what happens, in particular, when y is the integral of a linear equation with algebraic coefficients.

Thus, **we can always find a variable** z **of which the variable** x **and the integrals of such an equation are uniform functions**.

XII. Limit polygons

I say that two generating polygons R_0 and R'_0 of two Fuchsian groups G and G′ belong to the same class when they satisfy the following conditions:

1° The two polygons have the same number of sides.

2° If two sides of R_0 are equivalent, the corresponding sides of R'_0 are likewise equivalent. It follows that the vertices of the two polygons are partitioned into the same number of cycles, in such a way that each cycle of R_0 corresponds to a cycle of R'_0, and conversely.

3° The angle sums of corresponding cycles are the same.

The polygons of the same class depend upon a certain number of parameters. Suppose for example, to fix ideas, that they are completely defined* by three parameters x, y, z but that these three parameters are not independent but subject to a relation

(1) $$f(x,y,z) = 0$$

(*Of course, two polygons are not distinct when we can pass from one to the other by a linear transformation which leaves the fundamental circle invariant.)

If we regard x, y and z as the coordinates of a point in space, then each polygon in the class corresponds to a point on a surface S_1

defined by equation (1), so that the class itself corresponds to this surface or a portion of it.

In general, however, for the polygon R_0 to generate a Fuchsian group it is necessary that x, y and z satisfy not only equation (1), but also certain inequalities (2). So the class considered corresponds to only a part of the surface S_1, and this part is bounded by certain frontier curves whose equations are obtained by replacing inequality signs in (2) by equality.

We take an example. Let R_0 be a polygon of the second family and genus zero; it will be, for example, a hexagon abcdef. The sides ab and cd, cd and ed, ef and af, are equivalent, so that there are four cycles b, d, f and ace. The six vertices must satisfy the following relation:

$$-(a-b)(c-d)(e-f) = (b-c)(d-e)(f-a)$$

Let $\phi(\mu)$ be the cross ratio of μ with respect to the three points b, d and f, so that

$$\phi(b) = \infty, \quad \phi(d) = 0, \quad \phi(f) = 1$$

We put

$$\phi(a) = y, \quad \phi(c) = x, \quad \phi(e) = z$$

The polygon R_0 will be completely defined when we know x, y and z; but between these three parameters we have the following relation:

$$(1') \qquad x(1-z) = z(1-y)$$

This is the equation of a paraboloid.

However, x, y and z must also satisfy the inequalities

$$(2') \qquad x < 0 < z < 1 < y$$

so that we retain only a portion of the paraboloid (1′). This portion is bounded by the following six lines:

(3) $x = z = 0$

(4) $x = 0, \quad y = 1$

(5) $y = z = 1$

(6) $z = 1, \quad x = \infty$

(7) $x = y = \infty$

(8) $y = \infty, \quad z = 0$

These are the frontier curves of the used portion of the paraboloid; to each point of this used portion there corresponds **one** polygon of the class in question, and conversely.

Now suppose that, in the class considered, p parameters x_1, x_2, ..., x_p are needed to completely define the polygon R_0, and that these p parameters satisfy an equation (1) on the cne hand, and certain inequalities (2) on the other. If we regard x_1, x_2, ..., x_p as coordinates of a point in p—dimensional space, then this equation and these inequalities define a portion of surface, a manifold (**Mannigfaltigkeit**) as the Germans say. I let M_1 denote this (p—1)—dimensional manifold. It will in general be **open** and bounded by the (p—2)—dimensional frontier manifolds, since the x satisfy not only an equation but also the inequalities. These frontier manifolds are analogous to the frontier curves we spoke of above, and we obtain their equations by replacing inequality signs in (2) by equality.

Consider a point of the manifold M_1 which is infinitely close to one of the frontier manifolds. Corresponding to this point there is a polygon, certain elements of which are infinitesimal, and which I therefore call a **limit polygon**.

We shall first consider a class of polygons R_0, of genus zero, of the second family. The number of sides will be 2n, and each side of odd rank will be equivalent to the side whose rank is one greater (the side of rank $2p - 1$ will be equivalent to the side of rank 2p). Then if I let α_p denote the vertex which separates the sides of ranks p and $p - 1$, each vertex of odd index will form a cycle by itself, while the vertices of even index form a single cycle. Thus there are $n + 1$ cycles of vertices in all, which are on the fundamental circle.

To define such a polygon, it suffices to give 2n − 3 of these vertices, the other three being supposed fixed. Moreover, between these 2n − 3 parameters we have the relation

$$(1'') \qquad -(\alpha_1-\alpha_2)(\alpha_3-\alpha_4)\ldots(\alpha_{2n-1}-\alpha_{2n}) =$$

$$= (\alpha_{2n}-\alpha_1)(\alpha_2-\alpha_3)\ldots(\alpha_{2n-2}-\alpha_{2n-1})$$

and also the inequalities

$$(2'') \qquad \arg\alpha_1 < \arg\alpha_2 < \arg\alpha_3 < \ldots < \arg\alpha_{2n} < \arg\alpha_1 + 2\pi$$

On the other hand, I consider the Fuchsian functions engendered by R_0 and, among them, one in terms of which all others can be expressed rationally. I call it f(z). The singular points of the corresponding Fuchsian equation are then n + 1 in number; they are

$$f(\alpha_1),\ f(\alpha_3),\ \ldots,\ f(\alpha_{2n-1})$$

and

$$f(\alpha_2) = f(\alpha_4) = \ldots = f(\alpha_{2n})$$

I want to find the limit polygons of the class in question. To do this it is necessary to replace an inequality sign in (2″) by equality, whence

$$\arg\alpha_i = \arg\alpha_{i+1}; \qquad \alpha_i = \alpha_{i+1}$$

i.e., one of the sides of the polygon R_0 must become infinitely small. But by virtue of equation (1″), if a side of even rank becomes infinitely small, it is likewise so for a side of odd rank, and conversely. We therefore always have at least two sides which vanish. The three species of limit polygons are:

1° those in which two equivalent sides vanish;

2° those in which two inequivalent sides vanish;

3° those in which more than two sides vanish simultaneously.

Thus, in the case of the paraboloid (1′), the three frontier lines (3), (5) and (7) correspond to limit polygons of the first species; the lines (4), (6) and (8) to the polygons of the second species, and the points

$$x = z = 0, \quad y = 1; \quad x = 0, \quad y = z = 1; \quad \ldots$$

which simultaneously belong to two frontier lines, correspond to polygons of the third species.

The limit polygons of the third species are evidently particular cases of the first two. We leave them aside for the moment and show that those of the second species reduce to those of the first.

In fact, if we consider a polygon of the second species, one of its infinitesimal sides will be of even rank. I can assume that we have numbered the sides in such a way that it is the side $\alpha_1\alpha_2$ of rank 2. The other infinitesimal side will be, for example $\alpha_{2p}\alpha_{2p+1}$. These two sides will be separated by p − 1 pairs of equivalent sides $\alpha_2\alpha_3$ and $\alpha_3\alpha_4$, $\alpha_4\alpha_5$ and $\alpha_5\alpha_6$, ..., $\alpha_{2p-2}\alpha_{2p-1}$ and $\alpha_{2p-1}\alpha_{2p}$. We join α_{2p-1} and α_1 by an arc of a circle orthogonal to the fundamental circle, thus dividing the polygon R_0 into two others r_0 and r'_0; say with r_0 containing the vertex α_{2p} and r'_0 the vertex α_2. Let r''_0 be the transform of r_0 by the transformation which changes $\alpha_{2p-2}\alpha_{2p-1}$ into $\alpha_{2p}\alpha_{2p-1}$ and put

$$R'_0 = r_0 + r''_0$$

The polygon R'_0 will be piecewise equivalent to R_0, it will admit two infinitesimal sides, namely $\alpha_{2p}\alpha_{2p+1}$ and the transform of $\alpha_1\alpha_2$. These two sides will not be separated by more than p − 2 pairs of equivalent sides. Operating on R'_0 as we did on R_0, we obtain a polygon R''_0 on which the two infinitesimal sides are not separated by more than p − 3 pairs of equivalent sides and, continuing in this way, we finally arrive at a polygon S_0, piecewise equivalent to R_0, whose two infinitesimal sides are consecutive. All the polygons R_0, R'_0, R''_0, ..., S_0 have a common part, which will be r_0. The infinitesimal side $\alpha_{2p}\alpha_{2p+1}$ and its equivalent $\alpha_{2p+1}\alpha_{2p+2}$ then belong to S_0. The second infinitesimal side will be, for example, $\beta\alpha_{2p}$. We now join it to $\beta\alpha_{2p+1}$ by an arc of a circle orthogonal to the fundamental circle. The polygon S_0 will be divided into two parts, the triangle $\alpha_{2p}\beta\alpha_{2p+1} = s'_0$ and $s_0 = S_0 - s'_0$. Let γ be the transform of β by the transformation which changes $\alpha_{2p}\alpha_{2p+1}$ into $\alpha_{2p+2}\alpha_{2p+1}$ and let $\alpha_{2p+2}\gamma\alpha_{2p+1} = s''_0$ be the triangle to which s'_0 is sent by that transformation. The polygon $S'_0 = s_0 + s''_0$ will be piecewise equivalent to S_0, and it is easy to see that it has no more than two infinitesimal sides, namely $\beta\alpha_{2p+1}$ and $\gamma\alpha_{2p+1}$, which are equivalent.

It is true that the polygon S_0, though piecewise equivalent to R_0, does not belong to the same class, because the pairs of its equivalent sides are differently distributed. But it is easy to reduce the polygon S'_0, by suitable transformations, to a piecewise equivalent polygon S''_0 admitting equivalent infinitesimal sides $\beta\alpha_{2p+1}$ and $\gamma\alpha_{2p+1}$, like S'_0, and for which each side of odd rank is equivalent to the even side following it.

Thus we are enabled to concern ourselves mainly with limit polygons of the first species.

We consider such a polygon whose two equivalent sides are $\alpha_1\alpha_2$ and $\alpha_{2n}\alpha_1$. We distinguish two categories of the first species:

1° Those of the first category are such that

$$\frac{\alpha_1-\alpha_2}{\alpha_{2n}-\alpha_1} = 1$$

2° And those of the second category are such that

$$\frac{\alpha_1-\alpha_2}{\alpha_{2n}-\alpha_1} \neq 1$$

In the first case we have the following relations between the vertices of the polygon:

(9) $$\alpha_1 = \alpha_2 = \alpha_{2n}$$

(1[a]) $$-(\alpha_3-\alpha_4)\ldots(\alpha_{2n-1}-\alpha_1) = (\alpha_1-\alpha_3)\ldots(\alpha_{2n-2}-\alpha_{2n-1})$$

The polygon R_0 then reduces (by passing to the limit, where the infinitesimal sides vanish) to a polygon R'_0 which is completely analogous except that the number of its sides is $2n - 2$ instead of $2n$.

We shall say a few words about the groups G and G' generated by the two polygons R_0 and R'_0. If we let S_i be the parabolic transformation which changes $\alpha_{i-1}\alpha_i$ into $\alpha_{i+1}\alpha_i$ (where i is odd), then the group G will be generated by the n transformations

$$S_1, S_3, \ldots, S_{2n-1}$$

The product

$$S_1 S_3 S_5 \ldots S_{2n-1}$$

will be parabolic, with fixed point α_{2n}.

When R_0 reduces to R'_0, the transformations $S_3, \ldots, S_{2n-1}$ become $S'_3, \ldots, S'_{2n-1}$ and remain parabolic, but the transformation S_1 becomes illusory and no longer meaningful. The group G' is then generated by the $n - 1$ transformations

$$S'_3, \ldots, S'_{2n-1}$$

By virtue of the relation (1[a]), the product

$$S'_3 \ldots S'_{2n-1}$$

is parabolic.

Thus the vertices of even order of R'_0 form a parabolic cycle (third subcategory), and consequently R'_0 and its transforms by the different transformations in G' fill the whole fundamental circle (cf. Theory of Fuchsian Groups, Section VI).

Then, applying the second lemma, we see that the difference

$$f(\alpha_1) - f(\alpha_2)$$

tends to zero and that the type T, to which the Fuchsian equation engendered by R_0 belongs, reduces to a simpler type T′ admitting no more than n singular points.

Now suppose that two sides $\alpha_1\alpha_2$ and $\alpha_{2n}\alpha_1$ tend to zero in such a way that

$$\lim \frac{\alpha_1-\alpha_2}{\alpha_{2n}-\alpha_1} \neq 1$$

Relation (1^a) ceases to hold. The product

$$S'_3 S'_5 \cdots S'_{2n-1}$$

will be hyperbolic and not parabolic. The vertices of even order in R'_0 will form a hyperbolic cycle (fourth subcategory), and consequently the transforms of R'_0 by the transformations in G′ will not fill the whole fundamental circle, but only a portion bounded by infinitely many circles (cf. Theory of Fuchsian Groups, Section VI).

The second lemma is therefore inapplicable, which shows the profound difference between limit polygons of the first and second category.

We can construct an entirely analogous theory of limit polygons of the third species, which are nothing but particular cases of the first and second.

As a second example, we consider a polygon R_0 of genus zero whose sides are disposed as in the preceding example, but which belongs to the first family, and whose vertices, consequently, are all inside the fundamental circle, with all cycles being elliptic (first category).

We let

$$a_1, a_2, \ldots, a_{2n-1}$$

be the angles of this polygon at the vertices

$$\alpha_1, \alpha_2, \ldots, \alpha_{2n-1}$$

and let the sum of the other angles of the polygon be b. These various angles must be aliquot parts of 2π; I suppose them given, and the class to which R_0 belongs is then determined. In what follows we consider the L of the sides of this polygon and not their geometric length. In the preceding example, when we spoke of a side being infinitesimally small, we meant in geometric length; here, on the contrary, we mean its L (which is none other than length from the point of view of non–Euclidean geometry (cf. Theory of Fuchsian Groups, Section I; On Fuchsian Functions, Section I)).

This being so, how can we conceive of the polygon R_0 becoming a limit polygon? This can be conceived in three ways:

1° if a side becomes infinitely small;

2° if an angle vanishes;

3° if a side becomes infinitely large.

It is easy to see that the first two cases cannot occur without the third also being present; we therefore examine the third.

Suppose that a side of R_0, $\alpha_{2n}\alpha_1$ for example, becomes infinite. It will be the same for its equivalent $\alpha_1\alpha_2$. But the angle between these two sides, which is a_1, is given and finite. Hence the diagonal $\alpha_{2n}\alpha_2$ is infinite. But this diagonal is smaller than the sum of the other $2n - 2$ sides, hence one of these sides is infinite. If it is, for example, $\alpha_{2p-2}\alpha_{2p-1}$, then it will be the same for its equivalent, $\alpha_{2p-1}\alpha_{2p}$. We join the two vertices α_1, α_{2p-2} by an arc of a circle orthogonal to the fundamental circle. We have then divided R_0 into two partial polygons: r_0, which contains the vertex α_2, and r'_0, which contains the vertex α_{2n}. Now let r''_0 be the transform of r_0 by the transformation which changes $\alpha_1\alpha_2$ into $\alpha_1\alpha_{2n}$. The polygon $R'_0 = r'_0 + r''_0$, is piecewise equivalent to R_0 and, like it, has four infinite sides, however these four sides are consecutive. It follows that we can always assume that these four infinite sides of R_0 are $\alpha_{2n}\alpha_1$, $\alpha_1\alpha_2$, $\alpha_2\alpha_3$, $\alpha_3\alpha_4$; this is what we shall do.

We can also suppose that R_0 has more than four infinite sides, but this is only a particular case that we shall leave aside for the moment. We suppose then that the other $2n - 4$ sides are finite. In this case we can show without much difficulty that the diagonal $\alpha_1\alpha_3$ is finite. We let r_0 be the triangle $\alpha_1\alpha_2\alpha_3$ and let r'_0 be the rest of R_0. We let r''_0 be the triangle $\alpha_3\beta\alpha_4$ into which r_0 is transformed by the transformation which changes $\alpha_3\alpha_2$ into $\alpha_3\alpha_4$. The polygon $R'_0 = r'_0 + r''_0$ will be piecewise equivalent to R_0 and will have only two infinite sides $\alpha_1\alpha_{2n}$ and $\beta\alpha_4$, which are equivalent, while the two equivalent sides $\alpha_1\alpha_3$ and $\alpha_3\beta$ will be finite. The latter two sides have finite L, but their geometric length is infinitely small. Suppose that we pass to the limit and that these two sides vanish. The polygon R'_0 then reduces to a polygon R''_0 with no more than $2n - 2$ sides and consequently coincidence of three vertices α_1, α_3 and β. Among these sides, there will be $2n - 4$ no different from those of R_0; the two others are the sides $\alpha_{2n}\alpha_1$ and $\alpha_1\alpha_4$ which are equivalent. The vertex α_1, which is on the fundamental circle, forms a cycle on its own, which can be parabolic or hyperbolic.

If it is parabolic, which happens when the circle which passes through $\alpha_1\alpha_{2n}$ and α_4 is tangent to the fundamental circle, then the limit polygon is of the first category.

If, on the contrary, it is hyperbolic, then the limit polygon is of the second category.

The properties of these two categories are the same as in the preceding example. Thus if G is the group generated by R_0 then this group will be generated by n elliptic transformations

$$S_1, S_3, \ldots, S_{2n-1}$$

where S_i has α_i as fixed point. The group generated by

$$S_5, S_7, \ldots, S_{2n-1} \text{ and } S_1S_3$$

will be called G''. When R_0 reduces to R''_0, the transformations S_5, S_7, ..., S_{2n-1} tend to the elliptic transformations S'_5, S'_7, ..., S'_{2n-1}, and S_1S_3 tends to a certain transformation Σ. S_1 and S_3 become illusory. The group G' generated by R''_0 will be generated by

$$S'_5, S'_7, \ldots, S'_{2n-1} \quad \text{and} \quad \Sigma$$

and will be isomorphic to G''.

If the polygon is in the first category, Σ will be parabolic; the transforms of R''_0 by the group G' will fill the whole fundamental circle. If R_0 tends to R''_0 the transform of R'_0 by a transformation in G'' will tend to the transform of R''_0 by the corresponding transformation in G', and the area covered by the transforms of R'_0 by transformations in G which do not belong to G'' will tend to zero. Applying the second lemma and retaining the notation above, we see that the difference $f(\alpha_1) - f(\alpha_3)$ tends to zero at the same time.

If on the contrary the polygon is of the second category, then Σ is hyperbolic and the transforms of R''_0 do not fill the fundamental circle. The preceding deductions are then no longer possible and we cannot affirm that $f(\alpha_1) - f(\alpha_3)$ tends to zero.

We treat the case where there are more than four infinite sides similarly.

As a third example, we choose of polygon R_0 of genus p, of the first family, with 4p sides, opposite pairs equivalent, and vertices forming a single cycle with angle sum 2π.

Here again we can conceive of three ways in which the polygon can become a limit polygon:

1° if one of the sides vanishes (in L value, as always);

2° if angle vanishes;

3° if a side becomes infinite.

As in the preceding example, the first two cases do not occur without the third being present at the same time, so it is the third case which remains for us to examine.

As above, I let α_i be the vertex of rank i and I suppose that the side $\alpha_{4p}\alpha_1$ and its equivalent $\alpha_{2p}\alpha_{2p+1}$ become infinite. Let β_1 be the midpoint of $\alpha_{4p}\alpha_1$ (by which I mean that the L of $\alpha_{4p}\beta_1$ equals that of $\beta_1\alpha_1$), and let β_2 be the midpoint of $\alpha_{2p}\alpha_{2p+1}$. We join $\beta_1\beta_2$ by an arc of a circle orthogonal to the fundamental circle. This will be a secant of our polygon R_0 and **it will be finite**. This secant will divide the polygon R_0 into two others; r_0, which contains the vertex α_1, and r'_0, which contains the vertex α_{4p}. Now let S_p be the transformation which changes $\alpha_p\alpha_{p+1}$ into its equivalent, $\alpha_{3p}\alpha_{3p+1}$. This transformation will change

$$r_0 = \beta_1\alpha_1\alpha_2 \cdots \alpha_{2p-1}\alpha_{2p}\beta_2$$

into

$$r''_0 = \beta'_1\alpha'_1\alpha'_2 \cdots \alpha'_{2p-1}\alpha'_{2p}\beta'_2$$

where

$$\alpha'_p = \alpha_{3p+1}, \quad \alpha'_{p+1} = \alpha_{3p}$$

The polygon $R'_0 = r'_0 + r''_0$ will be piecewise equivalent to R_0; it will have $4p + 2$ vertices which occur in the following order

$$\beta'_1\alpha'_1\alpha'_2 \cdots \alpha'_p\alpha_{3p+2}\alpha_{3p+3} \cdots \alpha_{4p}\beta_1\beta_2\alpha_{2p+1}\alpha_{2p+2} \cdots$$

$$\cdots \alpha_{3p}\alpha'_{p+2}\alpha'_{p+3} \cdots \alpha'_{2p}\beta'_2$$

These $4p + 2$ vertices form two cycles which both have angle sums equal to 2π. Four sides are infinite, namely: $\beta'_1\alpha'_1$, $\alpha_{4p}\beta_1$, $\beta_2\alpha_{2p+1}$, $\alpha'_{2p}\beta'_2$. The L of the sides $\beta_1\beta_2$ and $\beta'_1\beta'_2$ is finite and their geometric lengths are infinitely small.

If we pass to the limit, these lengths will vanish absolutely and the vertices β_1 and β_2, β'_1 and β'_2, will coincide. The polygon R'_0 will then reduce to another polygon R''_0 with no more than 4p vertices, β_1 and β'_1 on the fundamental circle and $4p - 2$ others off it. Each of the vertices β_1 and β'_1 forms, by itself, a single cycle which can be parabolic or hyperbolic.

We now study the group G generated by the polygon R_0 or, which comes to the same thing, by R'_0. If we let S_i be the transformation which changes the side $\alpha_i\alpha_{i+1}$ of R_0 into its equivalent $\alpha_{2p+i}\alpha_{2p+i+1}$, then the group G will be generated by the 2p transformations

$$S_1, S_2, \ldots, S_{2p}$$

between which we have the relation

$$S_1S_2 \ldots S_{2p} = S_{2p}S_{2p-1} \ldots S_2S_1$$

What are the transformations in G which change each side of R'_0 into its equivalent? They are

S_p which changes $\beta_1\beta_2$ into $\beta'_1\beta'_2$
S_{2p} which changes $\alpha_{2p+1}\beta_2$ into $\alpha_{4p}\beta_1$
$S_p^{-1}S_{2p}S_p$ which changes $\beta'_2\alpha'_{2p}$ into $\beta'_1\alpha'_1$

and

$S_p^{-1}S_i$ which changes $\alpha'_i\alpha'_{i+1}$ into $\alpha_{2p+i}\alpha_{2p+i+1}$

When the polygon R'_0 reduces to R''_0, the transformation S_p becomes illusory and the transformations S_{2p}, $S_p^{-1}S_{2p}S_p$ and $S_p^{-1}S_i$ reduce to certain transformations Σ, Σ' and S'_i. These

transformations generate a group G' which is precisely that generated by R''_0, with Σ and Σ' being the transformations which respectively change $\alpha_{2p+1}\beta_1$ into $\alpha_{4p}\beta_1$ and $\alpha'_{2p}\beta'_1$ into $\alpha'_1\beta'_1$.

If the multipliers of Σ and Σ' are equal to 1, these two transformations are parabolic; the polygon R''_0 will then be said to be of first category and its transforms will fill the whole fundamental circle. We conclude, as in the two preceding examples, that the second lemma is applicable and consequently that the Fuchsian functions engendered by R'_0 tend to the Fuchsian functions engendered by R''_0 as R'_0 tends to R''.

We enter into some more details. Let x and y be two Fuchsian functions engendered by R'_0; between them there will be an algebraic relation

$$\phi(x,y) = 0$$

of genus p, because R'_0 is of genus p. If this relation is regarded as the equation of a curve of degree m, then this curve will have

$$\frac{(m-1)(m-2)}{2} - p$$

double points. When R'_0 reduces to R''_0, the genus of this curve will be diminished by one, because the polygon R''_0 is of genus $p - 1$, i.e., **the curve will acquire a new double point**. This double point will correspond in reality to two different analytic points (according as the double point is regarded as belonging to one or other of the two branches of the curve which pass through it) and consequently **to two really distinct points of the polygon** R''_0. These two points will be β_1 and β'_1 which are, as we know, on the fundamental circle. It follows that the two Fuchsian functions lim x and lim y cannot take, inside the fundamental circle, the two values corresponding to the new double point.

It cannot happen that the multiplier of one of Σ, Σ' equals 1 while that of the other does not, so that one of the transformations is parabolic and the other hyperbolic. In fact, we have

$$\text{multiplier } S_{2p} = \text{multiplier } S_p^{-1} S_{2p} S_p$$

whence

$$\lim \text{ multiplier } S_{2p} = \lim \text{ multiplier } S_p^{-1} S_{2p} S_p$$

and

$$\text{multiplier } \Sigma = \text{multiplier } \Sigma'$$

However, it can happen that both multipliers are unequal to one and both transformations are hyperbolic. In this case R''_0 is said to be of the second category. The second lemma is no longer applicable and the Fuchsian functions engendered by R'_0 do not tend to those engendered by R''_0.

We can similarly treat the case where more than two sides of R_0 become infinite.

The preceding examples suffice, I think, to show how the question of limit polygons must be treated in each case.

In the case of the second example, we say that the limit polygon is:

1° of the first species if there are no more than four infinite sides, and these are consecutive;

2° of the second species if there are no more than four infinite sides and these are not consecutive (we have seen that a polygon of the second species can always be reduced to one of the first);

3° of the third species if there are more than four infinite sides; we easily show that we can always reduce to the case where all the infinite sides are consecutive.

In the case of the third example we say that the limit polygon R_0 is:

1° of the first species if there are only two infinite sides;

2° of the second species if there are more than two infinite sides.

XIII. Reduced polygons

We have seen that different polygons in the same class correspond to different points in a manifold M_1. But a single Fuchsian group corresponds to an infinity of generating polygons and consequently to an infinity of points of M_1. In fact, the procedure of Section IX of Theory of Fuchsian Groups permits the polygon R_0 to be transformed into another R'_0, which is piecewise equivalent to R_0 and generates the same Fuchsian group. We have an infinity of transformations which change a polygon into a piecewise equivalent, and consequently a point of M_1 to another point of that manifold. These transformations form a group which I call Γ and which is evidently discontinuous.* (*Translator's Note: Now known as the Teichmüller modular group or the mapping class group.) The manifold M_1 is partitioned into an infinity of domains D_0, D_1, ..., D_i, ... in such a way that each transformation in Γ corresponds to one of these domains, and the latter domain is precisely the transform of D_0 by the given transformation. The subdivision of M_1 by means of the group Γ is entirely analogous to the subdivision of the (disc within the) fundamental circle into an infinity of polygons R_0, R_1, ... by means of an arbitrary Fuchsian group.

Suppose that the point xy in the xy plane represents the imaginary quantity $z = x + iy$, and consider a rectilinear parallelogram R_0 with vertices 0, z_1, z_2 and $z_1 + z_2$. We can subdivide the plane into an infinity of parallelograms equal to R_0 and thus generate the group $(z, z + \alpha z_1 + \beta z_2)$ where α and β are arbitrary integers. This group gives rise to the doubly periodic functions with periods z_1 and z_2. But this parallelogram can be replaced by another whose vertices are

$$0, \ \alpha z_1 + \beta z_2, \ \gamma z_1 + \delta z_2, \ (\alpha+\gamma)z_1 + (\beta+\delta)z_2$$

where α, β, γ, δ are integers such that $\alpha\delta - \beta\gamma = 1$. This second parallelogram is piecewise equivalent to the first; I call it R'_0. To define R_0, we have to give the ratio

$$\frac{z_2}{z_1} = \omega.$$

Different parallelograms correspond to different points of the ω plane, or rather the half plane above the real axis, which I call the **upper half plane***. (*Translator's Note: Poincaré actually calls it the **positive part of the plane**, but I have changed this to avoid confusion.) R'_0 corresponds to the point $\frac{\delta\omega+\gamma}{\beta\omega+\alpha}$, so that the upper half of the ω plane is partitioned into an infinity of domains which transform into each other by application of the transformations $\left(\omega, \frac{\delta\omega+\gamma}{\beta\omega+\alpha}\right)$. Corresponding to each doubly periodic function there is exactly one point of each of these domains. But, among all the parallelograms equivalent to R_0, there is one which is simpler than all the others; it is that for which the positive quadratic form

$$\left|\xi z_1 + \eta z_2\right|^2$$

where ξ and η are integer indeterminates, is a reduced form. We can then say that the parallelogram is itself reduced. We can define this reduced parallelogram in the following way:

1° it must be such that the imaginary part of ω is as large as possible;

2° and among those which correspond to this maximum of the imaginary part of ω, and which are infinite in number, it must be the one for which the absolute value of the real part of ω is as small as possible.

We then see easily that ω must satisfy the inequalities

$$-\frac{1}{2} < \text{real part of } \omega < \frac{1}{2}, \quad |\omega| > 1$$

The points ω which correspond to reduced parallelograms therefore make up a domain D_0 defined by these inequalities and bounded by two lines and a circle. This domain D_0 and its transforms by the different transformations in the group $\left(\omega, \frac{\alpha\omega+\beta}{\gamma\omega+\delta}\right)$ then fill the upper half of the ω plane.

We can operate in exactly the same way on the problem we

are concerned with, since it is entirely analogous. It is just that the parallelograms are replaced by generating polygons of a Fuchsian group, the doubly periodic functions by the Fuchsian functions, and the upper half of the plane by the manifold M_1.

Among the polygons equivalent to R_0, there is one which we regard as more simple than all the others and which we call the **reduced polygon**. Here is how we can define this reduced polygon. Let ϕ be a function of the coordinates of a point of the manifold M_1. I suppose that this function is always between 0 and 1, and that it attains the value zero only on the frontiers of the manifold M_1, i.e., at points corresponding to limit polygons. Apart from that, it can be chosen arbitrarily. That being so, different polygons piecewise equivalent to R_0 correspond to different points of M_1, and hence to different values of ϕ. The polygon we call **reduced** will be that corresponding to the greatest value of ϕ.

Under these circumstances, the points of M_1 which correspond to reduced polygons will fill a certain domain I call D_0. This domain and its transforms by the different transformations in Γ will fill the whole manifold M_1. The domain D_0 will be bounded by a certain number of manifolds $m_1, m_2, \ldots, m_q$, which will have $p - 1$ dimensions if I suppose that M_1 has p. Let us see the form of the equations of these frontier manifolds. Let α be a point of M_1, and αS the transform of α by a suitably chosen transformation S in the group Γ. The desired equations will be of the following form:

$$\phi(\alpha) = \phi(\alpha S) \tag{1}$$

where ϕ denotes the function defined above. If we consider a point α belonging to one of the frontier manifolds $m_1, m_2, \ldots, m_q$, for example, to that which is defined by equation (1), then this point will correspond to a reduced polygon, and so will the point αS which corresponds to a reduced polygon piecewise equivalent to the first and which also belongs to a frontier manifold. So it is that certain classes of quadratic forms contain two reduced forms, and likewise any point on a side of a generating polygon for a Fuchsian group corresponds to an **equivalent** point on the equivalent side. The manifolds m_1, m_2, ..., m_q are then divided into pairs of equivalent manifolds in the fashion of sides of a polygon R_0. These manifolds m are bounded by (p—2)—dimensional manifolds m′, the latter by (p—3)—dimensional manifolds m″, and so on. These manifolds m′, m″, etc. are partitioned into cycles in the same way as the vertices of a polygon R_0. These cycles can contain one, two or several members; if one of the cycles contains more than two, then we will have more than two reduced polygons piecewise equivalent to a given polygon.

Thus, in general we have no more than one reduced polygon piecewise equivalent to a given polygon, but in certain exceptional cases we can have two or more. This is exactly what happens with the quadratic forms. We can describe the situation in another way. In general, among the values of ϕ which correspond to an infinity of piecewise equivalent polygons, there will be one which is greater than all the others; but it can happen, exceptionally, that we have two or more which are equal, and greater than all the others.

We must now deal with a question of great importance; when is a limit polygon reduced? And we examine successively the three examples of the preceding Section.

First recall the polygon R_0 of the first example, and suppose that the two equivalent sides $\alpha_{2n}\alpha_1$ and $\alpha_1\alpha_2$ are infinitely small. We have seen that, leaving aside the exceptional case of limit polygons of the third species, all the limit polygons can be reduced to this case. We therefore neglect, as against finite quantities, quantities of the order of $\alpha_1\alpha_2$ and $\alpha_1\alpha_{2n}$.

I join $\alpha_{2n-1}\alpha_2$ (here, as in what follows, this means connecting the two points by an arc of a circle orthogonal to the fundamental circle). Thus I divide the polygon R_0 into two others:

$r_0 = \alpha_{2n-1}\alpha_{2n}\alpha_1\alpha_2$ and $r'_0 = R_0 - r_0$.

Transforming r_0 by the linear transformation which changes $\alpha_{2n}\alpha_{2n-1}$ into $\alpha_{2n-2}\alpha_{2n-1}$, we obtain a quadrilateral

$$r''_0 = \alpha_{2n-2}\alpha_{2n-1}\alpha_{1,1}\alpha_{2,1}$$

Now put

$$R_{0,1} = r'_0 + r''_0$$

The polygon $R_{0,1}$ will be piecewise equivalent to R_0 and will have vertices α_{2n-1}, α_{2n}, α_2, α_3, ..., α_{2n-3}, α_{2n-2}, $\alpha_{1,1}$, $\alpha_{2,1}$. We see that $R_{0,1}$ has a vertex infinitely near to each of the points

$$\alpha_2, \alpha_3, \ldots, \alpha_{2n-2}$$

and three vertices infinitely near to α_{2n-2}. As for the lengths of the infinitely small sides, they will be given by the relation

$$(2) \qquad \alpha_{1,1} - \alpha_{2,1} = (\alpha_1 - \alpha_2)\frac{(\alpha_{2n-2}-\alpha_{2n-1})^2}{(\alpha_{2n-1}-\alpha_{2n})^2}$$

I now join $\alpha_{2,1}\alpha_{2n-3}$ and divide the polygon $R_{0,1}$ into two others: $r_{0,1} = \alpha_{1,1}\alpha_{2,1}\alpha_{2n-2}\alpha_{2n-3}$ and $r'_{0,1} = R_{0,1} - r_{0,1}$. Now let $r''_{0,1} = \alpha_{1,2}\alpha_{2,2}\alpha_{2n-4}\alpha_{2n-3}$ be the transform of $r_{0,1}$ by the linear transformation which changes $\alpha_{2n-3}\alpha_{2n-2}$ into $\alpha_{2n-3}\alpha_{2n-4}$. Finally, let $R_{0,2} = r'_{0,1} + r''_{0,1}$. It is easy to see that $R_{0,2}$, which is piecewise equivalent to $R_{0,1}$ has a vertex infinitely near to each of

the points

$$\alpha_2, \alpha_3, \ldots, \alpha_{2n-2}$$

and three vertices α_{2n-4}, $\alpha_{2,1}$, $\alpha_{2,2}$ infinitely near to α_{2n-4}. Moreover, we have

$$(2') \qquad \alpha_{1,2} - \alpha_{2,2} = (\alpha_1-\alpha_2)\left[\frac{(\alpha_{2n-2}-\alpha_{2n-1})(\alpha_{2n-4}-\alpha_{2n-3})}{(\alpha_{2n}-\alpha_{2n-1})(\alpha_{2n-2}-\alpha_{2n-3})}\right]^2$$

We next operate on $R_{0,2}$ as we have on R_0 and $R_{0,1}$. I.e., we join $\alpha_{2,2}\alpha_{2n-5}$, let $r''_{0,2}$ be the transform of $r_{0,2}$ by the linear transformation which changes $\alpha_{2n-4}\alpha_{2n-5}$ into $\alpha_{2n-6}\alpha_{2n-5}$, and put

$$r_{0,2} = \alpha_{2n-4}\alpha_{2n-5}\alpha_{1,2}\alpha_{2,2}, \quad r'_{0,2} = R_{0,2} - r_{0,2}$$

$$R_{0,3} = r'_{0,2} + r''_{0,2}$$

We then operate on $R_{0,3}$ as on $R_{0,2}$, and so on.

It is easy to see that $R_{0,i}$ has a vertex infinitely near to each of the points

$$\alpha_2, \alpha_3, \ldots, \alpha_{2n-2}$$

and three infinitely near to α_{2n-2i}.

Consider in particular the polygon $R_{0,n-1}$. It will have three vertices infinitely near to α_2 and consequently to α_{2n}. They will be the vertices α_2, $\alpha_{1,n-1}$ and $\alpha_{2,n-1}$ (which should not be confused with α_{2n-1}). It follows from this that $R_{0,n-1}$ differs infinitely little from R_0. Moreover, these two polygons are piecewise equivalent, whence it follows that the transformation T which changes R_0 into $R_{0,n-1}$ belongs to the group Γ.

We put

$$H = \frac{(\alpha_2-\alpha_3)(\alpha_4-\alpha_5)\ldots(\alpha_{2n-2}-\alpha_{2n-1})}{(\alpha_3-\alpha_4)(\alpha_5-\alpha_6)\ldots(\alpha_{2n-1}-\alpha_{2n})}$$

We have the following relation, which is to the polygon $R_{0,n-1}$ what the relations (2) and (2′) are to the polygons $R_{0,1}$ and $R_{0,2}$:

$$\alpha_{1,n-1} - \alpha_{2,n-1} = (\alpha_1-\alpha_2)H^2$$

and likewise

$$\alpha_2 - \alpha_{1,n-1} = (\alpha_{2n}-\alpha_1)H^2$$

This being so, we recall our function ϕ and suppose that it behaves regularly, which is no problem since this function has been arbitrary until now. For a polygon with infinitely small sides, the function ϕ is infinitely small by hypothesis. We suppose that it is of the same order as $\alpha_1 - \alpha_2$. Neglecting second order infinitesimals, we can say that the three quantities ϕ, $\alpha_1 - \alpha_2$ and $\alpha_1 - \alpha_{2n}$ are proportional. For the polygon R_0 we then have

$$\phi = (\alpha_1-\alpha_2)F$$

where F depends only on the position of the points

$$\alpha_2, \alpha_3, \alpha_4, \ldots, \alpha_{2n-2}$$

and not on the lengths of the infinitely small sides $\alpha_1\alpha_2$ and $\alpha_1\alpha_{2n}$.

We let ϕ' and F' be the values of ϕ and F which correspond to $R_{0,n-1}$. We have

$$F' = F \text{ up to first order infinitesimals}$$

$$\phi' = \phi H^2 \text{ up to second order infinitesimals.}$$

Then if $H^2 > 1$, the transformation T transforms R_0 into a piecewise equivalent polygon corresponding to a greater value of ϕ. **The polygon R_0 is therefore not reduced.**

If, on the contrary, $H^2 < 1$, then the transformation inverse to T will transform R_0 into a piecewise equivalent polygon corresponding to a greater value of ϕ, so that R_0 is again not reduced.

There remains the case where $H^2 = 1$. But H is essentially negative. We therefore have $H = -1$. But this relation, which is none other than the relation (1^a) of the preceding section, says that the polygon R_0 is of the first category.

Thus, a limit polygon cannot be reduced unless it is of the first category.

We shall obtain the same result in the second and third examples.

We first treat the second example, supposing, to fix ideas, that

$$a_1 = a_3 = \dots = a_{2n-1}$$

In the second example, and also the third, I define the reduced polygon in the following way. Among the polygons piecewise equivalent to a given polygon, I take it to be the one whose longest side is as small as possible. If there are several which satisfy this condition, I choose the one among them whose second longest side is as small as possible, and so on. (Here, as in the second and third examples in the preceding section, when speaking of a side as large or small, I am referring to its L and not its geometric length.)

Here is a proposition which I shall merely state, and which I am going to apply to my purpose.

Let a triangle ABC be deformed in such a way that the three vertices tend respectively to three points A′, B′, C′, of which the first is on the fundamental circle and the other two inside. The two sides AB and AC increase indefinitely, while BC remains finite, but the difference between the two sides AB and AC tends to a finite limit λ:

$$\lim(AB-AC) = \lambda$$

This limit depends only on the position of the three points A′, B′, C′, not on the manner in which A, B, C tend to A′, B′, C′. Through the points B′ and C′ we can draw two circles tangent to the fundamental circle. Let M and N be the points of contact. We join MN by an arc of a circle orthogonal to the fundamental circle.* (*Translator's Note: This is the non-Euclidean perpendicular bisector of B′C′.) This circular arc will cut B′C′ at a point P such that the L of PB′ equals that of PC′. I suppose that B′ is to the left of MN and C′ to the right. The limit λ will be positive if A′ is to the right of MN, negative if A′ is to the left; it will be zero if A′ is at M or N.

We then consider the polygon R_0 whose sides $\alpha_{2n}\alpha_1$, $\alpha_1\alpha_2$, $\alpha_2\alpha_3$, $\alpha_3\alpha_4$ we are going to increase indefinitely, as in the preceding section. It will then happen that the two vertices α_{2n} and α_4 tend to two points B′ and C′ inside the fundamental circle, while α_1 tends to a point A′ on the circle itself. At the same time, the points α_2 and α_3 also tend to A′, since the sides $\alpha_1\alpha_2$ and $\alpha_2\alpha_3$, which have infinitely large L, have infinitely small geometric length.

We can find the points M and N on the fundamental circle defined above; we join MN, and I first suppose that A′, like B′, is to the left of MN. By virtue of the lemma I have stated above, we have

$$\lim \alpha_{2n} = B', \quad \lim \alpha_4 = C', \quad \lim \alpha_1 = \lim \alpha_2 = \lim \alpha_3 = A'$$

$$\alpha_2\alpha_4 > \alpha_2\alpha_{2n}, \quad \alpha_3\alpha_4 > \alpha_3\alpha_{2n}$$

In the two triangles $\alpha_2\alpha_3\alpha_4$, $\alpha_2,\alpha_1\alpha_{2n}$ we have

$$\alpha_2\alpha_3 = \alpha_3\alpha_4, \quad \alpha_1\alpha_2 = \alpha_1\alpha_{2n}$$

$$\text{angle } \alpha_2\alpha_1\alpha_{2n} = \text{angle } \alpha_2\alpha_3\alpha_4$$

and

$$\alpha_2\alpha_4 > \alpha_2\alpha_{2n}$$

whence

$$\alpha_3\alpha_4 > \alpha_1\alpha_2$$

The side $\alpha_3\alpha_4 = \alpha_2\alpha_3$ is then the longest side of R_0 because the others among $\alpha_1\alpha_2$, $\alpha_3\alpha_4$, $\alpha_2\alpha_3$, $\alpha_1\alpha_{2n}$ are finite.

I claim that we can find a polygon equivalent to R_0 whose longest side is smaller than $\alpha_3\alpha_4$.

By joining $\alpha_3\alpha_{2n}$, we divide the polygon R_0 into two others:

$$r_0 = \alpha_{2n}\alpha_1\alpha_2\alpha_3, \quad r'_0 = \alpha_3\alpha_4\alpha_5 \dots \alpha_{2n}$$

We transform r'_0 by the transformation which changes $\alpha_3\alpha_4$ into $\alpha_3\alpha_2$. We obtain

$$r''_0 = \alpha'_3\alpha'_4\alpha'_5 \dots \alpha'_{2n}$$

with

$$\alpha'_3 = \alpha_3, \quad \alpha'_4 = \alpha_2$$

The polygon $R'_0 = r_0 + r''_0$ is piecewise equivalent to R_0. It has $2n - 4$ sides equal to $2n - 4$ finite sides of R_0. It also has the two infinite sides

$$\alpha_1\alpha_2 = \alpha_1\alpha_{2n}$$

which belong to R_0, and the two infinite sides

$$\alpha_3\alpha_{2n} = \alpha_3\alpha'_{2n}$$

which do not belong to R_0. If $\alpha_3\alpha_{2n} > \alpha_1\alpha_2$, then $\alpha_3\alpha_{2n}$ is the longest side of R_0, and we have

$$\alpha_3\alpha_{2n} < \alpha_3\alpha_4$$

Q.E.D.

If $\alpha_3\alpha_{2n} < \alpha_1\alpha_2$ then $\alpha_1\alpha_2$ is the longest side of R'_0, and we have

$$\alpha_1\alpha_2 < \alpha_3\alpha_4$$

Q.E.D.

Thus, if the point A′ is to the left of MN, the polygon R_0 is not reduced. But since there is nothing to distinguish left from right, it is also not reduced if A′ is to the right of MN. Thus it is necessary for A′ to equal M or N. But then the circle A′B′C′ is tangent to the fundamental circle, which means that the limit polygon is of the first category.

Thus each reduced limit polygon is of the first category.

We move on to the third example, i.e., the polygon R_0 of 4p sides whose opposite sides are equivalent. We join the opposite vertices, and thus obtain 4p diagonals. I claim that if R_0 is reduced, then the longest of the 4p sides is smaller than the smallest of the 4p diagonals. I let δ be this smallest diagonal; if there is a side longer than δ, we divide the polygon R_0 into two others r_0 and r'_0 by cutting along the diagonal δ. We then let r''_0 be the transform of r_0 by the transformation which changes the longest side of R_0 into its conjugate. The polygon $R'_0 = r'_0 + r''_0$ will be piecewise equivalent to R_0, it will have its opposite sides equivalent, moreover, it will have all the sides of R_0 except the longest, which has been replaced by diagonal δ, whose length is less, by hypothesis. Thus the longest side of R'_0 will be smaller than that of R_0, and R_0 will not be reduced.* (*Translator's Note: Thus there is a finite set of cut—and—paste operations which can be used to reduce any polygon. Assuming there are also only finitely many reduced polygons, the mapping class group Γ is finitely generated.)

That being so, we suppose, as in the preceding section, that the two sides $\alpha_{4p}\alpha_1$ and $\alpha_{2p}\alpha_{2p+1}$ increase indefinitely. It will be the same for the diagonals $\alpha_{4p}\alpha_{2p}$ and $\alpha_1\alpha_{2p+1}$; all the other sides remain finite. Suppose that the vertices α_{4p} and α_{2p+1} tend to two points B′ and C′ inside the fundamental circle. The vertices α_1, α_2, ..., α_{2p} tend to the same point A′ on this circle, and the point A′ is also that approached by the vertices β_1 and β_2 of the polygon R'_0. This fact yields the following identity:

$$\lim(\alpha_1\alpha_{4p}-\alpha_1\alpha_{2p+1}) = \lim(\alpha_{2p}\alpha_{4p}-\alpha_{2p}\alpha_{2p+1})$$

But if the polygon R_0 is reduced we also have, since the sides are smaller than the diagonals,

$$\alpha_1\alpha_{4p} \leqslant \alpha_1\alpha_{2p+1}, \quad \alpha_{2p}\alpha_{4p} \leqslant \alpha_{2p}\alpha_{2p+1}$$

Then we have

$$\lim(\alpha_1\alpha_{4p}-\alpha_1\alpha_{2p+1}) = 0$$

We conclude, as in the preceding example, that the circle A′B′C′ is tangent to the fundamental circle and that the polygon R_0 is of the first category.

Thus every reduced limit polygon is of the first category.

All the preceding applies only to polygons of the first species.

We now move on to polygons of the second species which can, as we have seen, be reduced to limit polygons of the first species. Suppose then that P_0 is a polygon of the second species; this polygon will be piecewise equivalent to another, R_0, of the first species. If R_0 is not of the first category, then I claim that P_0 is not reduced.

In fact, consider the function ϕ defined above; this function, in the case of the second or third example could be, say, the reciprocal of the longest side of R_0.

The value of ϕ corresponding to either P_0 or R_0 will be infinitely small. Let a and b be the points of M_1 which correspond to

R_0 and to P_0; each of these points will be infinitely close to a frontier of the manifold M_1. Let c and d be the points of these frontiers which are approached by a and b respectively. The corrdinates of the point a are functions of those of the point b, because the corresponding polygons R_0 and P_0 are derived from each other by a definite rule. If a and b are very close to the frontier, as we are assuming, then the position of the point c will depend on that of the point d, in such a way that the coordinates of one of these points are continuous functions of those of the other. Now if the points c and d remain fixed then, **neglecting higher order infinitesimals**, the point a will approach c in such a way that when the infinitely small distance ac is halved, for example, the infinitely small distance bd will also be halved.

If we let ϕ_1 and ϕ_2 be the values of ϕ at the points a and b then we have, up to second order infinitesimals,

$$\phi_1 = \overline{ac}\ f_1(c), \qquad \phi_2 = \overline{bd}\ f_2(d)$$

where f_1 and f_2 are finite functions, in general, depending only on c and d.

Thus if the points c and d remain fixed, or vary infinitely little, and the distance ac diminishes in a certain ratio, then the two infinitely small values ϕ_1 and ϕ_2 diminish in the same ratio.

That being so, we have seen that if R_0 is not of the first category, then it is not reduced and consequently we can find a piecewise equivalent polygon R'_0 corresponding to a greater value of ϕ. Let P'_0 be the polygon of the second species derived from R'_0 the way P'_0 is derived from R_0, let a′, b′, c′, d′ be the points for R'_0 and P'_0 as a, b, c, d are for R_0 and P_0, and let ϕ'_1 and ϕ'_2 be the values of ϕ which correspond to R'_0 and P'_0. It is easy to check that the points c′ and d′ differ infinitely little from c and d; we therefore have

$$\frac{ac}{a'c'} = \frac{bd}{b'd'}$$

$$\phi'_1 = a'c'f_1(c') = a'c'f_1(c)$$

$$\phi'_2 = b'd'f_2(d') = b'd'f_2(d)$$

up to higher order infinitesimals and consequently

$$\frac{\phi_1}{\phi'_1} = \frac{\phi_2}{\phi'_2}$$

But we have

$$\phi'_1 > \phi_1$$

and hence

$$\phi'_2 > \phi_2$$

consequently the polygon P_0 will not be reduced.

Thus, **a limit polygon of the second species cannot be reduced unless it is piecewise equivalent to a polygon of the first species and of the first category.**

The case of polygons of the third species, which are particular instances of those of the first and second, is treated along the same lines. Here we shall study only a single example. In this way we fix ideas, and the reader is given a better grasp of the way in which we surmount the various difficulties than by direct study of the general case.

We choose the polygon R_0 of the second example and suppose that the six sides $\alpha_{2n}\alpha_1$, $\alpha_1\alpha_2$, $\alpha_2\alpha_3$, $\alpha_3\alpha_4$, $\alpha_4\alpha_5$ and $\alpha_5\alpha_6$ increase indefinitely.

We suppose that this polygon is reduced, and we shall show that it is of the first category, i.e., that the five vertices α_1, α_2, α_3, α_4, α_5 tend towards a single point on the fundamental circle, and in such a way that the circle $\alpha_{2n}\alpha_1\alpha_6$ is tangent to the fundamental circle.

To do this, we shall appeal to the following two principles.

1° If the polygon R_0 is reduced, then the sides issuing from a vertex of odd rank are smaller than the diagonal joining this vertex to any vertex of even rank. For example, let C be the side $\alpha_1\alpha_2$ and let D be a diagonal joining α_1 to any vertex of even rank. It is easy to construct a polygon R'_0 piecewise equivalent to R_0, with sides of the same lengths, except that C is replaced by D. If we have $C > D$ then the longest side of R_0 will be greater than the longest side of R'_0, and R_0 will not be reduced.

2° Suppose that the three vertices of a triangle ABC tend respectively to three points A′, B′, C′, the first of which is inside, and the other two on, the fundamental circle. **If the two points** B′ **and** C′ **are distinct**, then the three sides increase indefinitely, and the angle BAC will tend to a finite limit, so that the difference

$$AB + AC - BC$$

and, **consequently**, **the side** BC **will be greater than each of the other two**.

That being so, suppose that the two vertices α_{2n} and α_6 in the reduced polygon R_0 tend toward two fixed points B′ and C′ in the plane. The five vertices α_1, α_2, α_3, α_4, α_5 whose distance from α_6 is infinite, tend toward certain points on the fundamental circle.

If we have

$$\lim \alpha_4 \neq \lim \alpha_5$$

then we shall have, by virtue of the second principle,

$$\alpha_5\alpha_4 > \alpha_5\alpha_{2n}$$

which is contrary to the first principle.

Likewise, if we have

$$\lim \alpha_4 \neq \lim \alpha_3, \lim \alpha_3 \neq \lim \alpha_2 \text{ or } \lim \alpha_2 \neq \lim \alpha_1$$

we have

$$\alpha_3\alpha_4 > \alpha_3\alpha_{2n}, \quad \alpha_2\alpha_3 > \alpha_3\alpha_{2n} \quad \text{or} \quad \alpha_2\alpha_1 > \alpha_1\alpha_6$$

which is contrary to the first principle.

Thus the five vertices α_1, α_2, ..., α_5 tend toward a single point A′ on the fundamental circle.

We join the two points M and N defined above. If A′ is to the left of MN (i.e., on the same side as B′), then we have

$$\alpha_5\alpha_{2n} < \alpha_5\alpha_6$$

and if it is on the right

$$\alpha_1\alpha_6 < \alpha_1\alpha_{2n}$$

and these are contrary to the first principle.

Thus A′ is at M or N, i.e., the circle A′B′C′ is tangent to the fundamental circle, i.e., R_0 is of the first category.

Q.E.D.

XIV. The method of continuity

All the lemmas we have established now permit us to apply the method of continuity in all rigor.

We consider an infinity of types T belonging to the same class and defined by a certain number of parameters which are, if the general equation of the type is written

(1) $$\frac{d^2v}{dx^2} = \phi(x,y)v$$

(2) $$\psi(x,y) = 0 \ ,$$

the moduli of the relation (2) and the singular points of equation (1). Let q be the number of these parameters. These parameters define a point on a certain manifold M of q dimensions. I claim that this manifold is closed and without boundary. In fact, Riemann has shown that if a q—dimensional manifold is bounded by another manifold, then the latter must be of q − 1 dimensions. Thus a surface, for example, can only be **cut (Zerschnitten)** by a line and not by a point.

Now by varying the parameters of the type we cannot arrive at the boundary of M without attaining certain singular points of that manifold, corresponding to the case where the type T reduces to a simpler type T′ (cf. Section IX, p.396). But this can happen in only two ways.

1° Either, by two singular points of (1) coming into coincidence; but the necessary condition for this is **complex**, i.e., two **real** conditions;

2° or else, by decrease of the genus of relation (2) by one, i.e., by the curve represented by this relation acquiring a new double point. But here again two real conditions are necessary.

Thus in either case it is necessary to impose two conditions, and the singular points which satisfy these conditions form a manifold of q − 2 dimensions, which cannot serve as the boundary of M, since it has q. I let σ_1, σ_2, ... denote the (q—2)—dimensional manifolds defined by such conditions.

Now consider the domain D_0 defined in Section XIII. To each point of this domain there corresponds a polygon R_0 and consequently a Fuchsian equation. Consider the parameters of the type to which it belongs: they define a certain point of the manifold M. Thus to each point of D_0 there corresponds exactly one point of M. On the other hand, to each point μ of M there corresponds a Fuchsian type which contains a Fuchsian equation, corresponding to exactly one point S_0 of D_0. The exception to this is where the point δ lies on one of the manifolds m_1, m_2, ... which separate D_0 from the rest of the manifold M_1. In that case, in fact, we have two or more piecewise equivalent reduced polygons, so that the point μ corresponds to two or more points δ. Either a part or the whole of M will correspond to the domain D_0.

I claim that the first case does not occur. If it did, the manifold M would be divided into two parts, those which correspond to D_0 and those which do not, and the two parts would be separated from each other by a frontier which would have to have, by Riemann's theorem, $q - 1$ dimensions.

But what would happen to the point δ in D_0 when the corresponding point μ of M reached this hypothetical frontier manifold?

Could the Jacobian of the coordinates of μ with respect to those of δ vanish? This could never happen because the lemma of Section VII shows that no point μ corresponds to more than one point δ.

Could the point δ reach one of the manifolds m_1, m_2, ... which separate D_0 from the other parts of the manifold M_1? Suppose for example that δ crossed a manifold m_1 separating D_0 from another domain D_1. The point μ would then cross a certain manifold μ_1 corresponding to m_1. But since the point δ is now in the domain D_1, which again forms part of the manifold M_1, this point δ again corresponds to a polygon R_0 and consequently to a Fuchsian equation and a point of the manifold M. Thus the manifold μ_1 does not separate the Fuchsian types which contain Fuchsian equations from those which do not.

Finally, could the point δ reach the limit of the manifold M_1 itself? It would then mean that the corresponding polygon R_0 became a **limit polygon**. But since δ is inside D_0, the polygon R_0 is reduced and consequently **of the first category or piecewise equivalent to a polygon of the first category**. But the second lemma shows that when R_0 reduces to a limit polygon of the first category, the corresponding type T reduces to a simpler type T′, i.e., two singular points of (1) come into coincidence, or else the genus of (2) decreases by unity. Thus when the point δ, remaining inside D_0, tends toward one of the boundaries of m_1, the corresponding point μ will tend toward one of the manifolds σ_1, σ_2, ... But these manifolds have no more than $q - 2$ dimensions, hence they cannot divide M into two parts.

Thus M is not divided into two parts, one corresponding to D_0 and the other not corresponding to D_0.

Hence **every Fuchsian type contains a Fuchsian equation.**

XV. A particular application

We now make a particular application of the preceding principles. To do this we choose the simplest case which can arise, i.e., the first example of Section XII, and suppose that R_0 has no more than six sides. This is the case of the paraboloid (1′) of Section XII. This example will facilitate comprehension of the above principles.

In the case we are concerned with, the manifold M_1 is represented, as we have seen, by the paraboloid (1′). We shall see what becomes of the manifold M.

The Fuchsian equation engendered by R_0 admits four singular points, which are such that each of the four determinant equations has a double root.

Let α, β, γ, δ be these four singular points. Let $\psi(\mu)$ be the cross ratio of μ with respect to β, γ, δ so that

$$\psi(\beta) = 0, \quad \psi(\gamma) = 1, \quad \psi(\delta) = \infty$$

We put

$$\psi(\alpha) = X$$

The type containing the Fuchsian equation in question is completely defined when we know X. The manifold then reduces to the sphere or plane representing the complex quantity X.

When we permute the four singular points in all possible ways, we obtain the following six values for X:

$$\left[X,\ 1 - X,\ \frac{1}{X},\ \frac{1}{1-X},\ 1 - \frac{1}{X},\ \frac{X}{X-1}\right] \tag{1}$$

We can divide the X plane into six regions:

$$\rho_0 + \rho'_0,\ \rho_1 + \rho'_1,\ \rho_2 + \rho'_2,\ \rho_3 + \rho'_3,\ \rho_4 + \rho'_4,\ \rho_5 + \rho'_5$$

Figure 15 indicates this subdivision; the two circles and the lines shown there have the equations

$$|X| = 1,\ |1-X| = 1,\ \text{real part of } X = \tfrac{1}{2}$$

$$\text{imaginary part of } X = 0$$

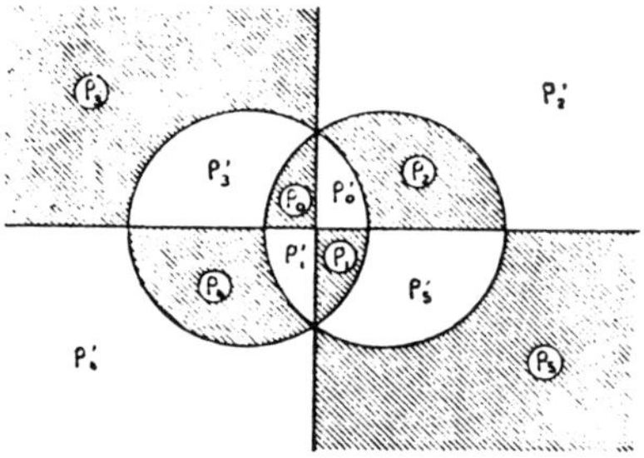

Figure 15

One easily sees that if X is inside ρ_0 and if X_0 is the complex conjugate of X then the quantities

$$1 - X,\ \frac{1}{1-X},\ \frac{X}{X-1},\ 1 - \frac{1}{X},\ \frac{1}{X},$$

$$1 - X_0,\ X_0,\ \frac{1}{X_0},\ 1 - \frac{1}{X_0},\ \frac{X_0}{X_0-1},\ \frac{1}{1-X_0}$$

respectively make up

$$\rho_1,\ \rho_2,\ \rho_3,\ \rho_4,\ \rho_5,\ \rho'_0,\ \rho'_1,\ \rho'_2,\ \rho'_3,\ \rho'_4,\ \rho'_5.$$

Now the paraboloid (1′) is subdivided into an infinity of domains D_0, D_1, ..., as described in Section XIII, i.e., in such a way that D_0 contains all the points of this paraboloid which correspond to reduced polygons R_0, and to each point of M there corresponds exactly one point of each of these domains. This is not all. Just as M is partitioned into twelve regions ρ and ρ', D_0 is partitioned into twelve partial regions corresponding to the twelve regions of M, and it will be the same for each of the domains D_1, D_2, ..., etc.

Figure 16 represents the paraboloid (1′) projected on to the xy plane; the line BCF has equation y = 1 and the line BAE, x = 0.

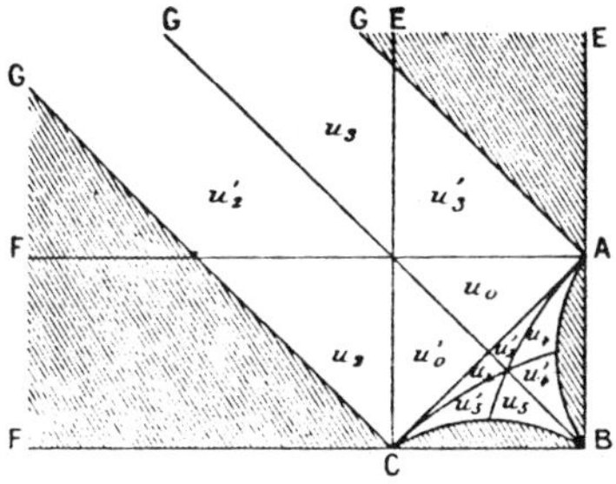

Figure 16

The line x = 0, y = 1 projects completely to the point B. The non–hatched part represents the projection of D_0. This projection is bounded by the lines CG and AG and by the parabolas AB and BC. It is subdivided into twelve regions by the lines AC, CE, AF, BG and by two curves.

These lines have the equations

$$CG,\ x = y = 0;\ BG,\ x + y = 1;\ AG,\ x + y = 2;$$

$$AC,\ x - y = -1;\ AF,\ y = 2;\ CE,\ x = -1$$

The two parabolas AB and BC are tangent to the lines AC and

BG. Finally, the other two curves drawn on this figure are arcs of conics. We see that the domain D_0 is divided into twelve regions u_i and u'_i corresponding respectively to the twelve regions ρ_i and ρ'_i

Here is how to proceed in showing that, when the point δ traverses the domain u_i or u'_i on M_1, the corresponding point μ traverses the region ρ_i or ρ'_i on the manifold M.

Let R'_0 be the polygon symmetric to R_0 in its diagonal be. I let a′, b′, c′, d′, e′, f′, be the vertices of R'_0; and in order to have the same cyclic order as R_0, a′ will be the reflection of c, c′ that of a, d′ that of f, f′ that of d, while b′ and e′ will be no different from b and e. Thanks to these conventions, we easily see that the new values of x, y, z are respectively

$$1 - y, \quad 1 - x, \quad 1 - z$$

Now let X′ be the value of X which corresponds to the polygon R'_0; we easily see that

$$X' = 1 - X_0$$

In particular, if the polygon R_0 is symmetric in be, so that it does not differ from R_0, then we have

$$x + y = 1, \quad z = \frac{1}{2}$$

$$X = 1 - X_0$$

Then if the point δ describes the line BG, the point μ will describe the line

$$\text{real part of } X = \frac{1}{2}$$

We similarly see that if we change R_0 into its reflection in the diagonal cf, then x, y, z will change into $\frac{1}{x}$, $\frac{1}{z}$, $\frac{1}{y}$, and X will change into $\frac{1}{X_0}$. We conclude that if R_0 is symmetric with respect to cf, i.e., if the point δ describes the line CE, then the point μ will describe the circle

$$|X| = 1$$

The polygon R_0 is piecewise equivalent to a polygon R'_0 obtained by joining bf and transforming the triangle baf by the transformation which changes ba into bc. We thus obtain a polygon bfedcf', where f' is the transform of f by the transformation in question. We take the reflection of R'_0 in the diagonal bd. This reflection can be regarded as derived from a certain polygon R''_0 in the same way that R'_0 was derived from R_0. We easily see that in changing R_0 and R''_0 we change x, y, z into

$$\frac{z^2}{x}, \quad 1 + \frac{z(z-1)}{x}, \quad z$$

and X into X_0.

Thus if the point δ describes the line AC, the point μ will describe the line

$$X = X_0, \quad \text{imaginary part of } X = 0$$

The following problem: **given a Fuchsian type, find the Fuchsian group which engenders an equation contained in that type**, is a very complicated problem I shall say a few words about later.

However, we can use the preceding to resolve the following special cases:

$$X = \tfrac{1}{2}, \quad X = -1, \quad X = 2, \quad X = \frac{1+\sqrt{-3}}{2}, \quad X = \frac{1-\sqrt{-3}}{2}$$

by simple symmetry considerations, and we find the respective parameters defining the generating polygon of the corresponding Fuchsian group to be:

$$x = -\frac{1}{2}, \quad y = \frac{3}{2}, \quad z = \frac{1}{2}$$

$$x = -1, \quad y = 3, \quad z = \frac{1}{3}$$

$$x = -2, \quad y = 2, \quad z = \frac{2}{3}$$

$$x = -1, \quad y = 2, \quad z = \frac{1}{2}$$

$$x = -\frac{1}{4}, \quad y = \frac{3}{4}, \quad z = \frac{1}{2}$$

I let S_1, S_2, S_3 be the operations under consideration and now show the changes they produce in x, y, z and X.

$$S_1 \quad (x, y, z; 1-y, 1-x, 1-z) \quad (X, 1-X_0)$$

$$S_2 \quad (x, y, z; \frac{1}{x}, \frac{1}{z}, \frac{1}{y}) \quad (X, \frac{1}{X_0})$$

$$S_3 \quad (x, y, z; \frac{z^2}{x}, 1 + \frac{z(z-1)}{x}, z) \quad (X, X_0)$$

Consider:

1° The group Γ_1 generated by these three transformations. If we transform the triangle u_0 by all elements of the group, then we obtain an infinity of analogous triangles which completely fill the manifold M_1.

2° The group Γ_2 consists of all products of an **even number** of transformations S_1, S_2 or S_3. This group is remarkable because it is isomorphic to the group

$$\left(z, \frac{\alpha z + \beta}{\gamma z + \delta}\right)$$

where α, β, γ, δ are integers such that $\alpha\delta - \beta\gamma = 1$.

3° The group Γ consisting of all transformations in the group Γ_1 which do not change X. If we apply all the transformations in Γ_1 to D_0, then we obtain a series of domains D_1, D_2, etc. which fill all the manifold M_1. The group Γ is isomorphic to the group of transformations

$$\left[z, \frac{\alpha z+\beta}{\gamma z+\delta}\right]$$

where α, β, γ, δ are integers and

$$\alpha\delta - \beta\gamma = 1, \quad \alpha \equiv \delta \equiv 1, \quad \beta \equiv \gamma \equiv 0, \quad (\text{mod } 2)$$

i.e., a group one encounters in the study of the modular function.

Each of the domains D_1, D_2, etc. is divided into twelve regions in the same way as D_0 is divided into twelve regions u_i and u'_i. But D_0 can also be regarded as divided into two triangles

$$T_0 = u_0 + u'_0 + u_2 + u'_2 + u_3 + u'_3$$

and

$$T'_0 = u_1 + u'_1 + u_4 + u'_4 + u_5 + u'_5$$

Similarly, the domains D_1, D_2, etc. can each be regarded as divided into two triangles, T_1 and T'_1, T_2 and T'_2, etc. This is the mode of subdivision we have adopted in Figures 17, 18 and 19, which represent portions of the paraboloid (1′) in perspective.

In Figure 17, the line BKL represents the generatrix $x = 0$, $y = 1$ which was represented only the point B in Figure 16. In the present

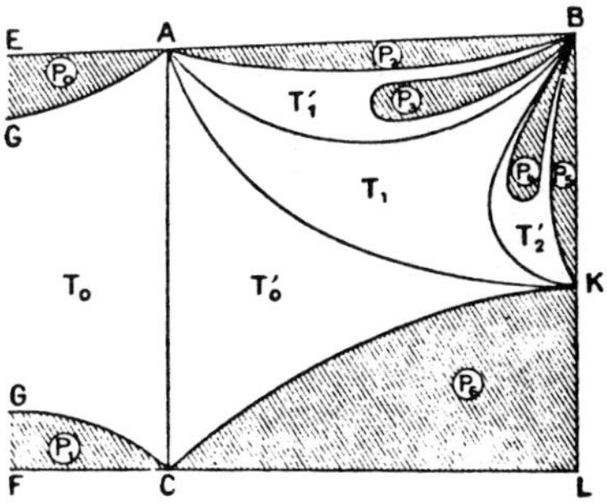

Figure 17

figure the five triangles T_0, T'_0, T_1, T'_1, T'_2 are shown not hatched, and the hatched portions P_0 to P_6 cover portions of the paraboloid where the subdivision is not shown. We also see that each of the two triangles T'_1, T'_2 have vertices which coincide at B.

Figure 18 shows the subdivision of the region $T'_1 + P_2 + P_3$ on a larger scale. The portion P_2 has been divided into a triangle T_3 and two hatched regions P_7 and P_8

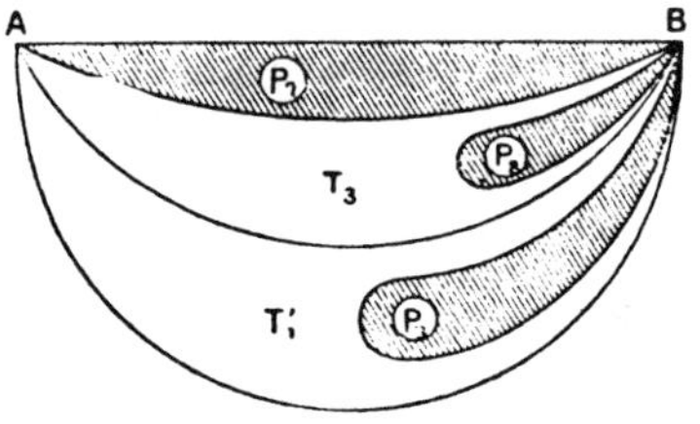

Figure 18

In all four figures 16, 17, 18, 19 the hatched regions are those where the subdivision is not shown.

Finally, Figure 19 represents the detail of the region P_3 which is divided into a triangle T_4 and two hatched regions P_9 and P_{10}. We see that the triangle T_4 has three vertices which coincide at B.

These figures suffice to give a complete account of the particulars of the subdivision. In fact, the region P_6 or CKL, is treated like the region AKB; the regions

P_0 and P_1 are treated like the region ABLC, except that a part of the figure is expelled to infinity. The regions P_5 and P_7 are treated like P_2; finally P_4, P_8, P_9 and P_{10}, are treated like P_3.

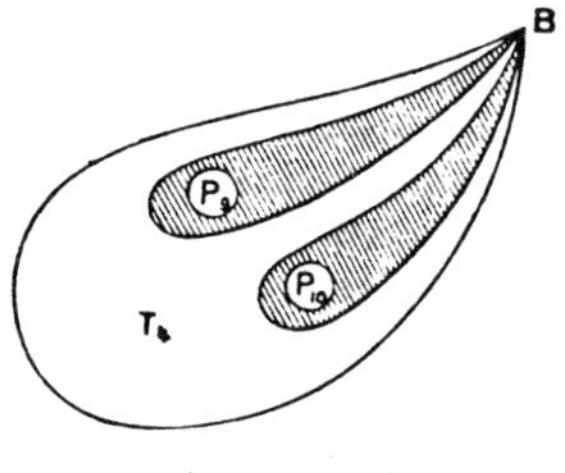

Figure 19

The points of the lines BE and FL correspond to polygons of the first species and, on these lines, the points A and C are the only ones which correspond to polygons of the first category; they are also the only ones in D_0.

On the line BL, which corresponds to polygons of the second

species, the point K is the only one in D_0. And in fact the transformation S_3 changes it into the point

$$x = y = \infty, \quad z = \frac{1}{2}$$

which corresponds to a polygon of the first category.

XVI. Theory of subgroups

We have shown that every Fuchsian type contains a Fuchsian equation; now it is necessary to find the Fuchsian equation within a given Fuchsian type, or else the corresponding Fuchsian group.

But before starting on this problem, it is necessary to say a few words on the subgroups of a given Fuchsian group.

These subgroups are of two kinds:

1° The subgroups of finite index, i.e., those for which all transformations in the principle group are obtainable by multiplying all the transformations in the subgroup by a finite number of transformations.

2° The subgroups of infinite index.

From another point of view, the subgroups are again of two kinds:

1° The subgroups which are distinguished* (*Translator's Note: i.e., normal), **ausgezeichnet**, as the Germans say, i.e., invariant under conjugation by all elements of the principle group.

2° The non–distinguished subgroups.

(Cf. KLEIN, **Mathematische Annalen, 17**; WALTHER DYCK, Ueber reguläre Riemannsche Fläche, Gruppentheoretischen Studien, **Mathematische Annalen, 17, 20.**)

The subgroups of finite index are none other than the Fuchsian groups. Let G be the principle group and let g be a subgroup. Let R_0 be the generating polygon of G, R_1, R_2, ... its transforms by the different transformations in the group. Let P_0 be the generating polygon of g. It is easy to see that P_0 is the union of a certain number of polygons R_0, R_1, R_2, ..., R_n.

We choose any n + 1 polygons from among the polygons

$$R_0, R_1, R_2, \dots$$

For example, let the n + 1 polygons by R_0, R_1, R_2, ..., R_n and suppose that their union is a single simply–connected piece. Let P_0 be this set. A certain number of sides of the polygon R remain free; they will be the sides of P_0. It remains to see how these sides are distributed in equivalent pairs. Let a_0b_0 be a side of R_0, $a'_0b'_0$ its equivalent; let a_1b_1, $a'_1b'_1$ be the homologous sides of R_1; let a_2b_2, $a'_2b'_2$ be those of R_2, etc. Among the n + 1 sides a_ib_i, there will be a certain number p which remain free; likewise, among the n + 1 sides $a'_ib'_i$ there will be p, i.e., precisely the same number, which remain free. We can therefore set each of the free sides a_ib_i equivalent to one of the free sides $a'_ib'_i$, and this in an arbitrary manner. Then all the sides of P_0 are found divided into equivalent pairs.

There is one more condition if P_0 is to give rise to a discontinuous group; it is that the division of vertices into cycles give an angle sum for each cycle which is an aliquot part of 2π. If this condition is satisfied, then P_0 will generate a Fuchsian group, and it will be a subgroup of G. Moreover, we obtain all subgroups of G of finite index in this way.

Now let

$$(1) \qquad X = F_1(z), \quad Y = F_2(z), \quad F_3(z), \dots$$

be the Fuchsian functions engendered by G and let

$$(2) \qquad x = f_1(z), \quad y = f_2(z), \quad f_3(z), \dots$$

be the Fuchsian functions engendered by g. The class of Fuchsian functions (2) contains the class of functions (1) as special cases.

First suppose that G and g are both of genus 0. Then all the functions (1) are expressible rationally in terms of X, all the functions (2) in terms of x. We conclude that X is a rational function to x. Moreover, it is easy to find the coefficients of this rational function when we know the values of X for the different vertices of R_0.

Now if G is of genus 0 and g is of genus p, then all the functions (1) are rational in X, and X is rational in x and y.

Finally, it can happen that neither G nor g is of genus 0. In that case, X and Y are rational in x and y, and the converse is not true.

It should be remarked that this theory is entirely analogous to the theory of transformation of elliptic functions. What, in fact, is this? It consists in studying the relations between the elliptic functions engendered by the group

(3) $$(z, m\omega_1 + n\omega_2) \quad (\omega_1 \omega_2 \text{ given periods; } m, n \text{ integers})$$

and the elliptic functions engendered by the group

(4) $$(z, m\Omega_1 + n\Omega_2)$$

where the new periods Ω_1, Ω_2 are chosen so that the second group is a subgroup of the first. We also remark that if we apply these principles to the Fuchsian functions engendered by the hypergeometric equation, then we recover without difficulty the results obtained by Goursat in his remarkable thesis (**Annales scientifiques de l'ecole Normale**, 1881).

In the particular case where g is a normal subgroup, the subdivision of P_0 into n + 1 polygons R is regular, likewise after we deform P_0 and paste together the equivalent sides to make a closed surface. Moreover, the relations between X, Y, x and y are of a particular nature. For example, if G and g are both of genus 0, then the group of the algebraic equation connecting X and x is only singly transitive.

We consider a particular example which will be useful to us later. Suppose that R_0 is a polygon

$$\alpha_1\alpha_2\alpha_3 \cdots \alpha_n\alpha_{n+1}\beta_n\beta_{n-1} \cdots \beta_3\beta_2$$

which is symmetric along the diagonal $\alpha_1\alpha_{n+1}$, so that β_i is the reflection of α_i. I also assume that the mirror-image sides are equivalent, that all the vertices are on the fundamental circle and that all the angles are zero. Let $X = F(z)$ be one of the Fuchsian functions engendered by R_0, in terms of which all the others are expressible rationally. We have been able to choose F(z) so that this function remains real on the perimeter of R_0 and on the diagonal $\alpha_1\alpha_{n+1}$. (Cf. the memoir, <u>On Fuchsian Functions</u>, **Acta Mathematica, I**, p. 232 and 272, this volume p.176 and 228.)

This function then gives a conformal mapping of the polygon

$$r_0 = \alpha_1\alpha_2\alpha_3 \cdots \alpha_n\alpha_{n+1}\alpha_1$$

onto the half plane. Now let R_1 be the transform of R_0 by the transformation which changes $\beta_i\beta_{i+1}$ into $\alpha_i\alpha_{i+1}$. This polygon will be the reflection of R_0 in $\alpha_i\alpha_{i+1}$. Now consider the polygon $P_0 = R_0 + R_1$ and take its equivalent pairs of sides to be those which are symmetric with respect to $\alpha_i\alpha_{i+1}$. Since all the cycles of this polygon are parabolic, it will engender a system of Fuchsian functions. Among them I choose one in terms of which all others are expressible rationally and which also remains real on the whole perimeter of R_0 and R_1.

I call it $x = f(z)$. This function gives a conformal mapping onto the half plane, not just of r_0 as in the preceding case, but of the whole polygon R_0.

To achieve the definition of x I suppose that

$$f(\alpha_1) = 0, \quad f(\alpha_{n+1}) = \infty$$

and I shall have

$$x = \pm\sqrt{\frac{X-F(\alpha_1)}{X-F(\alpha_{n+1})}}$$

We can take either the + sign or the — sign.

We now push this further. Let P_1 be the mirror image of P_0 in one of its sides, which I call C. Let $Q_0 = P_0 + P_1$ and take its equivalent pairs of sides to be those which are symmetric with respect to C. Let y be a Fuchsian function engendered by Q_0, in terms of which all the others are expressible rationally. I suppose that y is real along the perimeter of Q_0 and consequently along C.

Moreover, I put

$$y = \phi(z), \quad \phi(\alpha_i) = 0, \quad \phi(\alpha_{i+1}) = \infty$$

whence

$$y = \frac{x-f(\alpha_i)}{x-f(\alpha_{i+1})} = \sqrt{\frac{\sqrt{\frac{X-F(\alpha_1)}{X-F(\alpha_{n+1})}} - \sqrt{\frac{F(\alpha_i)-F(\alpha_1)}{F(\alpha_i)-F(\alpha_{n+1})}}}{\sqrt{\frac{X-F(\alpha_1)}{X-F(\alpha_{n+1})}} - \sqrt{\frac{F(\alpha_{i+1})-F(\alpha_1)}{F(\alpha_{i+1})-F(\alpha_{n+1})}}}}$$

This example shows the nature of the relations which exist between the Fuchsian functions engendered by a Fuchsian group G and those engendered by a subgroup of G.

We shall now say a few words about the normal subgroups of finite or infinite index. Here is how we can be led to consider such groups. Let

$$(5) \qquad \frac{d^p w}{dx^p} + \phi_{p-1}\frac{d^{p-1}w}{dx^{p-1}} + \ldots + \phi_0 w = 0$$

be a linear equation whose coefficients ϕ are rational in x. Suppose that the singular points of this equation are

$$a_1, a_2, \ldots, a_n$$

and are such that the roots of the determinant equation relative to a_i

are all multiples of $\frac{1}{K_i}$, where K_i is a positive integer. If we cannot find any integer such that the roots are multiples of $\frac{1}{K_i}$, then we take $K_i = \infty$.

By virtue of Sections VIII and XIV, we can find a Fuchsian equation

$$\frac{d^2v}{dx^2} = \psi(x)v \tag{6}$$

admitting the singular points

$$a_1, a_2, \ldots, a_n$$

and such that the difference between the roots of the determinant equation relative to a_i is $\frac{1}{K_i}$. This Fuchsian equation belongs to a Fuchsian type which I call T.

Let z be the ratio of the integrals; x will be a Fuchsian function of z and I call the group of that function G. The integrals

$$w_1, w_2, w_3, \ldots, w_p$$

of equation (5) are uniform functions of z. Let S be a transformation in G. When we apply this transformation to z, the integrals w become

$$w_1^1, w_2^1, w_3^1, \ldots, w_p^1$$

and it is clear that the w_i^1 are linear functions of the w_i. I let Σ be the linear transformation by which we pass from the w_i to the w_i^1, and I write in abbreviated form:

$$w(z.S) = [w(z)]\Sigma \tag{7}$$

This equation says that when z undergoes the transformation S, the w undergo the transformation Σ.

The transformations Σ form a group Γ which will be homomorphic to G, but this homomorphism can be one—to—one or many—to—one. If it is many—to—one there will be an infinity of transformations

$$(8) \qquad S_1, S_2, \ldots, S_p, \ldots$$

in the group G corresponding to the identity transformation in the group Γ. These transformations (8) form a subgroup g of the group G. I claim that this subgroup is normal. In fact, let s be a transformation in G which does not belong to g, and let σ be the corresponding transformation in the group Γ. Let S_1 be a transformation in g, then I claim that $s^{-1}S_1s$ will also belong to g. In fact we have

$$w(zS_1) = w(z)$$

$$w(zs^{-1}S_1) = w(zs^{-1})$$

$$w(zs) = [w(z)]\sigma$$

$$w(zs^{-1}S_1s) = [w(zs^{-1}S_1)]\sigma = [w(zs^{-1})]\sigma$$

$$w(z) = w(zs^{-1}s) = [w(zs^{-1})]\sigma$$

or finally

$$w(zs^{-1}S_1s) = w(z)$$

Q.E.D.

Thus we are led to consider the normal subgroups g of finite or infinite index and the functions w of z which are unchanged by the transformations in these subgroups. If equation (5) is integrable algebraically, then the group Σ of transformations is of finite order, and g is a subgroup of finite index. Thus the question of algebraic integration of linear equations reduces to that of normal subgroups of finite index and the transformation of Fuchsian functions. We shall be concerned with all this later.

When g is of infinite index, we can nevertheless find something analogous to the generating polygon of a Fuchsian group. In fact, let

(9) $$1, s_1, s_2, s_3, \ldots, s_p, \ldots$$

be infinitely many transformations in G such that if

(10) $$1, \sigma_1, \sigma_2, \sigma_3, \ldots, \sigma_p, \ldots$$

are the corresponding transformations in Γ then the table (10) contains each transformation in Γ exactly once.

Now let

$$R_0, R_1, R_2, R_3, \ldots, R_p, \ldots$$

be the polygons analogous to R_0 which correspond to the different transformations in the table (9). These polygons will form a region Q_0 which will be a sort of generating polygon for the group g.

In particular, if equation (5) is of second order and we let t be the ratio of the integrals, then t will be a uniform function of z, invariant under the transformations in g. Now suppose that (5) belongs to a type T and that x is a Kleinian function of t existing only inside a domain D bounded by a non–analytic curve, as considered in the Memoir on Kleinian Groups. Then the group g reduces to the identity and t cannot take a value inside D more than once. The relation between t and z then gives us a conformal mapping of D onto a circle.

This brings me to a very important point. I suppose that (5) is a Fuchsian equation belonging to a type T′ and that the type T is subordinate to T′. Then x is a Fuchsian function of t. Hence t is a uniform function of z invariant under g. It can take all values inside the fundamental circle, and no others. Moreover, it cannot take any of these values more than once inside Q_0, but it can take any of them infinitely often inside the fundamental circle.

It remains to examine what happens when the coefficients ϕ

of equation (5), instead of being rational in x, are rational in x and y, where these two variables are connected by the relation

$$\theta(x,y) = 0 \tag{11}$$

I consider equation (6)

$$\frac{d^2v}{dx^2} = \psi(x)v \tag{6}$$

where ψ is rational in x and has the singular points, not only of equation (5), but also of equation (11), i.e., the values of x for which $\frac{d\theta}{dy} = 0$. If b is a singular point of this species and if, in the neighborhood of this singular point, λ values of y are permuted in the well known way, then the difference between the roots of the determinant equation of (6) must be $\frac{1}{K}$, where K is a multiple of λ. To avoid any confusion we emphasize that each value of x corresponds to several points of the Riemann surface (11). If one of these points is a singular point of (5), in general it will not be the same for the others; however all these analytic points are then singular points of (6), since $\phi(x)$ is rational in x and independent of y and, consequently, it is not necessary to know which sheet of the Riemann surface the corresponding analytic point belongs to.

This being so, let G be the group of equation (6) and z the ratio of the integrals; x will be a Fuchsian function of z, y will be a uniform function of z, but it will not be invariant under all transformations in G. If I let g' be the group of all transformations in G which leave y invariant, then g' will be a subgroup of finite index in G, though it will not in general be normal (at least if the group of the algebraic equation (11) is not singly transitive).

If, on the other hand, I consider simultaneously the m values of y which can be extracted from equation (11), supposing it to be of degree m in y, and if I let g'' be the group of transformations in G which do not change **any** of these m values, then g'' will be a normal subgroup.

Finally, if we let g be the group of transformations in G which leave the integrals of equation (5) invariant, then this will be a subgroup of finite index in G and g'. As a subgroup of G it will not be normal, but as a subgroup of g', it will be.

We can consider the group g' to be generated by a polygon P'_0 formed from m transforms of R_0. The group g will be generated by a region Q_0 formed from infinitely many transforms of P'_0. As for g'', it will be generated by a polygon P''_0 formed from ω transforms of R_0, where ω is the order of the group of the algebraic equation (11). If the latter group is singly transitive, then $\omega = m$ and g' and g'' are the same.

XVII. Third problem. Symmetric types

We have seen in Section X how easily it can be shown that each symmetric Fuchsian type contains a Fuchsian equation. Now we shall show how one can calculate, to arbitrary precision, the coefficients of this equation.

Consider a symmetric type T:

$$\frac{d^2 v}{dx^2} = \psi(x)v \tag{1}$$

where $\psi(x)$ is rational in x, all singular points of the equation are real and all determinant equations have a double root. Let

$$a_1, a_2, \ldots, a_n$$

be the real singular points in question. Instead of assuming that these singular points are real, we assume that they are all on the circle with center 0 and radius 1. In fact, it is easy to pass from one case to the other by a linear change of the variable x.

The problem is to determine the arbitrary coefficients in equation (1) so that this equation becomes Fuchsian. Suppose the problem is solved. Let z be the ratio of the integrals. The variable x will be a Fuchsian function of z and the generating polygon R_0 of that function will be symmetric and will have all its vertices on the fundamental circle. It will be divided into two mirror image polygons r_0 and r'_0 by one of its diagonals. I assume that the center of the fundamental circle is inside r_0 and I let b be a certain **real** point also inside r_0. The function

$$x = f(z)$$

will be one of the Fuchsian functions in terms of which all others are expressible rationally. I define it by the following conditions

$$|x| = 1 \quad \text{when } z \text{ is on the perimeter of } r_0,$$

$$x = 0 \quad \text{for} \quad z = 0$$

$$\arg x = 0 \quad \text{for} \quad z = b$$

The function $x = f(z)$ then gives us a conformal mapping of the polygon r_0 onto the circle with center 0 and radius 1. This being so, let R_1, R_2, ... be the different transforms of R_0 by the group this polygon generates. These polygons fill the whole fundamental circle, and each of them is subdivided symmetrically into two parts r_1 and r'_1, r_2 and r'_2, etc. We consider a certain number of polygons r_i and r'_k whose union is a single simply connected piece. Let P_0 be the polygon formed by their union. Consider the function

$$y = \theta(z)$$

which gives the conformal mapping of P_0 onto the circle with center 0 and radius 1. To define y, I suppose that

$$\theta(0) = 0, \qquad \arg \theta(b) = 0$$

This function $\theta(z)$ will be a Fuchsian function whose group g will be a subgroup of the group G of the function f(z). It follows that x is a function $\phi(y)$ which is rational in y. Moreover, it is easy (and we shall return to this point a little later) to find the coefficients of the function $\phi(y)$ when we know the singular points a and the relative positions of the different polygons r_i and r'_k. Hence the coefficients of the rational function $\phi(y)$ can be regarded as known.

By taking a sufficiently large number of polygons r_i and r'_k, we can always choose these polygons so that the circle with center 0 and radius ρ is entirely inside P_0, no matter how large ρ is, provided, of course, that this radius remains less than 1. We then have the following inequality:

$$|z| < |y| < \frac{|z|}{\rho}$$

We can then choose a radius ρ sufficiently close to 1, and hence a polygon P_0 sufficiently large, that the difference between $|z|$ and $|y|$ is as small as we please. In other words, when ρ tends to unity we have

$$\lim |y| = |z|$$

But this is not all. Let

$$z = \xi + i\eta, \qquad u = \log\left|\frac{y}{z}\right|, \qquad v = \arg \frac{y}{z}$$

Suppose that z is a point such that

$$|z| < \rho_1 < \rho < 1$$

where ρ_1 is a suitably chosen positive quantity. Inside the circle of radius ρ we have

$$0 < u < \log \frac{1}{\rho}$$

We conclude that, inside the circle of radius ρ_1,

$$\left|\frac{du}{d\xi}\right| < \frac{4 \log \left(\frac{1}{\rho}\right)}{\left(1 - \frac{\rho_1}{\rho}\right)^2} > \left|\frac{du}{d\eta}\right|$$

But

$$\frac{du}{d\xi} = -\frac{dv}{d\eta}, \quad \frac{du}{d\eta} = \frac{dv}{d\xi}$$

We conclude that $\frac{dv}{d\xi}$ and $\frac{dv}{d\eta}$ tend **uniformly** to zero as ρ tends to 1. But for $z = b$ we have $v = 0$; we can then write

$$v = \int_b^z \left[\frac{dv}{d\xi} d\xi + \frac{dv}{d\eta} d\eta\right]$$

whence we conclude

$$\lim v = 0 \text{ for } \lim \rho = 1$$

and

$$\lim(u + iv) = 0, \quad \lim \frac{y}{z} = 1, \quad \lim y = z$$

I shall now show that, for any value of z, the limit of the rational function $\phi(z)$ (which is none other than $\phi(y)$, with y replaced by z) is precisely the Fuchsian function f(z). We have

$$y = \theta(z)$$

Likewise, put

$$z = \theta(z_1)$$

I claim that

$$\lim z_1 = z$$

In fact, when z is inside the circle with center 0 and radius

ρ_1 we have

$$\left|\frac{dy}{dz}\right| > 1 - \frac{\log(\frac{1}{\rho})\sqrt{32}}{(1 - \frac{\rho_1}{\rho})^2}$$

I abbreviate the right hand side of this inequality by K. Now we have

$$|z| < |y|$$

and consequently

$$|z_1| < |z| < \rho_1$$

It then follows that

$$|z_1 - z| < |y - z| \frac{1}{K}$$

But

$$\lim |y-z| = 0 \quad \text{hence} \quad \lim |z-z_1| = 0$$

Moreover, we have

$$\phi(z) = f(z_1)$$

But f(z) is a continuous function and, since the difference $z - z_1$ tends to zero, we have

$$\lim f(z_1) = f(z)$$

or

$$\lim \phi(z) = f(z)$$

Q.E.D.

We similarly show that the limit of the nth derivative of $\phi(z)$ is the nth derivative of $f(z)$.

It is a matter of profiting from the preceding to calculate the remaining arbitrary coefficients in equation (1). Let p be the number of these coefficients, which we call the coefficients c for short. Let v_1 and v_2 be two integrals of equation (1), defined as follows:

$$\text{for } x = 0, \quad v_1 = 0, \quad \frac{dv_1}{dx} = 1$$

$$\text{for } x = 0, \quad v_2 = 0, \quad \frac{dv_2}{dx} = 0$$

The ratio $\frac{v_1}{v_2}$ is then a well defined function of x. We calculate the coefficients of

$$x^2, x^3, x^4, \ldots, x^{p+3}$$

in the expansion of that function. These p + 2 coefficients are rational functions of the coefficients c and the singular points a. But we have

$$\frac{v_1}{v_2} = \frac{\alpha z}{z+\gamma} \tag{2}$$

where α and γ are two constants we do not know yet. In the identity (2) we have a relation between $x = f(z)$ and z. From it we can express the first p + 2 coefficients in the expansion of f(z) in powers of z as rational functions of α, γ, the c and the a. Conversely, we then have the c, α and γ as rational functions of the a and the first p + 2 coefficients in the expansion of f(z). But the a are given to us, the coefficients in the expansion of f(z) can be regarded as known, since they are the limits of the coefficients in the expansion of $\phi(z)$, and the coefficients c can therefore be calculated with arbitrary precision.

It may be proposed that, instead of calculating the coefficients of equations (1), we should find the group of that equation directly.

Here is how we can proceed.

Suppose that x describes a closed positive contour around the singular point a_i; z will change into

$$\frac{\alpha_i z+\beta_i}{\gamma_i z+\delta_i}$$

where the transformation

$$S_i = \left[z, \frac{\alpha_i z+\beta_i}{\gamma_i z+\delta_i}\right]$$

is parabolic. At the same time y will change into y_i. If the initial value of z is 0, its final value will be $\frac{\beta_i}{\delta_i} = \lambda_i$; if the initial value of y is 0, its final value will be called μ_i. Knowledge of λ_i suffices to determine the transformation S_i; but when we know the rational function $\phi(y)$ the determination of μ_i is merely an algebraic question. Moreover, we have

$$\lambda_i = \lim \mu_i$$

We thus determine all the transformations S_i and consequently the group G. We also have an algebraic relation between the λ_i which facilitates the calculation.

It now remains to enter more deeply into the question and to see how we can effectively calculate the coefficients of the rational function $\phi(y)$. Naturally, the calculation depends upon the choice of polygons r_i and r'_k which make up P_0. But the choice is itself a little arbitrary and we are forced, in order to fix ideas, to choose certain particular examples. Of course, these are only examples, and we do not want to claim that our choice is the simplest or best. It is probable that experience and practice would lead to a different choice.

Let $\alpha_1, \alpha_2, \ldots, \alpha_n$ be the vertices of r_0 corresponding to the singular points $a_1, a_2, \ldots, a_n$ respectively. We consider the polygon

P_0^1 formed by r_0 and its mirror image in one of its sides, $\alpha_1\alpha_n$ for example. We let y_1 be the function y generated by this polygon P_0^1, and let z_1 be a function related to y_1 be a suitable linear relation whose coefficients we shall determine later. We then have

$$z_1 = \sqrt{\frac{x-a_1}{x-a_n}}, \quad y_1 = \frac{A_1 z_1 + B_1}{C_1 z_1 + D_1}$$

Now consider a polygon P_0^2 formed from P_0^1 and its reflection in one of its sides, C. I let y_2 and z_2 be the functions which are to P_0^2 what y_1 and z_1 are to P_0^1, and let b and c be the values of z_1 at the extremities of the side C. We then have

$$z_2 = \sqrt{\frac{z_1-c}{z_1-d}}, \quad y_2 = \frac{A_2 z_2 + B_2}{C_2 z_2 + D_2}$$

We proceed with P_0^2 as we have with P_0^1 and so on **ad infinitum**. Thus we successively calculate the values of

$$z_1, z_2, \ldots, z_i, \ldots$$

To pass from z_i to y_i it is necessary to know the coefficients of the linear relation

$$y_i = \frac{A_i z_i + B_i}{C_i z_i + D_i}$$

We shall determine these coefficients in such a way that y_i vanishes with x, has its absolute value equal to 1 at the same time as x, and is real for a certain real value of x.

The mode of generation of the polygons P_0^1, P_0^2, ... remains somewhat arbitrary. To form P_0^i from P_0^{i-1} we annex to the latter polygon its mirror image in one of its sides, C; but how do we choose the side C? It will be convenient to choose the side of P_0^{i-1} whose geometric length is greatest.

I repeat that I am only giving an example here, and it is not my intention to recommend this mode of generating polygons P_0 above any other. On the contrary, I believe that it will be more advantageous to arrange this so that the group of the Fuchsian function $\theta(z)$ is always a normal subgroup of G.

Guided in this way, the calculation of the coefficients c in equation (1) does not have to be pursued for very long in order to give a very close approximation. But in the majority of applications we can be content with a rougher approximation. In fact, if the coefficients c, instead of being exactly the values which yield the Fuchsian equation, are merely sufficiently close to these values, then the variable x is not a Fuchsian function of the ratio z of the integrals, but it is a Kleinian function, which suffices in almost all cases.

Here is how we can apply this to the case of non–symmetric types. Let

$$\frac{d^2v}{dx^2} = \psi(x,y)v \tag{3}$$

$$\theta(x,y) = 0 \tag{4}$$

be an equation belonging to a Fuchsian type T. This type contains a Fuchsian equation and, if I let t be the ratio of the integrals of this Fuchsian equation, the variables x and y will be Fuchsian functions of t. But we do not propose, for the moment, to find the coefficients of this Fuchsian equation itself; we want only to find a Fuchsian equation in the type T′ subordinate to T. To do this, let

$$a_1,\ a_2,\ \ldots,\ a_n \tag{5}$$

be the table to singular points of equation (3) and those of relation (4). In Section XI we have seen that there is always a Fuchsian equation

$$\frac{d^2v}{dx^2} = F(x)v \tag{6}$$

where F(x) is rational and has singularities which include the n given quantities (5), together with k other non–given quantities b_1, b_2, ...,

b_k. Moreover, all the determinant equations have double roots.

But if we refer to what was said in Section XI, we see that we have formed this equation (6) with the aid of another Fuchsian equation whose singular points are all real and which, consequently, belongs to a symmetric type. Moreover, we have seen that the coefficients of a Fuchsian equation contained in a given symmetric type can be regarded as known. It will then be the same for those of equation (6).

This being so, let z be the ratio of the integrals of (6); x will be a Fuchsian function of z, with group G. If g' is the group of transformations in G which leave y invariant (cf. the preceding section), then y will be a Fuchsian function of z with group g'. Thus the group g' will engender a Fuchsian equation which is none other than

$$\frac{d^2v}{dx^2} = F(x)v \tag{6}$$

with the relation

$$\theta(x,y) = 0 \tag{4}$$

This new Fuchsian equation (which I regard as different from equation (6) unaccompanied by relation (4)) belongs to a type T' subordinate to T and we have seen that its coefficients are known. The problem proposed is therefore resolved.

XVIII. Third problem; general case

We recall equations (3), (4) and (6) of the preceding section. We let R_0 be the generating polygon of G, and let P_0 be that of g'. The group g of transformations in g' which leave the integrals of equation (3) invariant is a normal subgroup of g'. It is well defined and generated by a region Q_0 which plays the role of a generating polygon, except that it has infinitely many sides. These sides are equivalent in pairs in such a way that each side is the transform of its equivalent by one of the transformations in the group g. Two points on equivalent sides are said to be **corresponding** when one goes to the other by such a transformation.

Now it is a question of determining the remaining arbitrary coefficients in equation (3) in such a way that this equation becomes Fuchsian. Let t be the ratio of the integrals of equation (3), supposed Fuchsian, and let z be the ratio of the integrals of equation (6). **It is a question of determining** t **as a function of** z.

I assume, in order to determine t completely, that it vanishes at the same time as z. Then t will be a uniform function of z, invariant under transformations in g, and such that

$$|t| < 1$$

In addition, t can take all values whose absolute value is less than 1.

This being so, I claim that we have the following identity:

$$\log \left|\frac{1}{t}\right| = \sum_i \log \left|\frac{\gamma_i z + \delta_i}{\alpha_i z + \beta_i}\right| \tag{7}$$

In this identity, the transformations

$$\left(z, \frac{\alpha_i z + \beta_i}{\gamma_i z + \delta_i}\right)$$

are the transformations in g and the summation is over all elements in this group. Now consider an infinity of circles

$$C_1, C_2, \ldots, C_n, \ldots$$

with centers at 0 and radii $R_1, R_2, \ldots, R_n, \ldots$ increasing with n and tending to 1 as n increases indefinitely.

Among the terms of the series (7) there will be a certain number p which become infinite inside the circle C_n. I denote their sum by θ_n. It is a matter of showing that

$$\lim \theta_n = \log \left|\frac{1}{t}\right| \quad \text{for} \quad n = \infty$$

I remark, in addition, that $\log \left|\frac{1}{t}\right|$, the general term

$$\log \left|\frac{\gamma_i z+\delta_i}{\alpha_i z+\beta_i}\right|$$

of the series (7), and consequently θ_n, are essentially **real** and **positive**.

Let

$$z = \xi + i\eta$$

and let u_n be a real function of ξ and η which satisfies the equation

$$\frac{d^2u_n}{d\xi^2} + \frac{d^2u_n}{d\eta^2} = 0 \tag{8}$$

inside C_n, which is zero along the circumference of that circle and which, finally, is such that the difference

$$\log \left|\frac{1}{t}\right| - u_n$$

is holomorphic inside that same circle. We have inequalities

$$0 < u_n < \log \left|\frac{1}{t}\right|$$

and also

$$u_n < u_{n+1}$$

so we conclude that u_n is increasing with n but less than $\log \left|\frac{1}{t}\right|$, which means that u_n tends to a finite limit, which I call u, as n tends to ∞.

In a recent **Bulletin de la Société mathematique de France** (Paris, Gauthier–Villars, 1883) I have inserted a small work entitled

Sur un théorème de la théorie générale des fonctions, in which I show that if y is any non–uniform function of x then we can always find a variable z such that x and y are uniform functions of z. By repeating the argument of that work we see that u is an analytic function of ξ and η and that we can find a second function v of ξ and η such that u + iv is a monogenic analytic function of z. Moreover, we have

$$\log \left|\frac{1}{t}\right| \geqslant u$$

It is a question of showing that the sign which holds is = and not >. To do this, we study the function u more closely. I claim that this function is invariant under transformations in g. In fact, let

$$S = \left(z, \frac{\alpha z+\beta}{\gamma z+\delta}\right)$$

be one of these transformations and put

$$u'_n(z) = u_n\left(\frac{\alpha z+\beta}{\gamma z+\delta}\right)$$

There is another way to define the function u'_n. In fact, let C'_n be the circle which is the image of C_n under the transformation inverse to S. Like u_n, the function u'_n will satisfy equation (8), it will be zero along C'_n and the difference

$$\log \left|\frac{1}{t}\right| - u'_n$$

will be holomorphic inside that circle.

We then have an infinity of circles

$$C'_1, C'_2, \ldots, C'_n, \ldots$$

and an infinity of functions

$$u'_1, u'_2, \ldots, u'_n, \ldots$$

arising from these circles in the same way that the u_i arise from the circles C_i. Moreover we have

$$\lim u'_n(z) = u\left(\frac{\alpha z+\beta}{\gamma z+\delta}\right) \quad \text{for} \quad n = \infty$$

This being so, whatever the size of n, we can find a circle C'_p entirely outside C_n, and a circle C_q entirely outside C'_p. We then have

$$u_n(z) < u'_p(z) < u_q(z)$$

But for $n = \infty$ we have

$$p = q = \infty$$

Then

$$\lim u_n = \lim u_q = u$$

hence

$$\lim u'_p(z) = u(z)$$

and finally

$$u(z) = u\left(\frac{\alpha z+\beta}{\gamma z+\delta}\right)$$

Q.E.D.

We conclude that

$$v\left(\frac{\alpha z+\beta}{\gamma z+\delta}\right) = v(z) + \text{const}$$

Now consider the function

$$V = \log\left|\frac{1}{t}\right| - u$$

in the t plane and put

$$t = \xi_1 + i\eta_1$$

The function V will be a uniform function of t. In fact, when t describes an arbitrary contour and returns to its initial value, z will return to its initial value or else undergo one of the transformations in the group g. In either case, the function u, and hence V, will be unaltered.

Moreover, V satisfies the equation

$$(8') \qquad \frac{d^2V}{d\xi_1^2} + \frac{d^2V}{d\eta_1^2} = 0$$

We consider the circle in the t plane

$$\xi_1^2 + \eta_1^2 = \rho^2$$

where ρ is a constant smaller than 1.

The function V will be holomorphic* inside this circle. (*In fact there can be no doubt about this except for certain isolated points corresponding to vertices of R_0 on the fundamental circle. But we know that if a function V is holomorphic inside a circle, except for certain isolated points for which we know nothing, if it satisfies equation (8′), and is uniform and bounded between two given limits, then the function is also holomorphic at the isolated points in question.) Moreover, **along the circumference of this circle** it will always be between zero and $\log \frac{1}{\epsilon}$. Then, by virtue of the well known property of equation (8′), we have

$$0 \leqslant V \leqslant \log \frac{1}{\rho}$$

inside this same circle. But π can be chosen as close as we please to unity, so we necessarily have

$$V = 0$$

or

$$\log \left|\frac{1}{t}\right| = u$$

Q.E.D.

On the other hand, we have the following easily proved inequalities:

$$\log \left|\frac{1}{t}\right| > \theta_n > u_n$$

But for $n = \infty$, $\lim u_n = \log \left|\frac{1}{t}\right|$, hence also

$$\lim \theta_n = \log \left|\frac{1}{t}\right|$$

which proves the identity

$$(7) \qquad \log \left|\frac{1}{t}\right| = \Sigma \log \left|\frac{\gamma_i z+\delta_i}{\alpha_i z+\beta_i}\right|$$

giving t as a function of z.

Thus, assuming the existence of the Fuchsian equation (3) and consequently the function t, we have shown that the series (7) is convergent and that the sum of this series is precisely $\log \left|\frac{1}{t}\right|$. Conversely, assuming the convergence of the series (7), we can show that the sum of this series defines the logarithm of the absolute value of the ratio of the integrals of a Fuchsian equation of type T. Consequently, one shows that this type contains a Fuchsian equation. But this proof would be very long and also unnecessary.

We can nevertheless use it for something I shall say a few words about. Suppose T is subordinate to a type T″. Then T′, which is subordinate to T, will be subordinate to T″. Suppose that we know that T″ contains a Fuchsian equation; let (3′) be that equation and let t_1 be the ratio of its integrals. Let g_1 be the group of transformations in G which leave t_1 invariant; g will be a subgroup of g_1. From the preceding, we have

$$(7') \qquad \log \left|\frac{1}{t_1}\right| = \Sigma \log \left|\frac{c_i z+d_i}{a_i z+b_i}\right|$$

where

$$\left[z, \frac{a_i z+b_i}{c_i z+d_i}\right]$$

is the general transformation in the group g_1. But the series (7) can be obtained by suppressing certain terms of the series (7′). It is therefore convergent and consequently the type T contains a Fuchsian equation.

Thus each type subordinate to a type which contains a Fuchsian equation itself contains a Fuchsian equation. In a great number of cases, this principle allows the method of continuity to be dispensed with.

XIX. Reflections on the convergence of the preceding series

Here is how we can give a direct proof of the convergence of the series (7) in the preceding section.

In Section I of the memoir On Fuchsian Functions we have shown the convergence of the following series

$$\Sigma \left|\gamma_i z + \delta_i\right|^{-4}$$

where

$$\left[z, \frac{\alpha_i z+\beta_i}{\gamma_i z+\delta_i}\right]$$

is the general transformation in a Fuchsian group G. On the other hand, still in the case of a Fuchsian group, the series

(9) $$\Sigma \left|\gamma_i z + \delta_i\right|^{-2}$$

is not convergent. However, it can converge if we have, rather than the Fuchsian group itself, a subgroup of infinite index, for example the group g we are concerned with here.

I claim that if the series (9) is convergent, then so is the series (7). In fact, let

$$\rho_i = \left| \frac{\alpha_i z+\beta_i}{\gamma_i z+\delta_i} \right|$$

and we have

$$\left| \gamma_i z + \delta_i \right|^{-2} = \frac{1-\rho_i^{\,2}}{1-|z|^2}$$

and

$$\frac{|\gamma_i z+\delta_i|^{-2}}{\log \frac{1}{\rho_i}} = \frac{\rho_i-1}{\log \rho_i} \, \frac{\rho_i+1}{1-|z|^2}$$

When I increases indefinitely, ρ_i tends to unity and we have

$$\lim \frac{|\gamma_i z+\delta_i|^{-2}}{\log \frac{1}{\rho_i}} = \frac{2}{1-|z|^2}$$

Thus the ratio of a term of the series (9) to the corresponding term of the series (7) tends to a finite limit as the order of the term increases indefinitely.

Hence the two series converge or diverge together.

That being so, let z be any point inside Q_0. We consider a circle C whose S is

$$\frac{\pi}{4}(e^{2R} + e^{-2R} - 2) \tag{10}$$

and which has the point z in its interior. Let F(R) be the S of that part of the circle inside Q_0. Now let f(R) be the smallest value taken by F(R) as the circle C varies with its S remaining constant and always containing the point z. Then f(R) will be a continuous function of R which increases indefinitely with R.

Let N be the number of points

$$\frac{\alpha_i z+\beta_i}{\gamma_i z+\delta_i}$$

inside the circle with center 0 whose S precisely equals (10). We have

$$N < \frac{\pi}{4f(R)}\left[e^{2R} + e^{-2R} - 2\right]$$

Now let R_1, R_2, ..., R_n, ... be a series of values of R; positive, and increasing indefinitely with n. Let K_n be the circle with center O whose R is R_n. Let U_n be the sum of all the terms of the series (9) which correspond to points

$$\frac{\alpha_i z+\beta_i}{\gamma_i z+\delta_i}$$

outside K_{n-1} and inside K_n.

The series (9) can be replaced by the series

$$U_1 + U_2 + \dots + U_n + \dots \tag{11}$$

where we have

$$U_n < K \frac{e^{2R_n}+e^{-2R_n}+2}{f(R_n)(e^{2R_{n-1}}+e^{-2R_{n-1}}+2)} < K' \frac{e^{2(R_n-R_{n-1})}}{f(R_n)}$$

where K and K′ are suitably chosen constants.

The series (7), (9) and (11) will then be convergent if we can choose the R_n in such a way that the series

(12) $$\sum \frac{e^{2(R_n - R_{n-1})}}{f(R_n)}$$

is convergent. But this is what happens if the integral

(13) $$\int_a^\infty \frac{dR}{f(R)}$$

has a finite value.

This then is a sufficient condition for the convergence of the series (7). Thus it is necessary to show that the integral (13) is finite, which can be done by studying the form of the region Q_0.

Once it has been carried out, this proof will dispense with applications of the method of continuity in all cases.

XX. Resume

In this memoir, after showing how to calculate the parameters of the group of a given linear equation, I have given some properties of these parameters as functions of the coefficients of the equation, and, conversely, properties of these coefficients as functions of the parameters of the group.

I have then arrived at another problem. Consider an equation of the following form:

(1) $$\frac{d^2v}{dx^2} = \phi(x,y)v$$

(2) $$\theta(x,y) = 0$$

where ϕ and θ are rational in x and y. I suppose that relation (2) is given, as well as the singular points of equation (1) and the determinant equation relative to each of them, but that all other coefficients of equation (1) remain arbitrary. Moreover, I suppose that the difference between the roots of each determinant equation is zero or an aliquot part of unity. I let z be the ratio of the integrals and I consider x as a function of z.

I show that we can dispose these remaining arbitrary coefficients, **in exactly one way**, so that x becomes a Fuchsian function of z, existing only inside a circle which I have shown how to calculate from the coefficients.

The same coefficients can be disposed **in infinitely many ways** so as to make x a Kleinian function of z **not existing over the whole plane**. Finally, we can again dispose of these coefficients, **in exactly one way**, so that x is a Fuchsian or Kleinian function of z, **existing over the whole plane**. This latter fact has not been proved, to do so it is necessary to apply the method of continuity.

Now consider an arbitrary linear equation

$$\frac{d^m v}{dx^m} = \sum \phi_p \frac{d^p v}{dx^p}, \qquad \theta(x,y) = 0$$

where θ and the ϕ are rational in x and y. Let $v_1, v_2, \ldots, v_m$ be the integrals of this equation. We can find a variable z such that z is the ratio of the integrals of a second order equation whose coefficients are rational in x and y, and so that $v_1, v_2, \ldots, v_m$ as well as x and y are uniform functions of z. Moreover, it can be arranged that x, as a function of z is rational, or doubly periodic, or Fuchsian existing only inside a circle, or Kleinian not existing over the whole plane, or finally Fuchsian or Kleinian existing over the whole plane. We leave aside the latter case.

Then, leaving this last case aside, we can choose the variable z in infinitely many ways to satisfy all the conditions stated above. I let $z_1, z_2, \ldots, z_n$ be the different variables z of this kind obtained. Among all these variables, we can choose exactly one which is simpler than all the others and which I call z_1. In general, x will be a Fuchsian function of z_1 (existing only inside a circle), but in certain particular cases it could be a rational or doubly periodic function. In all cases, z_1 will be a uniform function of the other variables $z_2, \ldots, z_n, \ldots$, which explains why x, y and the v, which are uniform in z_1, are also uniform in $z_2, z_3, \ldots, z_n, \ldots$

Thus we can express the integrals of a linear equation with

algebraic coefficients as uniform functions of a variable z.

It remains to study the properties of these uniform functions and to expand them in series. This is what I shall do in the next memoir, which will be the last of this series.* (*Translator's Note: Mémoire sur les Fonctions Zéta Fuchsiennes, **Acta Mathematica**, **5**(1884), 209–278.)